COMPUTING WITH MAPLE

CHAPMAN & HALL/CRC MATHEMATICS

OTHER CHAPMAN & HALL/CRC MATHEMATICS TEXTS:

Functions of Two Variables, Second edition
S. Dineen

Network Optimization
V. K. Balakrishnan

Sets, Functions, and Logic: A foundation course in mathematics, Second edition
K. Devlin

Algebraic Numbers and Algebraic Functions
P. M. Cohn

Dynamical Systems: Differential equations, maps, and chaotic behaviour
D. K. Arrowsmith and C. M. Place

Control and Optimization
B. D. Craven

Elements of Linear Algebra
P. M. Cohn

Error-Correcting Codes
D. J. Bayliss

Introduction to Calculus of Variations
U-Brechtken-Mandershneid

Integration Theory
W. Filter and K. Weber

Algebraic Combinatorics
C. D. Godsil

An Introduction to Abstract Analysis-PB
W. A. Light

The Dynamic Cosmos
M. Madsen

Algorithms for Approximation II
J. C. Mason and M. G. Cox

Introduction to Combinatorics
A. Slomson

Galois Theory
I. N. Stewart

Elements of Algebraic Coding Theory
L. R. Vermani

Linear Algebra: A geometric approach
E. Sernesi

A Concise Introduction to Pure Mathematics
M. W. Liebeck

Geometry of Curves
J. W. Rutter

Full information on the complete range of Chapman & Hall/CRC Mathematics books is available from the publishers.

COMPUTING WITH MAPLE

FRANCIS WRIGHT

CHAPMAN & HALL/CRC

A CRC Press Company

Boca Raton London New York Washington, D.C.

Library of Congress Cataloging-in-Publication Data

Wright, Francis J.
Computing with Maple /Francis J. Wright.
p. cm.
Includes bibliographical references and index.
ISBN 1-58488-236-0
1. Algebra—Data processing. 2. Maple (Computer file) I. Title.
II. Series

QA155.7.E4 W75 2001
512′.0285′5369—dc21 2001037510

Direct all inquiries to CRC Press LLC, 2000 N.W. Corporate Blvd., Boca Raton, Florida 33431.

Visit the CRC Press Web site at www.crcpress.com

International Standard Book Number 1-58488-236-0
Library of Congress Card Number 2001037510
Printed in the United States of America 2 3 4 5 6 7 8 9 0
Printed on acid-free paper

Contents

Preface **xiii**

1 Introduction **1**
1.1 Learning computing 1
1.2 Background references 3
1.2.1 Computer programming in general 4
1.2.2 Maple in particular 4
1.2.3 Algorithms in general 5
1.2.4 Computer algebra algorithms 5
1.2.5 Numerical algorithms 5
1.2.6 Online references 6
1.3 Fundamental terminology 6
1.4 Maple interfaces 8
1.5 The Maple worksheet interface 11
1.5.1 The online help system 11
1.5.2 Balloon Help 13
1.5.3 Use of the mouse 13
1.5.4 Useful keyboard shortcuts 15
1.5.5 New lines and execution groups 15
1.5.6 Opening and executing worksheets 17
1.5.7 Application and document windows 17
1.5.8 Averting disaster 17
1.5.9 My worksheet shrank! 19
1.5.10 Don't work in the dark 19
1.5.11 Compatibility with Maple V 19
1.6 Exercises: the worksheet interface 20

2 Review of Basics **23**
2.1 Maple syntax: special symbols 23
2.1.1 Variables, assignments, equations, procedures, and mappings 24
2.1.2 Quotes 25
2.1.3 Brackets 28

2.1.4 Mathematical symbols 30
2.1.5 Programming symbols 33
2.2 Assignment, evaluation, substitution, and simplification 41
2.2.1 Assignment and evaluation 42
2.2.2 Evaluation and substitution 46
2.2.3 Multiple substitutions 48
2.2.4 Simplification 49
2.3 Common errors 51
2.3.1 Juxtaposition 51
2.3.2 Upper and lower case letters 52
2.3.3 The exponential function and the letter e 53
2.3.4 Assignments and equations 54
2.4 Sequences, lists, and sets 55
2.4.1 Parallel assignment 55
2.4.2 Functions and their argument sequences 57
2.4.3 Automatic simplification; inert functions 57
2.4.4 Functions and operators 58
2.4.5 Operations on sequences, lists, and sets: `op` 59
2.4.6 Operations on elements of sequences, lists, and sets: `op`, `[ ]` 60
2.4.7 The empty sequence: `NULL` 63
2.4.8 Constructing sequences, lists, and sets 64
2.4.9 Relatives of sequences: sums and products 66
2.5 Characters and strings 68
2.6 Exercises 70

3 Two-Dimensional Plotting **73**
3.1 Simple plotting 73
3.1.1 Plotting graphs of functions 73
3.1.2 Plotting graphs of "strictly numerical" functions 76
3.1.3 Plot ranges 77
3.1.4 Multiple plots 78
3.1.5 Parametric plotting 79
3.1.6 Plotting points 81
3.1.7 Plot options 82
3.1.8 Plotting discontinuous functions 84
3.1.9 Displaying and manipulating plots 85
3.2 The `plots` package and animation 88
3.2.1 Aside: packages and the `with` command 88
3.2.2 Simple animation: `animate` 89
3.2.3 Combining and animating plots: `display` 90
3.2.4 Changing default plotting options: `setoptions` 93
3.2.5 Annotating plots: `textplot` 93
3.2.6 A simple geometrical example 94
3.3 A graphical investigation of the cardioid 94

3.3.1 The basic geometry 95
3.3.2 The geometry in more detail 97
3.3.3 Animating the geometry 98
3.4 Exporting plots (and worksheets) 100
3.5 Exercises: the cycloid 101
3.6 Appendix: code to plot the figures 103

4 Three-Dimensional Plotting 105
4.1 The `plot3d` function 106
4.1.1 Graphs of real-valued functions of two real variables . . 106
4.1.2 Parametric plotting of surfaces 107
4.1.3 Multiple three-dimensional plots 110
4.2 Three-dimensional plotting options 110
4.3 Interacting with three-dimensional plots 112
4.4 Using the `plots` package 112
4.4.1 Setting default options and plotting text: `setoptions3d`, `textplot3d` 112
4.4.2 Combining and animating three-dimensional plots: `animate3d`, `display3d` 112
4.4.3 Plotting curves in space: `spacecurve` 113
4.4.4 Plotting contours: `contourplot3d`, `contourplot` 114
4.4.5 Plotting gradient fields: `gradplot` 117
4.4.6 Implicit plotting: `implicitplot` 118
4.5 Three-dimensional bouncing ball animation 119
4.6 Exporting three-dimensional plots 121
4.7 Exercises 123
4.8 Appendix: code to plot the figures 125

5 Numerical and Semi-Numerical Computation 127
5.1 Number systems 128
5.1.1 Number systems in mathematics 128
5.1.2 Number systems in Maple 131
5.1.3 Number systems in other languages 139
5.2 Floating-point computation in Maple 139
5.2.1 Floating-point evaluation: `evalf`, `evalhf` 140
5.2.2 Solution of algebraic equations: `solve`, `fsolve` 142
5.2.3 Numerical integration (quadrature): `evalf/Int` 146
5.2.4 Solution of differential equations: `dsolve` 148
5.3 Multiple roots 150
5.4 Aliases and algebraic numbers 151
5.4.1 Alias 151
5.4.2 Algebraic numbers 152
5.4.3 Implicit dependence 153
5.5 Computing with modular integers 154
5.6 Exercises 156

5.7 Appendix: code to plot the figures 158

6 **Numerical Linear Algebra** **159**
6.1 Linear algebra in Maple 159
6.2 Numerical linear algebra 160
6.3 Interactive linear algebra 161
6.4 Viewing large matrices and vectors 164
6.5 Using the LinearAlgebra package 166
6.6 Solving systems of linear equations 167
6.7 Special matrices 169
6.8 Eigenvalue problems 172
6.9 Ill-conditioned problems 176
6.10 Under- and over-determined problems 180
6.10.1 Under-determined problems 180
6.10.2 Over-determined problems 181
6.10.3 Solution by singular value decomposition 182
6.10.4 Least squares solution 184
6.11 Exercises 185

7 **Logic and Control Structures** **187**
7.1 Relational operators: <, <=, >, >=, =, <> 187
7.2 Boolean algebra 191
7.3 Boolean operators in Maple: and, or, not 191
7.3.1 Truth tables in Maple 192
7.3.2 Boolean expressions 193
7.4 Conditional execution: if 194
7.4.1 Simple conditional execution: if 194
7.4.2 Alternative execution: else 195
7.4.3 Multiple alternative execution: elif 196
7.5 Piecewise-defined functions 198
7.5.1 Piecewise-defined functions using if 198
7.5.2 Piecewise-defined functions using piecewise 201
7.5.3 Continuity class 202
7.5.4 Conditional evaluation: the ‘if‘ function 204
7.6 Loops: do, while, for 204
7.6.1 The general controlled loop: while 205
7.6.2 Nested control structures 206
7.6.3 Iterating over an integer range: for ... to 207
7.6.4 Iterating over a structure: for ... in 210
7.6.5 Other ways to control loops: break, next 211
7.6.6 The for ... to-loop implemented as a while-loop . . . 212
7.6.7 Double loops 212
7.7 Exercises 216

8 Procedures and Recursion **219**
8.1 Defining and using procedures . . . 220
8.1.1 Returning versus printing values . . . 220
8.2 General procedure definition syntax . . . 221
8.3 Variable scope . . . 222
8.4 Procedure arguments . . . 225
8.5 Special identifiers in procedures . . . 227
8.5.1 Variable numbers of arguments: `nargs`, `args` . . . 227
8.5.2 Returning unevaluated: `procname` . . . 229
8.6 Terminating procedure execution . . . 231
8.7 Examples: integer factorization . . . 232
8.7.1 Aside: `ifactor` . . . 236
8.8 Displaying Maple procedures . . . 237
8.9 Procedure options and remember tables . . . 238
8.9.1 Remember tables . . . 239
8.10 More about variable scope . . . 241
8.10.1 Lexical scoping . . . 241
8.10.2 Returning local variables . . . 242
8.11 Recursion . . . 244
8.11.1 The factorial function . . . 244
8.11.2 Recursion and stacks . . . 245
8.11.3 Avoiding infinite recursion . . . 246
8.11.4 Recursive summation . . . 247
8.11.5 Recursive power computation . . . 248
8.12 Procedures that output plots . . . 250
8.13 Recursive plotting . . . 253
8.14 Animating the Koch snowflake fractal . . . 255
8.14.1 The snowflake construction . . . 256
8.14.2 Basic geometry . . . 256
8.14.3 Constructing a single side . . . 257
8.14.4 Animating the "growth" of a snowflake . . . 258
8.14.5 Code to plot the snowflake construction . . . 259
8.15 Program design . . . 259
8.15.1 Understand the problem . . . 260
8.15.2 Design the algorithm . . . 261
8.15.3 Write the program . . . 261
8.15.4 Test the program . . . 262
8.16 Exercises . . . 263

9 Operators and Functions **267**
9.1 Operators . . . 267
9.1.1 Syntax . . . 267
9.1.2 Inert operators . . . 268
9.1.3 Active operators . . . 270
9.2 Converting between expressions and mappings . . . 273

9.3 Composition of mappings and functions . . . 274
9.4 The derivative operator D . . . 275
9.4.1 Partial derivatives . . . 278
9.4.2 Specifying conditions on differential equations . . . 279
9.5 Set operations . . . 279
9.5.1 Union . . . 279
9.5.2 Intersection . . . 280
9.5.3 Difference . . . 281
9.5.4 Symmetric difference . . . 281
9.5.5 Membership . . . 282
9.5.6 Set algebra . . . 284
9.5.7 Power sets . . . 284
9.6 Examples of recursive integer and polynomial functions . . . 286
9.6.1 Euclidean division . . . 286
9.6.2 Greatest common divisors . . . 287
9.6.3 Digits and coefficients . . . 288
9.7 Exercises . . . 291

10 Data Types **295**
10.1 Primitive types: `whattype` . . . 296
10.2 Use of type information: error checking, polymorphism . . . 298
10.3 Alternative types: {*type1, type2, ...*} . . . 298
10.4 The main Maple data types . . . 299
10.4.1 Numerical types . . . 299
10.4.2 Algebraic types . . . 300
10.4.3 Boolean types . . . 300
10.4.4 Structural types . . . 300
10.4.5 Programming types . . . 301
10.5 Type testing: `type` and `::` . . . 301
10.5.1 Error-checking of procedure arguments using `::` . . . 301
10.5.2 General type testing: `type` or `::` . . . 302
10.5.3 Quoting type specifications . . . 303
10.6 Boolean combinations of types: `And`, `Or`, `Not` . . . 304
10.7 Structured or nested data types . . . 305
10.7.1 Generic structured types . . . 306
10.7.2 Specific structured types . . . 307
10.7.3 Type-testing commuting operators . . . 308
10.7.4 Type-testing sequences . . . 310
10.8 Special procedure-argument types . . . 311
10.9 Examples of polymorphic procedures . . . 313
10.9.1 Active relational operators . . . 314
10.9.2 A more general maximum function . . . 315
10.9.3 A more general (rational) factorial function . . . 318
10.9.4 A more general digit-sequence function . . . 320
10.10 Defining new types . . . 321

10.11 Exercises 323

11 Conventional Programming 329
11.1 Operations on structures 330
11.1.1 Mapping over structures: `map` 330
11.1.2 Mapping on other arguments: `map2` 331
11.1.3 How is `map` defined? 332
11.1.4 Mapping over arrays and tables 334
11.1.5 Selecting from structures: `select`, `remove` 335
11.2 Implementing vector and matrix algebra 337
11.2.1 Vector addition 338
11.2.2 Scalar product 339
11.2.3 Matrix multiplication 341
11.3 Data processing in Maple 342
11.3.1 Outputting to a file 342
11.3.2 Inputting from a file 343
11.3.3 Elementary statistics 344
11.3.4 Other methods of input and output 346
11.3.5 Aside: offline processing 346
11.4 Interactive programs in Maple 347
11.5 Error handling: `try`, `catch` 351
11.6 Debugging 354
11.6.1 `infolevel` 356
11.6.2 `printlevel` 357
11.6.3 `debug` and `trace` 359
11.6.4 Interactive debugging 360
11.7 Exercises 364

12 Algebraic Programming 369
12.1 Univariate polynomial algorithms 369
12.1.1 Squarefree factorization 369
12.1.2 Sturm sequences 373
12.1.3 Finding zeros by interval halving 377
12.2 Multivariate polynomial algorithms 381
12.2.1 Generalized polynomial division 381
12.2.2 Gröbner bases 388
12.3 Exercises 397

13 Spreadsheets 399
13.1 Introduction 400
13.2 Copying and moving cells 402
13.3 Working with sequences 405
13.4 Tabulating data 406
13.5 Programming spreadsheets 410
13.6 Case study: the simple pendulum 412

13.6.1 Exact solutions for small and large amplitude 413
13.6.2 Approximate numerical solution 415
13.6.3 Numerical computation of the period 418
13.6.4 Numerical computation of the amplitude 421
13.7 Exercises 423

14 Text Processing 425
14.1 Text files 426
14.2 Counting lines, words, and characters 428
14.2.1 Counting words and characters within a line 428
14.2.2 Counting lines, words, and characters in a file 433
14.3 Text formatting 436
14.4 A Markov chain algorithm 440
14.5 Exercises 444

15 Object Orientation and Modules 445
15.1 Introduction to object orientation 446
15.2 Block matrices 447
15.3 Modules as records 448
15.4 Modules in more detail 450
15.5 Module constructors 451
15.6 Modules as objects 455
15.7 Data encapsulation 457
15.8 Accessing block matrix elements 463
15.9 Square and inverse block matrices 464
15.10 Modules and types 466
15.11 Block matrix multiplication 467
15.12 Limitations of modules as objects 469
15.13 Operator overloading 470
15.14 Using a private Maple library 477
15.15 Modules as packages 481
15.16 Adding online help 485
15.17 Exercises 487

Bibliography 489

Index 491

Preface

This is a book about computer programming that uses Maple™ for its concrete realization and so discusses both general concepts and technical details of Maple.[1] Current versions of Maple encompass most key aspects of modern programming languages: control structures, data types, numerics, graphics, spreadsheets, text processing, and object oriented programming, all of which are covered in this book. However, Maple is primarily a system for mathematical computation, which is also where my own interests mainly lie, so there is an emphasis on mathematical examples and a chapter devoted to algebraic programming.

Level

It is probably fair to describe this book as an "intermediate" course in Maple. It could be used to learn Maple from scratch, but the learning curve would be fairly steep for a reader who had no previous computing experience. Much of the material that I would expect to be covered in a first course is compressed into the first two chapters. By selecting between about half and two thirds of the material, it should be possible to use this as a textbook for a one-semester course on programming for mathematically oriented students pitched at some intermediate level, the precise level being determined by whether the easier material that comes earlier or the harder material that comes later is selected.

Maple coverage

In this book I focus on Maple 6 and throughout I use the new syntax introduced with Maple 6. I mention the major incompatibilities with recent earlier versions of Maple, but I do not (intentionally) use any obsolete syntax or facilities. My intention in writing this book was primarily to regard Maple 6 as the start of a new direction in Maple development and, as is probably inevitable, Maple 7 was announced while I was making the final revisions to

[1]Maple is a registered trademark of Waterloo Maple, Inc., 57 Erb St. W., Waterloo, Canada N2L 6C2, www.maplesoft.com.

the book. However, based on the announcement that I have seen, I anticipate that most, if not all, of this book will still apply to Maple 7, but I will make any necessary updates available via the web site for this book; see below.

I think it would be impossible to do justice to the whole of Maple in one book and I have not tried to do so. In particular, I mention very few of the many Maple packages and I hardly mention the "assume" facility. I have omitted in part or in total many advanced technical details that are covered in the *Maple 6 Programming Guide* [22] and the online help, such as Maple's internal data representations, advanced input and output facilities, very low and very high level graphics programming (I cover only the middle ground), and the new interface to external functions written in C.

Background

This book is based on about eight years' experience using and teaching Maple. It was developed from the lecture notes for a second Maple-based course for first-year mathematics students at Queen Mary, University of London, that I designed during the academic year 1993–94 and have taught ever since. It is primarily intended for joint mathematics and computer science students, and part of its aim is to provide an introduction to computer science for mathematics students.

The lecture notes on which this book is based were revised through at least four releases of Maple. However, this book is completely rewritten for Maple 6 and about half of the material is new.

A book developed from the first Maple-based course for first-year mathematics students at Queen Mary has just been published as *Experimental Mathematics with Maple* by Franco Vivaldi [27]. It provides a gentler introduction to Maple basics than I provide in Chapter 2 and an introduction to the mathematical background behind some of my examples, although it has more emphasis on mathematics and less on programming and does not use the new syntax that I use exclusively.

Structure of the book

This book falls into two parts: the first part comprises Chapters 1 to 6 and concerns basic Maple facilities that are accessible for interactive use; the second part comprises Chapters 7 to 15 and concerns programming in Maple. (Spreadsheets are difficult to classify; I regard them as an interactive programming facility and I have put them toward the end of the second part.) The book is mainly intended to be read sequentially. The chapters in the first part form pairs: Chapters 1 and 2 introduce Maple's user interfaces and basic syntax, Chapters 3 and 4 describe plotting in two and three dimensions, and Chapters 5 and 6 introduce numerical computing and numerical linear algebra. In the second part, Chapters 7 to 10 form a sequence that builds up the techniques of computer programming using Maple and then Chapters 11 and 12 consolidate these programming techniques. The last three chapters are largely independent of each other and cover spreadsheets, text processing, and modules. The discussion of the use of Maple modules for object oriented programming and building packages and libraries is left until last because it involves probably the most sophisticated programming concepts and certainly the largest single body of code discussed in the book, which is too large to maintain comfortably within a worksheet.

Web site

There is a web site for this book at[2]

http://www.maths.qmw.ac.uk/~fjw/CwM/

Every chapter ends with a number of exercises, the incisiveness of many of which has been honed on my students over the years. The web site for the book provides model solutions in the form of Maple 6 worksheets, contact information, and supplementary material such as some of the more entertaining graphics developed in the book (two- and three-dimensional animations and a very simple example of "virtual reality", exported in standard web formats as explained in the text). It will provide any updates necessary for Maple 7 and it may also provide further discussion, examples, and corrections of errors — please let me know if you find any!

Production of the book

This book was written using various versions of Microsoft Windows. Maple is supposed to be largely independent of the platform on which it is run, but there are some inescapable differences, such as the number of buttons on a standard mouse: one on a Macintosh, two on Windows, and three on UNIX/Linux (which makes Windows a nice compromise). I have tried to deal with the "platform" issue in an even-handed way, but readers may still detect some Windows bias.

Each chapter of this book was written as a single Maple 6 worksheet consisting of text regions interleaved with execution groups, and containing all the necessary mathematical symbolism and diagrams. Only in the final stage of preparation was each worksheet exported as LaTeX [20] and then re-edited to correct and improve the formatting, but I have consciously tried to preserve the worksheet structure as closely as possible. Hence, all the Maple examples are essentially "live" and so should reflect how Maple 6 actually behaves. Before exporting each worksheet, I re-executed it from beginning to

[2] At some future date, the "qmw" in this URL may change to "qmul" to reflect the recent change of working title of the College to "Queen Mary, University of London".

end after restarting Maple, so each chapter is "WYSIWIG" (What You See Is What I Got) — more or less. At least, this defines the precise environment in which every Maple example was run and is the reason for occasional gratuitous Maple statements such as `unassign` and `restart`.

However, the LaTeX that Maple generates does not always correspond precisely with what is displayed in the worksheet. Hence, in particular, the Maple input examples generally do not show execution group delimiters, except in Chapter 1, where I reconstructed them by hand, and in one of the bitmap figures in Chapter 13 that I captured from the screen. I discuss execution group delimiters further near the end of Chapter 1.

The typography of the Maple output examples is an approximation to that used in the worksheets, whereas the typography of the mathematics that is part of the discussion is slightly different. In particular, Maple uses a roman font for *all* identifiers when they are applied as functions, whereas normal mathematical typography does this only for standard mathematical functions such as "log". Hence, in Maple output examples you will see typography such as $\mathrm{f}(x)$, whereas in the discussion I normally use $f(x)$.

A few Maple output examples introduce a label %1 for a common subexpression, which is defined immediately after it is used. This was not the case in the actual worksheets and probably occurred because I had to set a narrower page width for the LaTeX export than I was using for the worksheets. Also, quite a lot of minor editing was required to meet the requirements of publishing a book rather than worksheets, in particular the constraint of fitting the material into discrete pages having a fixed and fairly small size!

The example plots are shown as they appear in the worksheets; more precisely, in the form in which they were exported (automatically) by Maple as encapsulated PostScript files during export of the worksheets as LaTeX, which automatically converted all the colours to black. In the case of animations, Maple exports only the first frame, regardless of which frame is currently displayed in the worksheet. The first frame is often the least interesting, so I have cheated slightly and manually constructed and exported a more interesting frame in some cases. Nevertheless, to see actual animation, I encourage you to enter and run the code for yourself. (Or look at the animations on the web site.) A few figures are part of the discussion rather than Maple examples. Most of these were also produced using Maple, in which case the code is included (as extra examples, usually in an appendix) in each chapter in which such figures appear.

Chapter 13 is exceptional in that all the figures are bitmaps. The spreadsheets shown were all "captured" from the Maple worksheet display — I have sacrificed the typographical display quality of the tables in favour of showing how spreadsheets really appear in Maple. For variety, I constructed the two diagrams in this chapter as Microsoft Word Picture objects embedded in the Maple worksheet, which I later output directly from Word.

The worksheets contain a few non-ASCII characters (via Windows Character Map) and some embedded Microsoft Equation Editor objects to provide mathematical notation that is not directly available in Maple. Care is required when typesetting the former in LaTeX and unfortunately Maple silently discards embedded objects when exporting as LaTeX so it was necessary to process non-ASCII characters and embedded objects by hand.

[A cautionary tale: When I was previewing a draft of this book, one figure appeared completely blank, for no obvious reason. It was Figure 13 in Chapter 13 and the day was May 13. Had it been a Friday I would have been suspicious, but in fact it was a Sunday! The following day I discovered that this and related problems were caused by the white space surrounding the bitmap figures, which needed to be clipped.]

Acknowledgements

For the final editing of the LaTeX version of this book I used GNU Emacs[3] extensively and several other GNU[4] utilities, many of them from the Cygwin[5] package. To produce the final PDF file I used MiKTeX[6] and GhostScript[7]. I thank the original developers of all this excellent free software and those who ported it to, and continue to develop it for, Windows.

I thank the following for helping to make this book possible: Marcelo Rebouças for inviting me to spend a sabbatical year from 1992 to 1993 at the Brazilian Center for Research in Physics (Centro Brasileiro de Pesquisas Físicas, CBPF), Rio de Janeiro, Brazil, where I developed and taught my first Maple course; my colleagues Franco Vivaldi and Leonard Soicher for many interesting discussions on Maple, computational mathematics, and algorithms; my colleague David Arrowsmith for professional support and encouragement; Waterloo Maple, Inc., for providing me with a personal copy of Maple 6 through their author support programme; and last but not least my family — Heather, Kathryn, and Madelyn — for allowing me the time to write this book.

Francis Wright
Queen Mary, University of London
June 2001

[3]http://www.gnu.org/software/emacs/
[4]http://www.gnu.org/
[5]http://sources.redhat.com/cygwin/
[6]http://www.miktex.org/
[7]http://www.cs.wisc.edu/~ghost/

Chapter 1

Introduction

This chapter introduces computing and computer programming in general and Maple in particular, although without going beyond the Maple user interface. The first two sections provide general advice about how to learn computing and suggest some background reading. The next section introduces some of the unavoidable jargon of the subject and warns about ambiguous terminology that arises because Maple spans such a wide range of application areas. Section 1.4 introduces the Maple worksheet and command-line interfaces. I recommend using the worksheet interface, and the last section covers important practical details that you should be aware of before trying to use it seriously.

1.1 Learning computing

This book is intended to provide an introduction to computing, which is about making a computer perform tasks that you want done but don't want to do yourself. Computing as a discipline embraces art, science, and engineering. (Some might include magic, and the term *wizard* is often used to refer both to people and more recently to programs, although not in this book.) Computer programming is necessary when the computer does not yet know how to perform the required task. In that case, you have to provide some of the intelligence required to perform the task, but the more intelligence already built into the computer the less you have to provide. This "built-in" intelligence comes in the form of "interfaces" to make it easier to communicate with the computer, "languages" for describing what you want the computer to do, and "libraries" of solutions to subtasks that someone has already programmed. All these facilities are provided by Maple.

Computing is an essentially practical activity and some aspects can be learnt only by doing them. However, it is wise to be guided by the experience that has been developed over the last 50 or so years. There is a great deal of relevant theoretical background in, for example, languages and grammars, algorithms and complexity, and the psychology of human–computer interactions. Most of the important concepts in computing are quite general and independent of the details of any particular computing machinery or programming language. However, they would be useless without some concrete system to which to apply them. In this book I will discuss the use of Maple, specifically Maple 6, the latest version at the time of writing. When I refer to *Maple* I shall mean Maple 6 unless I specifically indicate some other version.

So, my advice is to read this book with a computer running Maple close at hand. Use Maple as a computing laboratory and experiment. Try things that you think should work. If they do, think about ways to do them better. If they don't, think about why what you tried may not be feasible. But trial-and-error should only be a vehicle for discovering and understanding features of Maple; it is not a sound programming strategy! Once you have understood computer programming in general and Maple in particular, you should be able to write correct programs on paper or type programs into Maple and have them execute correctly the first time (usually).

How should you learn computing and programming? You need several things. You need motivation. The motivation might be that you want the credit for passing a course, but it's much more fun to have a specific task in mind that you want the computer to perform for you. For example, you might like the idea of using a computer to produce music or works of art (Chapters 3 and 4 cover computer graphics), or you might want to produce a game to be played on the computer (a very simple game is suggested as an exercise at the end of Chapter 11) or a facility to help you play other games or solve puzzles. (As a student, I had fun and learnt a lot by writing a very simple interactive program to play noughts-and-crosses, sometimes called tic-tac-toe, which had nothing at all to do with the numerical analysis course for which the computing facility had been provided!) Such motivations may seem impossibly grandiose for a beginner, but the way to approach any large task is to break it down into smaller and smaller tasks until you get down to a scale that is manageable. Aim to develop small programs to solve small subtasks independently, test them to make sure they work correctly alone, and then put the pieces together. This is a good approach to learning how to program, because you can focus on the facilities that you need for each subtask one at a time.

You need good examples of how to perform tasks that are similar to what you want to do, which you can modify to do what you want. To do so you have to understand the original example to some extent, and the act of modifying it helps you to remember what you have learnt. Just typing examples from a book into a computer is also a useful way to learn, *provided* you think about what you are typing. There are lots of Maple examples in this book and lots more in the Maple online help. An interactive programming environment is very helpful when learning because it gives you instant feedback — instant reporting of errors when you get it wrong and instant gratification when you get it right! Maple is normally used interactively in just this way, although it can also be run non-interactively when appropriate, such as for solving a difficult problem once the program has been developed and tested on simpler examples.

Learning is an iterative process, meaning that it proceeds via a sequence of small steps, and learning computing and programming is no exception. It is useful to have some idea of what might be possible, and a rapid superficial read through a book (such as this one) describing a computing system or programming language should give you a feel for its capabilities. Then think about your problem, or more precisely about the small subtask that you are currently working on. Then sketch out a program that you think ought to work. Then why not try it? The worst that can happen is that you will get some complaints from your computing environment. You can learn a lot from a good interactive programming environment such as Maple by just experimenting. Then go back to the documentation and read the details of the facilities that you are trying to use, either to check that you have got them right or to correct what you have got wrong. You also need to think about what you are doing. Put yourself in the place of the computer: you can execute simple programs in much the same way that the computer can; if the instructions don't make sense to you then they are unlikely to make sense to the computer, in which case you may need to rethink the logic of what you are trying to do.

It is important to become familiar with the full documentation for any computing system or programming language, although it can be a little intimidating at first. If you like analogies then think of this book as a motorway map that you are using as the basis for a "grand tour". Every so often along the tour you will want to stop off to explore sites of interest with the help of a detailed local guide, which for this "Maple grand tour" is the Maple online help. If I use any Maple facilities without explanation in this book then I assume you will look them up in the online help.

1.2 Background references

The discipline of computer programming is largely independent of any particular programming language. A well-rounded programmer will be familiar with a number of languages and be able to choose the appropriate language for any particular task. In this book, I make occasional reference to other languages, in particular to what I regard as currently the most important mainstream languages, namely C, C++, and Java. I recommend that you at least read about some other languages. This book touches on a lot of aspects of computing that I cannot pursue to any depth. I give here a brief annotated bibliography of the books and other information sources that I have found most useful and most influential. I will refer to many of these books again in later chapters where they are specifically relevant.

1.2.1 Computer programming in general

Anything with Brian Kernighan's name on it is worth reading. In particular, these three books have been influential in forming my view of computing and my programming style:

Software Tools by Kernighan and Plauger [15]

The Elements of Programming Style by Kernighan and Plauger [16]

The Practice of Programming by Kernighan and Pike [14]

The next three books are my primary references for what I regard as currently the three most important general-purpose programming languages:

The C Programming Language by Kernighan and Ritchie [17]

The C++ Programming Language by Bjarne Stroustrup [26]

Java in a Nutshell: A Desktop Quick Reference by David Flanagan [6]

The following are books that I have found useful as references for a couple of other important symbolic programming languages. Both languages predate Maple and both have been used for work in Artificial Intelligence. This was one of the original motivations for developing computer algebra systems, of which Maple is one of the later examples. I believe that LISP in particular has influenced the design of Maple.

LISP by Winston and Horn [28]

Prolog Programming for Artificial Intelligence by Ivan Bratko [2]

1.2.2 Maple in particular

I learnt Maple originally by reading the books distributed with the early releases of Maple V, reading the online help, and experimenting, and I have kept up to date by doing the same with subsequent releases. These are the two books currently distributed with Maple and I recommend you read them:

Maple 6 Programming Guide by Monagan *et al.* [22]

Maple 6 Learning Guide by Heal *et al.* [11]

Here are a couple of other books that I have found useful and recommended to my students in the past, although they describe Maple V and so are now a little out of date:

The Maple Handbook by Darren Redfern [24]

Introduction to Maple by André Heck [12]

1.2.3 Algorithms in general

Like Kernighan, anything with Donald Knuth's name on it is worth reading. In particular, his book on *Seminumerical Algorithms* provides essential background to several of the algorithms that I discuss.

The Art of Computer Programming by Donald Knuth:
Vol. 1, *Fundamental Algorithms* [18]
Vol. 2, *Seminumerical Algorithms* [19]

The following was one of the first books on algorithms that I read and it has played a crucial role in shaping my approach to algorithms:

Data Structures and Algorithms by Aho *et al.* [1]

1.2.4 Computer algebra algorithms

A lot of good books have been written on computer algebra and I have just picked the two most recent ones. The first is very recent and I regard it as currently the definitive text on computer algebra algorithms. It is remarkably good value for money, includes some attractive colour printing, and I find the historical sketches that begin each of the five main parts very enticing.

Modern Computer Algebra by von zur Gathen and Gerhard [7]

The second is the definitive background text for Maple because it was written by a subgroup of the principal Maple developers, and while most computer algebra theory applies to most computer algebra systems, this book clearly reflects the Maple approach.

Algorithms for Computer Algebra by Geddes *et al.* [8]

1.2.5 Numerical algorithms

Numerical analysis is a relatively old subject and there is not a great deal of new material other than the very esoteric. However, when it was first published, *Numerical Recipes* did present a fresh approach. It has a very practical orientation but also presents enough simple theory to confirm that the algorithms are not entirely magic. It has been repackaged using other languages, such as C, but I prefer the original version, which uses FORTRAN 77 (with slight cosmetic modifications) to implement all the recipes, with a translation into Pascal in an appendix. Although I doubt that I would use much of the code verbatim, it is very useful to have it because it makes the whole book very concrete and well-tested computer code is necessarily unambiguous.

Numerical Recipes: The Art of Scientific Computing by Press *et al.* [23]

The following is one of several good, solid, intermediate-level classical numerical analysis texts that I like:

Introduction to Numerical Analysis by Stoer and Bulirsch [25]

1.2.6 Online references

There are at least two active Internet discussion groups that focus on Maple. They can be useful as a source of occasional information from the Maple developers or for asking specific questions, and other people's questions and the answers to them often help to illuminate the murky corners of Maple.

The Maple User Group (MUG) — see
http://www.maplesoft.com/support/discussion.html

While writing, I find the following online reference works useful, mainly for checking background details. As a general encyclopaedia, *Encarta* is surprisingly good for computing topics.

Concise Oxford Dictionary (Ninth Edition) CD-ROM,
Oxford University Press (1997)

Oxford Reference Shelf CD-ROM, Oxford University Press (1994)

Microsoft® Encarta® Encyclopedia CD-ROM, Microsoft Corporation

For more obscure information I have usually found that a simple web search finds what I need.

1.3 Fundamental terminology

A Maple *identifier* or *symbol* is a letter or underscore followed by zero or more letters, digits, or underscores. Alternatively, it is any run of characters enclosed in *backward* quotes. Lower and upper case letters are distinct. There is no practical limit on the length of an identifier. Identifiers that begin with an underscore or begin or end with a tilde (in the latter case the identifier must be quoted) have special significance and should not be used for general identifiers.

Maple is principally a system for mathematical computation and so unavoidably uses both mathematical and computational jargon. Moreover, the applications of Maple span a wide range of mathematics, science, and engineering. Unfortunately, the various cultures occasionally disagree on terminology so the meanings of a few terms in Maple depend on the context. Here are a couple of important ambiguities.

Pure mathematicians and programmers normally use the term *function* to mean an *operation* with particular properties whereas applied mathematicians tend to use it to mean an *expression* representing the result of applying an operation. The Maple *data-type system* (see Chapter 10) uses the latter meaning. When there might be any ambiguity, I will try to use the term *mapping* or *abstract function* for the former and *function* for the latter.

The term *name* is often used in computing to mean an identifier or symbol, which in a programming language is the simplest type of reference to an object.

The Maple documentation uses this meaning in the context of "name quotes", which it uses as a synonym for the backward quotes that turn any string of characters into a symbol, but the Maple data-type *name* is defined to be either a symbol or an indexed symbol.

Here are a few important terms, in alphabetical order, and what they mean in this book:

active — an active mapping is either a procedure or an identifier that has a procedure assigned to it.

apply — make a mapping or abstract function operate on some object(s); e.g., $f(x)$ is the mapping f applied to the argument x.

argument — the object that a mapping or abstract function operates on; e.g., x in $f(x)$.

body — the executable code contained in some programming construct, as in *loop body, procedure body, module body.*

code — part or all of a computer program (because it *encodes* an algorithm).

default — pre-selected option adopted when no alternative is specified by the user.

evaluate — replace assigned variables by their values and run the programs associated with functions, procedures, commands, etc.

execute — run a program, make something happen.

execution group — a worksheet segment that is marked by a long square bracket down the left side; all the code in an execution group is executed at the same time.

expression — any code that represents a *value* and could be assigned to a variable.

identifier — a symbol that can be used to identify objects such as variables and procedures.

inert — an inert mapping is an identifier that has no assigned value and is used as a mapping.

mapping — an abstract function not applied to any argument; e.g., $(x, y) \to x^2 + y^2$, or, in the case of an inert mapping, a symbol (that has been or will be applied to arguments to construct an expression of function type).

sequence — expressions separated by commas or statements separated by semicolons (or colons).

simple mapping — a mapping or abstract function defined with one argument; e.g., $x \to x^2$.

simplify — reduce an expression to a simpler form, where *simpler* has a technical meaning that does not necessarily agree with its colloquial meaning.

statement — the smallest executable instruction; in Maple, any expression is also a statement.

1.4 Maple interfaces

Maple 6 runs on 32-bit versions of Microsoft Windows (i.e., Windows 95 and later), various versions of UNIX including Linux, and the Power Macintosh. The differences among the various platforms are minor. Maple has two distinct interfaces: *worksheet* and *command-line.* The worksheet version provides a Graphical User Interface (GUI), which is the kind of modern friendly interface that users of personal computers will probably expect. The command-line version is much more "bare bones" and less friendly, although it was the only interface available until about 1990. Both interfaces have their advantages and disadvantages. The Maple worksheet interface uses the native GUI provided by Windows or the Macintosh; on UNIX-like platforms it uses the X Window System, provided it is installed.

The worksheet interface is most appropriate for the beginning user and for simple interactive use of Maple, and hence is what I will primarily describe in this book, although much of Maple is independent of the interface used to access it. I prefer the worksheet interface and use it almost exclusively. However, the worksheet interface uses more computing resources than does the command-line interface because of the overhead of the GUI. Using Maple with the worksheet interface is very similar to using a word processor. A Maple worksheet is like a word-processor document, and text, Maple input, and Maple output can be interleaved and placed almost anywhere. The output can include two- or three-dimensional plots (the latter represented as projections into two-dimensional space).

The GUI supports a mouse, menus and toolbars, multiple fonts, typeset-quality display of symbolic mathematics, interaction with plots and arrays, spreadsheets, and multiple worksheets. In fact, the Maple worksheet interface can be "driven" almost entirely by using the mouse to select items from menus and palettes, which may be a very convenient way to solve simple mathematical problems. However, I do not normally use Maple this way myself, and this "scientific calculator" approach is not the view of computing that I will expound in this book, although it does tie in with the use of spreadsheets, which is the topic of Chapter 13.

On Windows, other Windows "objects" can be *embedded* in Maple worksheets, and many types of files can be *linked* into Maple worksheets using "Object Linking and Embedding" via the *OLE Object...* item on the Insert menu. (A linked object is a reference to an external file whereas an embedded object is stored within the worksheet.) I find it particularly useful to be able to embed mathematics generated with Microsoft Equation Editor (which comes with Microsoft Word but may not be installed by default), because some mathematical notation is currently difficult or impossible to compose using Maple itself (e.g., set notation and general inequalities). Some of the mathematical notation in the Maple draft of this book consists of embedded Equation Editor objects.

Some technical diagrams are best constructed using a programming interface. Most of the figures in this book were generated as Maple plots and the code for all these figures is included as appendices in the relevant chapters. However, some diagrams are easier to generate another way, such as interactively using a drawing or painting application. (A *drawing* is represented as a collection of distinct editable objects whereas a *painting* is represented as a single bitmap.) Such pictures, originating from any convenient source, can be embedded (or linked) into a worksheet. I composed one of the figures in Chapter 13 as a Microsoft Word drawing, which was then embedded in the Maple draft of this book. (As an experiment, I also successfully embedded both a new Microsoft Paint bitmap and a digital photograph stored as a JPEG file in a worksheet.)

The command-line version is perhaps most appropriate for non-interactive use. It is entirely possible for another program rather than a human user to communicate with Maple. For example, a human user could access Maple remotely via a web browser, in which case it would be the web server at the remote site that communicated directly with Maple, probably via at least one intermediary program implementing the Common Gateway Interface (CGI). Alternatively, the command-line version allows you to use a different interactive interface. For example, Emacs is a free and very powerful editor, which can act as a configurable user interface to other programs. (GNU Emacs 21 is available for all platforms that support Maple 6.) The command-line version of Maple has the advantage that it leaves more memory for the Maple kernel since very little is taken up by the interface, so it may allow you to solve a slightly larger problem than you could with a GUI version.

The Maple input and output in a worksheet is best regarded as a picture of part of the abstract computational space maintained by the Maple kernel. It shows you some of the input to the kernel and some of the output from it. But the relationship is not one-to-one because you can delete information from the worksheet without deleting it from the kernel, and vice versa. Moreover, the spatial order of information in the worksheet may have little relation to the temporal order of information flow to and from the kernel. When you save a worksheet you do not also save the state of the kernel, so when you open a new worksheet the kernel knows nothing about its contents. You must execute it, or at least those parts that are relevant, before you can make any use of the information in the worksheet, although you can manipulate plots without replotting them. The power of the worksheet interface can easily confuse a beginning user. In this respect, the command-line interface may be slightly less confusing, in that the spatial arrangement of input and output is in one-to-one correspondence with the temporal input to and output from the kernel.

There are actually two slightly different versions of the worksheet interface. By default, there is only one kernel and all worksheets correspond to (probably completely different) views of that one kernel. Except on the Macintosh, there is also a "parallel server" version in which each worksheet has its own independent kernel. This is more or less equivalent to running more than one copy of Maple at the same time, which you can also do, but the latter probably requires more memory. However, it can be useful to have more than one worksheet accessing the same kernel since it allows you to make small experiments on your live data without disturbing the main worksheet, and to perform independent computations it is often sufficient just to ensure that you use different variables. So, for maximum flexibility, provided you have enough memory, I recommend running the GUI version of Maple in single-server mode, and if you need another Maple kernel then start another copy of Maple (which is the only choice on a Macintosh).

The Maple worksheet interface provides facilities for inputting and editing program text, which is often called *source code.* Indeed, it provides a very convenient interface for maintaining and testing small programs together with their documentation, and a worksheet (or part of it) can easily be saved to the Maple library and/or exported to the Maple online help system (as explained in Chapter 15). However, for larger programming projects a programmer would normally want to use a source code editor or development environment that provides more automated support for software engineering. For example, a source code editor should support automatic indentation of lines of code depending on the current control structure, matching of brackets, syntactic highlighting to distinguish keywords, strings, etc., and many other programming conveniences.

A good development environment will support both editing and running of programs. For example, Emacs can be configured to provide full support for editing and running Maple source code. Maple supports a simple text input format that is completely different from the worksheet interface. Maple text files can be generated using any convenient editor and can be read into Maple using either the command-line or the worksheet interface. A Maple text file can optionally be incorporated into a worksheet. Note that Maple text input is *read* (using the `read` command) whereas Maple worksheets are *opened* (using the *Open...* item on the File menu or the corresponding toolbar button provided by the GUI). These issues are explored further in Chapter 15. Maple programs written (or saved or exported) as text files can also be run "offline", as explained in Chapter 11. Software developers often use some kind of revision control system (e.g., RCS or CVS, both of which are free and well supported by Emacs), which keeps track of the development of a program efficiently by storing only the differences between successive revisions. This works well for line-based source code text files but may not work so well for worksheet files.

I wrote this book as a set of Maple worksheets using various versions of Microsoft Windows, so that is the platform that I have primarily in mind and on which I tested my examples. There should be very few aspects of the use of Maple that differ among platforms, and I have tried to cover most of those in the next section.

1.5 The Maple worksheet interface

This section consists of practical advice on how best to use the Maple worksheet interface. It is based on my experience using it myself and helping students learn to use it. Maple worksheets can be opened and saved only via the File menu or the toolbar provided by the GUI; there are no Maple commands for these purposes. (By contrast, Maple text and binary files can be read and saved only via Maple commands.)

It is well worth familiarizing yourself with the user interface provided by the platform(s) that you use, e.g., by reading the general platform documentation, taking the "online tour", or using whatever is available. To a large extent, Maple will work like other similar applications on a given platform. For example, mouse functions and some keyboard shortcuts are fairly standard across applications (and to some extent across platforms). In particular, I recommend learning how to "copy, cut, and paste" as a fast way of editing; copying and pasting text is not only faster but more accurate (and more fun) than re-typing it!

One particularly useful feature of the worksheet interface is the undo/redo facility. It can be accessed via the pair of buttons on the toolbar with arrows arcing respectively anti-clockwise and clockwise or via the top two items in the Edit menu. Operations, such as typing, deleting, or moving text, can be undone in "chunks" back to some earlier state of the worksheet, although Maple does not keep a full undo history. If you undo further than you want then you can redo or "undo the undo". This facility is similar but not identical to the undo/redo facilities in other editors, such as Microsoft Word and Emacs. It can save a lot of frustration if you make a small mistake or simply change your mind. But it is not infallible and I do not recommend relying too heavily on it. It is certainly no replacement for regular and/or automatic saving and for a secure file-backup strategy!

1.5.1 The online help system

The definitive Maple documentation is available online within Maple and there are various ways to access it. Maple is a very big system and a great deal of documentation is available online (which I use a lot). The two printed books currently distributed with Maple are very useful additional references but they do not cover the whole system and I certainly do not intend to either.

My hope is that once you have read and digested this book you will be able to proceed with assistance only from the online help.

The online help can be accessed by executing a special Maple command consisting of a question mark followed by a topic. Maple looks up the topic and either displays help on it or responds that it cannot find help on that topic. A topic can be qualified by following it with a comma, slash, or opening square bracket and then a subtopic. Spaces are optional. For example, this command gives help about the help system itself:

```
> ?help
```

This command gives general help about integration:

```
>  ? int
```

whereas this gives help specifically about numerical integration:

```
> ? int, numeric
```

This way of accessing help works with both the worksheet and command-line interfaces. With the worksheet interface, it opens a new document window whereas with the command-line interface the help text is output immediately after the command input line and scrolls up the screen or window.

With the worksheet interface, the ? command provides a quick way to access help if you know the appropriate topic, but there are several other friendlier ways to access help via the Help menu. The second entry in this menu provides *context-sensitive* help. When there is no appropriate context it is de-activated and reads *Help on Context.* When the text cursor is within a word or immediately after some non-word text character, the menu item changes and *Context* is replaced by the word or character. If there is any ambiguity, you can select (by *double*-clicking or dragging the mouse) a word or character, which will be used as the context. Selecting this menu item will then cause the help system to look for help on the *Context* as topic. This is a very quick way to access help relating to a word or special symbol in a worksheet.

If you are less certain of the precise topic about which you want help then you can use the *Topic Search...* menu item in the Help menu. Type slowly into the *Topic* input box and Maple will list matching topics. The more you type the fewer topics match. Often, a couple of letters produce a short enough list that you can quickly look through it and select the topic you want with the mouse. (As usual, double clicking a topic is equivalent to clicking the topic and then clicking the *OK* button.) If you have no idea of the appropriate help topic then you can use the *Full Text Search...* item in the Help menu. Type a subject into the *Word(s)* input box and then click on the *Search* button. Maple will list help topics in decreasing order of likely relevance, from which you can select one as before. There are other useful items in the Help menu.

Most worksheets accessed via the help system are special help worksheets, which are similar to normal worksheets except that they have a yellow background, they are read-only, and they have a help-browser section at the top. The relative sizes of the help text and browser sections can be changed by dragging the horizontal dividing line up and down. The browser displays up to five columns, which represent a hierarchy of help topics with up to five levels. Clicking on a topic will either show its sub-topics in the next column or its help text below if it has no sub-topics. The browser provides a complete overview of the entire help system and one way to navigate it. Help worksheets also contain hyperlinks that can be used to navigate through the help system; this works very much like a web browser.

1.5.2 Balloon Help

If you select the *Balloon Help* option in the Help menu then you can find out what any toolbar button or menu item does by placing the mouse pointer over it, and a cartoon-style text bubble will appear. This can be very useful until you are *completely* familiar with Maple.

1.5.3 Use of the mouse

One of the more obvious differences among Macintosh, Windows, and UNIX-like platforms is that normally the mice have, respectively, one, two, and three buttons. I will focus on the middle ground: Windows with a two-button mouse. The principles are the same on the other platforms, although it is well worth familiarizing yourself with precisely how the mouse support works on the platform(s) you use.

Mouse support in Maple is similar to that in other applications that use a GUI. But remember that it is the text cursor, which usually appears as a vertical bar, that indicates where keyboard actions take effect, not usually the position of the mouse pointer, although the mouse can be used to position the text cursor. Mouse actions take effect at the tip of the arrow-shaped normal-select mouse pointer or at the centre of the I-shaped text-select mouse pointer. Using the *left* mouse button (or the only button on a Macintosh):

- a single click positions the text cursor at the mouse pointer;
- a double click selects the word nearest to the mouse pointer;
- a triple click selects the whole paragraph of text, input, or output nearest to the mouse pointer.

Dragging the mouse (i.e., holding the *left* or only mouse button down while you move the mouse) selects arbitrary partial and/or complete lines of text as usual. Dragging after double or triple clicking changes the end-point of the selection to the position of the mouse pointer; it does not continue to select by words or paragraphs.

Once text has been selected, it can be deleted, cut, copied, or moved. Selected text is *deleted* when either of the *Delete* or *Backspace* keys is pressed, and ***deleted text is completely lost*** (except that the deletion can be undone). Selected text is *cut*, meaning that it is moved from the worksheet to the clipboard, by pressing the cut button on the toolbar (marked with a pair of scissors). Selected text is *copied* to the clipboard, leaving the original in the worksheet intact, by pressing the copy button on the toolbar (marked with two overlapping pages of text). Text stored on the clipboard can be *pasted* at the text cursor (which can be anywhere in the worksheet) by pressing the paste button on the toolbar (marked with a page of text on a clipboard). Clipboard operations can also be effected via the Edit menu or the shortcut keys shown on the Edit menu, and text can be shared with other worksheets and *other applications* via the clipboard, which is maintained by the underlying GUI and not by Maple itself.

Selected text can also be moved or copied by simply dragging it with the mouse to its new location; by default it is moved but if the *Control* (or *Command*) key is held down when the mouse button is released after the drag then it is copied. However, this method of manipulating text is unreliable in Maple because Maple does not show clearly where the text will go. (Some other applications show a text cursor to indicate the target position, but unfortunately Maple does not.) So, if there is any ambiguity use cut or copy followed by paste, which is unambiguous because it takes place at the text cursor. If a mouse drag operation puts the text in the wrong place, then I recommend undoing the operation and using cut or copy followed by paste instead.

These worksheet manipulations apply not only to normal text but also to Maple input and output, including complete mathematical expressions and parts of expressions (provided *Standard Math Notation* is selected on the Options menu), plots, and to some extent spreadsheets. Using the GUI to copy parts of Maple output expressions is not a good way to develop a formal computation, but it can be extremely convenient for interactive experimentation! If you copy mathematical output to a context where Maple input is required then the format is automatically converted, either to exactly the right format or to something close. Moreover, a chunk of Maple input, output, and text can be copied together from a Maple worksheet and pasted into another application, such as a text or e-mail editor, when it is automatically converted to text representation, which is extremely convenient for discussing mathematics in general (and Maple in particular) by e-mail!

Clicking the *right* mouse button causes a context menu to "pop up" that displays useful operations that relate to the location of the mouse pointer (i.e., to the context). On a Macintosh, a right mouse-button click can be emulated by holding the *Option* key and clicking the mouse button, so if you are using a Macintosh please interpret *right*-clicking to mean *Option*-clicking. I will add this proviso in parentheses sometimes (but not every time) when I refer to *right*-clicking in the rest of this book. If you are using the X Window System

Table 1.1: Useful keyboard shortcuts. (⇑ denotes the *Shift* key and *Cmd* denotes the Macintosh *Command* key.)

Function	Macintosh	Windows	UNIX
New Line	⇑-*Return*	⇑-*Return*	⇑-*Return*
Copy Selection to Clipboard	*Cmd-C*	*Ctrl-C*	*Ctrl-C*
Cut Selection to Clipboard	*Cmd-X*	*Ctrl-X*	*Ctrl-X*
Paste from Clipboard	*Cmd-V*	*Ctrl-V*	*Ctrl-V*
Delete Entire Object		*Ctrl-Delete*	*Ctrl-Delete*
Undo Last Action	*Cmd-Z*	*Ctrl-Z*	
Redo Last Action	*Cmd-Y*	*Ctrl-Y*	
Find	*Cmd-F*	*Ctrl-F*	⇑-*Ctrl-F*
Beginning of Line	*Cmd-Home*	*Home*	*Home*
End of Line	*Cmd-End*	*End*	*End*
Top of Worksheet	*Home*	*Ctrl-Home*	*Ctrl-Home*
Bottom of Worksheet	*End*	*Ctrl-End*	*Ctrl-End*
Next (Previous) Input	(⇑-) *Tab*	(⇑-) *Tab*	(⇑-) *Tab*
Next (Previous) Worksheet		(⇑-) *Ctrl-Tab*	
Split/Join Groups		*F3*/*F4*	*F3*/*F4*
Toggle Input/Text Mode		*F5*	*F5*

on a UNIX-like platform then clicking the middle mouse-button pastes from the clipboard into the document at the mouse pointer. Three-button mice used with Windows can usually be configured (via the Mouse Control Panel applet) so that the middle button performs some useful function, such as to paste from the clipboard. (However, this may prevent proper use of the middle button with an X Window server and with programs such as Emacs that have been ported from an X Window environment.) The wheel on a wheel mouse can also (somewhat uncomfortably) be used as a middle mouse button.

1.5.4 Useful keyboard shortcuts

The functions shown in Table 1.1 are either particularly useful or available only via the keyboard. For other keyboard shortcuts see the menus or the online help topic `worksheet,reference`. If the GUI is the X Window System on a UNIX-like platform then the precise details may depend on the window manager being used.

1.5.5 New lines and execution groups

Pressing the *Return* (or *Enter*) key starts a *new paragraph* if the cursor (the vertical bar that shows where new text goes) is within text or *executes* the *whole* of the current execution group and starts a new execution group if the cursor is *anywhere* within Maple command input. Holding down either of the *Shift* keys and pressing *Return* (or *Enter*) starts a new line in both text and Maple input, without starting a new paragraph or executing any code. (This is also the case in Microsoft Word, for example.) The difference becomes clear if display of invisible (non-printing) characters is turned on (via the ¶ toolbar button or the View menu).

It is important to format (i.e., lay out) Maple input tidily and consistently. Ideally, lines of code should not be allowed to wrap automatically onto subsequent lines (although this is fine for text), but instead should be broken at suitable points using *Shift-Return*. Each execution group corresponds to a "paragraph" of Maple code. Executing a group of code produces a new execution group below, whereas breaking a line of code does not. Execution groups can be joined (or split) via the Edit menu and inserted via the "[>" toolbar button (or the Insert menu). It is important that all *related* Maple code is in the *same* execution group, so that it is always all executed together (e.g., all lines of a loop including its initialization and final display statements). This can be achieved either by breaking lines between input statements using *Shift-Return* (rather than executing them using just *Return*), or by joining execution groups after executing them. (It is the continuity of the group range delimiter that is important, not the number of ">" prompts.)

The extent of an execution group is delimited by the vertical line at the left side of the screen like this:

```
[ > # This is a single execution group.
[   # This is a second line of input.
```

For normal use of Maple, I recommend that you do not delete execution group delimiters or disable display of execution group ranges.

However, I will show execution group delimiters only in this chapter and not in any subsequent chapters, partly because of the technical difficulty of showing long execution group delimiters and partly because they would look strange in a book. Execution groups will nevertheless be indicated unambiguously. An execution group will always begin with a single Maple prompt at the start of a line of Maple code and contain no further such prompts; it will end immediately after the last subsequent line of Maple code with no prompt or after the following output, if there is any.

The similar vertical lines with a ⊞ or ⊟ at the top delimit sections and sub-sections, which have no bearing on actual Maple computation. They are there to provide document structure and can be expanded and collapsed by clicking on the ⊞ or ⊟. These *widgets* are fairly standard in hierarchical views of data structures, such as file systems.

1.5.6 Opening and executing worksheets

When you open a worksheet it is *not automatically executed*, so that Maple is not in the same state that it was in when the worksheet was saved. You can re-execute the whole worksheet (or a selected region) by selecting the *Execute* item in the Edit menu. Executing the whole worksheet will put Maple back into a state similar to when the worksheet was saved. You can also execute each execution group in the worksheet individually (in whatever order you want). A common problem is to find that some computation does not work as expected because execution groups earlier in the worksheet have not been executed since the worksheet was opened.

You can also execute all the code in an execution group by putting the text cursor somewhere within the Maple input text (it does not need to be at the end) and pressing the *Return* or *Enter* key, or by *right*-clicking (*Option*-clicking on a Macintosh) the mouse over the code and selecting *Execute* from the pop-up context menu.

1.5.7 Application and document windows

Unless you have a fairly powerful computer, do not run any unnecessary applications together with Maple and do not run more than one copy of Maple at a time! Windows have three forms: *maximized* to fill the largest possible screen area, *minimized* as an icon or taskbar button, or an intermediate form in which windows can overlap each other. Unless you are lucky enough to have a very large display, it is normally best to maximize the Maple application window and also (if applicable) maximize the worksheet document window within Maple.

On Windows, Maple can open multiple worksheets within the main application window. It is normally best either to keep all worksheets maximized, and to switch among them by using the Window menu or *Ctrl-Tab* (which cycles through windows like *Alt-Tab* cycles through running applications), or to *tile* at most two worksheets (by using the Window menu). You might, for example, have two tiled worksheets, one an online help document and one a worksheet that you are developing. But beware that by default all worksheets opened within the same Maple application share the same kernel, so that they are equivalent to different pieces of the same worksheet. (On Windows and UNIX-like platforms it is also possible to start Maple in "parallel server" mode, which runs separate kernels, equivalent to running each worksheet in a separate Maple application.)

1.5.8 Averting disaster

Maple is a large and complicated piece of software and the GUI version in particular is a little fragile; it is not too difficult to perform some operation that Maple does not expect, which can cause the program to attempt an illegal operation and be unable to continue. This is sometimes called a program *crash*. There is also the possibility of a network failure (which matters if your files are stored remotely on a file server). If this happens, you will probably lose all the work that you have done since you last saved any open worksheets, so ***save each worksheet frequently!***

An easy way to protect yourself is to use the Maple *AutoSave* facility. This is controlled by a dialogue box accessed from the bottom item of the Options menu. You can turn it on and off (it is off by default) and select how frequently Maple should autosave your worksheet when it is on, the default interval being 3 minutes. The autosave file has "`_MAS`" appended to its filename (before the "`.mws`" extension), and it disappears when you explicitly save the worksheet but (one hopes!) remains if Maple crashes.

Maple may also crash if it runs out of memory, which may be more of a problem on a networked system than on a stand-alone system (because the network software takes up some of the memory). You can save memory by avoiding running unnecessary applications, closing unnecessary Maple document windows, clearing (unassigning) unnecessary variables, and deleting any unnecessary worksheet sections. (Just deleting parts of a worksheet does not clear variables.) Avoid creating large worksheets — use several smaller worksheets instead, and paste them together when they are finished. Alternatively, issuing the Maple command `restart` between unrelated computations is almost as effective — it will clear the memory used by the Maple kernel but preserve the worksheet.

There appears to be *only one way* to interrupt GUI versions of Maple and stop a computation, which is to click the *STOP* button on the toolbar. This may be necessary if you inadvertently start an "infinite" computation. You may have to wait a while for Maple to actually stop. Hence, do not turn off display of the toolbar — if you do then you will have no safe way at all to interrupt Maple!

If all else fails and your computer stops responding to the keyboard and mouse then try using the operating system facilities to interrupt Maple. For example, on Windows you can hold the *Alt* key and a *Ctrl* key down while you press the *Delete* key, which should give you access to the Windows Task Manager (although there may be better ways to do this depending on what version of Windows you are running, such as *right*-clicking the taskbar). From the Task Manager you should be able to terminate Maple. But you will lose any unsaved data and on older versions of Windows there is a risk that you will interrupt Windows itself and effectively restart the computer. If all else fails then either press the reset button on the computer, or switch the power off, wait briefly, and then switch it back on. (This may upset the operating system, but it should be able to recover.)

1.5.9 My worksheet shrank!

If everything within a Maple window suddenly shrinks in size then you probably accidentally changed the *Zoom Factor*. You can reset it by clicking on one of the "magnifying glass" buttons on the toolbar (the leftmost button resets the standard 100% magnification) or via the *Zoom Factor* sub-menu of the View menu. The problem arises because the keyboard shortcut *Ctrl-digit*, where *digit* is a digit between 0 and 6 inclusive, sets a particular magnification, and in particular *Ctrl*-0 and *Ctrl*-1 set magnifications less than 100%. These keyboard shortcuts are surprisingly easy to hit accidentally, e.g., when trying to press the right cursor key together with *Ctrl* to move to the right by a word!

1.5.10 Don't work in the dark

In most cases, while you are developing a Maple worksheet you should terminate all statements with a semicolon so that you can see the result of executing them. Once a section of code is working correctly, you might want to tidy it up by replacing top-level semicolons by colons and re-executing it. Alternatively, you can remove output using the *Remove Output* item at the bottom of the Edit menu.

1.5.11 Compatibility with Maple V

Maple V was the first version of Maple to support a worksheet interface and Maple 6 can open worksheets saved by all versions of Maple V at least as far back as Release 3 (and maybe earlier). However, while worksheets saved using Maple V Release 4 or 5 can (optionally) be automatically converted to Maple 6 syntax when they are opened, worksheets saved using earlier releases must be converted by hand. Once the syntax is correct, it is necessary or at least advisable in all cases to re-execute the worksheet, because output can be lost or mangled in the conversion. Occasionally further conversion by hand may be necessary to take account of more subtle changes to the Maple language. (In particular, it is essential to update the old use of double quote as the ditto operator to `%`. Maple may give a warning that this conversion may be necessary, but does not actually do it.) It is not possible to save worksheets in old formats. It may be possible to open Maple 6 worksheets in recent versions of Maple V, although any necessary syntax conversion will have to be done by hand.

The worksheet file format changed between Maple V Releases 3 and 4. Because of this, worksheet files for Maple V Release 4 and later versions have filename extension "`.mws`" whereas worksheet files for earlier releases had filename extension "`.ms`". Maple V Release 4 and later versions up to (at least) Maple 6 will open "`.ms`" files, whereas Maple V Release 3 and earlier releases cannot use "`.mws`" files at all.

1.6 Exercises: the worksheet interface

For these exercises, make sure that execution group ranges are displayed, i.e., that the *Show Group Ranges* option in the View menu is ticked. (In fact, I recommend keeping the Maple worksheet interface set this way if you intend to use it for any serious programming.)

Cut and paste

Enter, but do not execute, the following line of Maple:

```
[ > x + y;
```

Copy the statement and paste it back, then edit the + sign (only) in the copy to give the following line of Maple, and execute it by pressing the *Return* (or *Enter*) key:

```
[ > x + y;  x - y;
```

Practice using both the mouse and the cursor keys (the four arrow keys), and the Edit toolbar buttons, menu, and shortcut keys (listed to the right of the entries in the Edit menu). Learning to use these facilities now will save time later. Explore the context menus available by clicking the *right* mouse button (*Option*-clicking on a Macintosh) in different regions.

Do not type any new text other than spaces and newlines (*Shift-Return*) in the rest of this exercise. *Edit* the previous input so that each statement is on a separate line but both lines are still in the same execution group with only one prompt (>), *exactly as follows*, and execute the group again:

```
[ > x + y;
[   x - y;
```

Now split the execution group (see the Edit menu) so that each statement is in a separate group on its own, *exactly as follows*, and execute both groups one after the other. Note where the output appears now.

```
[ > x + y;

[ > x - y;
```

Finally, rejoin the two execution groups and then rejoin the two lines into a single text line, as it started out.

Execution groups

Enter and execute the following *two* execution groups, one after the other. Remember to end each line except the last in the second group with *Shift-Return* and then press *Return* or use the pop-up mouse context menu (*right-* or *Option*-click) to execute each group.

```
[ > i := 1:

[ > while i <= 5 do  # print the integers 1..5
      print(i);
      i := i + 1
    end do:
```

Now re-execute just the second group (e.g., try *right-* or *Option*-clicking the mouse somewhere within the input text) and notice that there is no output, because the loop initialization is missing (from the execution group) so that i already fails the test from the previous time the loop was executed. Join the two groups and re-execute the joined group. Check that now it always works correctly.

Finally, delete the whole execution group. You can do this either by selecting the group range delimiter by clicking it once with the mouse and then pressing *Ctrl-Delete* (or using the Edit menu, the only choice on a Macintosh), or by selecting the whole group, either by *double*-clicking the group range delimiter or by dragging the mouse diagonally from top left to bottom right within the group, and then pressing *Delete* (or *Backspace*).

Chapter 2

Review of Basics

This chapter provides a fairly rapid review of the basics that one needs to begin working with Maple. Computer languages have very precise rules about what constitutes valid input, which is the *syntax* of the language. A lot of the syntax revolves around the use of *special symbols* and I use this in the first section as a key to the main syntax of Maple. There are issues concerning the use of variables that are peculiar to computer algebra systems such as Maple, which I discuss in the second section. Experience of teaching Maple over a number of years has made me aware of a number of very common errors that catch inexperienced users, and I warn the reader about some of these in Section 2.3. Languages for symbolic and algebraic computing normally provide one or more list-like data structures that are nearly as fundamental as the primitive unstructured data types; Maple provides sequences, lists, and sets, which are distinct but closely related, and I discuss their use in Section 2.4. In the last section I introduce the use of literal or textual data. Typically, text plays a secondary role in high-level mathematical computation, but it plays a fundamental role at a lower level and in other application areas, and in fact it is possible to use Maple for quite sophisticated text processing, which is the subject of Chapter 14.

2.1 Maple syntax: special symbols

This section provides a rapid introduction to, or review of, Maple syntax from the somewhat unconventional viewpoint of special symbols. These are the symbols on a computer keyboard that are neither letters nor numbers; they include normal punctuation and also a few symbols that are not normally used in ordinary text. It is impossible to explain any one topic fully without reference to others, so a presentation of Maple cannot be linear and cannot avoid some forward references. The purpose of this section is to start the ball rolling. Most of the concepts introduced here will be explained in more detail later.

Maple uses all three types of quotes, all three types of brackets, and almost all of the other special symbols available on a standard ASCII keyboard, and it is important not to confuse their meanings. Unfortunately, the location of the symbols on a keyboard is not fixed; I will describe a common layout on full-size PC keyboards in the U.K. Moreover, the appearance of quotes depends on the font used to display them, so they may not appear on your screen exactly as printed in this book. If you have occasion to write Maple code by hand it is particularly important to write all characters carefully and distinctly, to distinguish upper and lower case (i.e., capital and small) letters, to distinguish letters from digits, and to distinguish the different types of quotes and brackets.

2.1.1 Variables, assignments, equations, procedures, and mappings

Perhaps the most fundamental operation in computer programming, as opposed to simple interactive "calculator-mode" computing, is giving names to objects for later use. Conventional low-level computation works with fixed locations in memory that can store different data. This is modelled in a programming language by analogy with mathematical terminology and the memory locations are referred to as *variables.* Storing data in a memory location is modelled as *assigning* a *value* to a variable, which replaces any data previously stored in the corresponding memory location. This low-level programming language terminology has been carried forward to higher level symbolic programming languages, even though the underlying operation may be much more complicated, because it is still a good abstract model of computation. However, in a symbolic context, assigning a value to a variables is also referred to as *binding* a variable to a value; the role reversal may reflect more accurately the underlying operation and the fact that now data values are more fundamental than variables.

In Maple, assignment is represented by the infix operator := (colon equal), which is used in the form:

```
variable := value
```

It cannot be reversed and the flow of information is always right to left. It may be useful to think of an analogy between the assignment operator and an arrow in which the colon represents the arrow head, and in fact in descriptions of algorithms it is quite common to use the notation "*variable* $\leftarrow$ *value*".

Many other programming languages (C, C++, Java) use just an equal sign to denote assignment, which is perhaps less satisfactory because it obscures the asymmetry of assignment. Maple uses the equal sign alone as an operator to construct an equation, which is simply an expression or data structure that in itself has no operational significance. There is clearly scope for confusion over assignments and equations and I will return to this important topic later.

The simplest class of *procedure* or *mapping* can be defined using an *arrow* syntax like that used in mathematics, which in Maple is constructed from the two special characters "-" and ">"; e.g., this is a squaring mapping:

```
> x -> x^2;
```

$$x \to x^2$$

This is the easiest way to write a simple function; internally it is a procedure:

```
> lprint(%):

  proc (x) options operator, arrow; x^2 end proc
```

It is simply an *expression* that represents an *anonymous* procedure and, just as for any other expression, a procedure or mapping is given a name by assigning it to a variable, e.g.,

```
> sqr := x -> x^2;
```

$$sqr := x \to x^2$$

2.1.2 Quotes

Forward quotes (')

The forward quote or apostrophe (') is usually to the right of the semicolon key. In Maple it is used to prevent one evaluation. *Evaluation* means replacing any assigned variables by their values and then *executing* any procedures (mappings) in an expression. When an expression enclosed in forward quotes is evaluated the effect is to remove the quotes but perform no further evaluation. Forward quotes can be "nested", as follows:

```
> a := b:  a, 'a', ''a'';
```

$$b,\ a,\ 'a'$$

```
> sin(Pi), 'sin'(Pi);
```

$$0,\ \sin(\pi)$$

```
> 'sin(Pi)', eval('sin(Pi)');
```

$$\sin(\pi),\ 0$$

The most common use of forward quotes is to undo or clear an assignment, by assigning the unevaluated name of a variable to itself, like this:

```
> a := b:
  a;  a := 'a';  a;
```

$$b$$
$$a := a$$
$$a$$

Backward quotes (`)

The backward quote or *backquote* (`) is usually at the top left of the keyboard. It is used in Maple to remove any special meaning or syntax from characters, words, or symbols and to make Maple treat them like letters. More formally, Maple treats *anything* enclosed in backward quotes as an *identifier* or *symbol*:

```
> `1 2 3`;  # an identifier
```

$$\mathit{1\ 2\ 3}$$

```
> 123;  # a number
```

$$123$$

```
> 1 2 3;  # an error!
```

```
Error, unexpected number
```

The above example shows how to include space characters in an identifier. It is quite common in mathematics to add a prime to the name of a variable to indicate a different but related variable. To do this in Maple it is necessary to enclose the whole identifier in backquotes every time it is used:

```
> `x'` := x;
```

$$x' := x$$

Note that backquotes have no effect on evaluation: to unevaluate an identifier that requires backquotes it is necessary to use both kinds of quotes with the backward quotes *inside* the forward quotes:

```
> `x'` := '`x'`';
```

$$x' := x'$$

Putting backquotes around an identifier that does not need them is not an error, it is just a waste of effort because Maple ignores them:

```
> `x` = x;
```

$$x = x$$

However, it is necessary to put backquotes around some special identifiers that are Maple *keywords* in order to use them outside their normal context, e.g.,

```
> and;
    Error, reserved word `and` unexpected
> `and`;
```

$$and$$

Double quotes (")

The double quote (") is usually on the "2" key (*Shift*-2). It is used in Maple (and most other programming languages) to construct *strings*. A string is a *text constant* consisting of anything enclosed within double quotes. (Note that a double quote is a single character that is not the same as two single quotes, even though the two may look similar in some fonts.) Common uses of strings in Maple are as plot titles and filenames, e.g.,

```
> title = "Plot of a very interesting graph.";
```

title = "Plot of a very interesting graph."

```
> file := "E:/Program Files/Maple 6/intro.mws";
```

file := "E:/Program Files/Maple 6/intro.mws"

```
> file := "E:\\Program Files\\Maple 6\\intro.mws";
```

file := "E:\\Program Files\\Maple 6\\intro.mws"

At first sight, there is little difference between the effect of backward quotes and double quotes and, in Maple, identifiers can currently be used almost anywhere that strings can. However, this is mainly for backward compatibility with versions of Maple up to Maple V Release 4, in which identifiers were used as strings and there was no distinct string type. The main difference is that identifiers can be used as variables whereas strings are constants just as numbers are constants. By definition, the value of a constant cannot be changed, so constants cannot have new values assigned to them. It is therefore much more reliable to use a string than an identifier to represent constant text; for example, it is quite likely that an identifier such as `x` will have a value assigned to it, to which it will evaluate by default, whereas the string `"x"` *cannot* evaluate to anything other than itself.

```
> "a string" := 123;
    Error, invalid left hand side of assignment
```

```
> 123 := 456;
```

```
Error, invalid left hand side of assignment
```

```
> `an identifier` := 123;
```

$$an\ identifier := 123$$

Another difference is that when strings are output automatically (or by `print` or `lprint`) they are surrounded by double quotes whereas identifiers are not surrounded by any quotes, which often looks neater and is probably the only situation in which use of an identifier to represent constant text might be justified. (It is the only situation in which I will do so, occasionally, in this book.)

```
> "a string", `another identifier`;
```

$$\text{"a string"},\ another\ identifier$$

Compatibility note: In versions of Maple up to Maple V Release 4, the double quote was used to represent the *ditto* operator which is now represented by the special character `%`.

2.1.3 Brackets

There are three different kinds of brackets — all of which consist of pairs of mirror-image symbols — that are used somewhat interchangeably for enclosing text in both prose and mathematics. ***However, in Maple they have distinct meanings that cannot be interchanged and it is a common error to forget this.***

Round brackets `( )`

Round brackets or *parentheses* `()` are usually on the "9" and "0" keys (*Shift*-9 and *Shift*-0). They are used for grouping expressions and to cause function (mapping) application. An expression within parentheses is evaluated and the result treated as a single object within a larger expression, e.g.,

```
> (a+b)*c <> a+b*c;
```

$$(a+b)\,c \neq a + b\,c$$

An expression of the form $f(x)$ causes the value of the variable f to be *applied* as a function or mapping to the value (in general) of the variable x (which may be an expression sequence), e.g.,

```
> f := g:  x := a,b:  f(x);
```

$$\mathrm{g}(a, b)$$

```
> unassign('f','x');
```

There is no other syntax for either expression grouping or function application in Maple, and *no other type of brackets can be used* for either purpose.

Square brackets `[ ]`

Square brackets `[ ]` are usually to the right of the "P" key. They are used *only* for constructing lists, e.g.,

```
> L := [a,b,c];
```

$$L := [a, b, c]$$

and for *indexing* or *subscripting*, which provides a way to *access* or *select* elements of structured objects (tables, arrays, lists, expression sequences, sets, and strings), e.g.,

```
> L[2];
```

$$b$$

```
> (a,b,c)[2];
```

$$b$$

```
> "abc"[2];
```

"b"

If the object that is indexed or subscripted is a *name* then the result is a new indexed or subscripted name that remains symbolic, e.g.,

```
> A[i];
```

$$A_i$$

```
> %[j];
```

$$A_{i\,j}$$

Some Maple functions accept subscripts to *qualify* them, as is common in mathematics; e.g., `log[b]` denotes logarithm to the base b thus:

```
> 'log[b](x)':  '%' = %;
```

$$\log_b(x) = \frac{\ln(x)}{\ln(b)}$$

Similarly, `D[i,j,...]` denotes partial derivation with respect to the ith, jth, ... arguments (see Chapter 9) and `evalf[d]` denotes approximate numerical evaluation to d significant figures of accuracy (see Chapter 5).

Compatibility note: In versions of Maple prior to Maple 6 the precision for `evalf` was supplied as an optional argument, so the function-call `evalf[d](x)` was written as `evalf(x,d)`. Maple 6 still accepts the old syntax for backward compatibility but the subscript syntax is recommended, because in Maple 6 `evalf` accepts a *sequence* of arguments to be numerically evaluated, which makes the old form ambiguous.

Curly brackets { }

Curly brackets or *braces* {} are usually to the right of the "P" key (*Shift*-[and *Shift*-]). They are used *only* for constructing sets, e.g.,

```
> S := {a,b,c};
```

$$S := \{a, b, c\}$$

(However, note that sets have a slightly special meaning when used in the context of the Maple type system, as we will see in Chapter 10.)

2.1.4 Mathematical symbols

Algebraic (+, -, *, /, ^)

The symbols +, -, *, /, ^ represent algebraic operators meaning "add", "subtract", "multiply", "divide", and "raise to the power", e.g.,

```
> a*x^2 + x/b - c;
```

$$a\,x^2 + \frac{x}{b} - c$$

The combination ** is a synonym for ^, but no other combination of these symbols is defined and so would be an error. If the expression that you are trying to construct is well defined then the error can be avoided by appropriate use of parentheses, e.g.,

```
> x^-2;
```

```
Error, '-' unexpected
```

```
> x^(-2);
```

$$\frac{1}{x^2}$$

Relational (=, <, >)

The symbols =, <, > represent relational operators meaning "equal to", "less than", and "greater than". They can be combined into the symbol pairs <>, <=, >= that represent relational operators meaning "not equal to", "less than or equal to", and "greater than or equal to", e.g.,

```
> 2*3 = 6;  evalb(%);
```

$$6 = 6$$

true

```
> 6 <= 5;  evalb(%);
```

$$6 \leq 5$$

false

No other combinations of relational operators are defined.

The meanings of these algebraic and relational operator symbols are fairly standard among programming languages, with the exception of = and <>; in C, for example, these are represented by == and !=, respectively. In Maple, all of these symbols just build expressions, the interpretation of which requires further instructions. For example, the determination of the *truth* of an equation or inequality may require "Boolean evaluation" by applying the function `evalb`, as illustrated above, although when relational expressions are used in programming contexts where a Boolean value is required then Boolean evaluation is automatic. Alternatively, *solving* an equation requires application of a suitable function such as `solve`. (There is a family of evaluation functions with names of the form eval*x* and a family of equation solving functions with names of the form *x*solve, where *x* represents various letters — see the online help for full details.)

Exclamation (!)

The exclamation sign (!) is used as in mathematics to indicate the factorial of a non-negative (or symbolic) integer, e.g.,

```
> 5!;
```

$$120$$

```
> n!;
```

$$n!$$

```
> (-2)!;
```

```
Error, the argument to factorial should be non-negative
```

It is the only algebraic operator that appears *after* its operand; Maple is unusual in supporting any such *postfix* algebraic operators at all. (However, the brackets [...] denoting selection and (...) denoting function application could also be regarded as postfix operators.)

At (@)

The "at" symbol (@) represents the functional composition operator, which is normally represented in mathematics as a small circle, e.g., $f \circ g$. Maple supports fairly sophisticated operator algebra, e.g.,

```
> (f@g)(x);
```

$$\mathrm{f}(\mathrm{g}(x))$$

A doubled "at" symbol (@@) represents repeated self-composition ("powering in operator space"); e.g., differentiating the sine function twice gives

```
> (D@@2)(sin);
```

$$-\sin$$

Dot (.)

The dot, period, point, or full-stop (.) is used in decimal fractions to separate the integer part from the fractional part, exactly as in normal mathematical usage. When this format is used in Maple it implies floating-point representation, which is normally approximate, e.g.,

```
> 1.234;
```

$$1.234$$

```
> 1/3:  % = evalf(%);
```

$$\frac{1}{3} = .3333333333$$

(Incidentally, the above example illustrates a useful trick for constructing an equation with a symbolic expression on the left and its value on the right; variations on this theme are possible using the many different eval*x* functions or the value function as appropriate to perform the evaluation.)

However, the dot is also often used in mathematics to indicate some kind of product; indeed, it is so commonly used to indicate the scalar product of two vectors that this is often called a "dot product". Maple also supports the use of the dot to indicate a product, which in general is non-commutative, meaning that it matters which way round it is performed:

```
> a.b <> b.a;
```

$$a \,.\, b \neq b \,.\, a$$

whereas the * operator is commutative:

```
> a*b = b*a;
```

$$a\,b = a\,b$$

If one of the multiplicands is a number then there may be ambiguity between the two meanings of dot. If the dot is preceded immediately (with no space) by an integer then it is part of a floating-point number; otherwise, it denotes multiplication. But multiplication by a number is always commutative and represented internally using *:

```
> 4 .x = x.4;
```

$$4\,x = 4\,x$$

```
> whattype(rhs(%));
```

$$*$$

The space after the first 4 in this example is necessary to avoid the first dot being interpreted as part of the floating-point number "4."; spaces elsewhere are optional.

Compatibility note: In previous versions of Maple, the dot represented the concatenation operator, which is now represented by `||`.

2.1.5 Programming symbols

I will present these symbols in the order in which they appear on my keyboard; your keyboard may be different!

Dollar ($)

The dollar symbol ($) represents the sequence or repetition *operator*. It has been largely superseded by the sequence *function* `seq`, although $ still provides a very convenient shorthand in a number of special situations, especially for constructing multiple derivatives, e.g.,

```
> diff(f(x), x$3);
```

$$\frac{\partial^3}{\partial x^3}\,\mathrm{f}(x)$$

However, I recommend avoiding use of any other currency symbol (unless you work in financial mathematics), since no currency symbol other than $ has any defined meaning in Maple. Other currency symbols will probably be treated like letters, but they may make your worksheets non-portable.

Percent (%)

The percent symbol (%) represents the *ditto* operator used to recall the value of the last expression evaluated by Maple, e.g.,

```
> Pi:  % = evalf(%);
```

$$\pi = 3.141592654$$

Here, "last" means "last in time", not "last in space", so that % may not give the value immediately before the expression containing % in a worksheet. If in doubt then re-execute the whole sequence of statements that lead to the value of %. Also, null values are ignored.

The ditto operator can be doubled or tripled to refer to the value of the expression 2 or 3 steps (but no further) back in time, e.g.,

```
> %%/4:  % = evalf(%);
```

$$\frac{1}{4}\pi = .7853981635$$

```
> a: b: c: %%%;
```

$$a$$

But beware that the error message arising from misuse of the ditto operator may not be very helpful:

```
> %%%%;
```

```
Error, missing operator or `;`
```

Ampersand (&)

The ampersand symbol (&) is used to indicate *operator* (as opposed to function) identifiers; e.g., this is the inert Maple non-commuting multiplication operator:

```
> a &* b <> b &* a;
```

$$a \mathbin{\&*} b \neq b \mathbin{\&*} a$$

and here is a new inert operator that I have just invented:

```
> a &splat b;
```

$$a \mathbin{\&\mathrm{splat}} b$$

Underscore (_)

The underscore or underline symbol (_) is treated as a letter and is allowed anywhere in an identifier without any need for backquotes. It can be used to make identifiers more readable, e.g.,

```
> x_prime;
```

$$x\_prime$$

However, an underscore at the start of an identifier is best avoided because Maple generates such identifiers automatically for its own purposes to represent unbound global variables, e.g.,

```
> solve(tan(x) + x);
```

$$\mathrm{RootOf}(\tan(\_Z) + \_Z)$$

Semicolon (;) and colon (:)

These symbols are used as statement terminators and separators. A top-level statement must be *terminated* in order to be executed; if it is terminated with a semicolon then any value(s) associated with it are displayed, whereas if it is terminated with a colon then (by default) they are not. Statements in statement sequences must be *separated* from each other. It makes little difference whether they are separated by semicolons or colons, but I recommend using semicolons for consistency with most other modern languages. Statement sequences are normally terminated by the `end` closing delimiter of a control structure and do not *need* to be terminated by a semicolon or colon. Unnecessary statement separators generate empty statements and so are allowed anywhere that a statement is allowed. One place where they are not allowed is between the closing parenthesis following `proc` and a declaration such as `local`.

A double colon is used to bind variables to data types, e.g., `x::numeric`.

Tilde (~)

The tilde symbol (~) should not generally be used for input; Maple uses it by default in output to indicate variables about which assumptions have been made, e.g.,

```
> x;
```

$$x$$

```
> assume(x, real);
```

```
> x;
```

$$x^{\sim}$$

```
> about(x);
  Originally x, renamed x~:
    is assumed to be: real
```

```
> x := 'x':  about(x);
  x:
    nothing known about this object
```

However, references to spreadsheet cells in Maple must begin with a tilde to distinguish them from other identifiers (see Chapter 13).

Hash (#)

The hash mark or sharp symbol (#) is used to introduce a comment within Maple program text or *source code*. The comment extends from the # character to the end of the line (ignoring any automatic wrapping of long text lines onto subsequent lines so as to fit within a window). Comments are ignored by Maple itself and are there for the benefit of other readers. When Maple code is saved in its compressed binary format, as the whole Maple library is, comments are not saved. This is the reason why no comments are visible if you look at the definition of a Maple library procedure.

```
> `This text is input`;  # but this is ignored
```

This text is input

Comments can also be included in Maple worksheets by using text regions, but comments introduced by the # character are the only way to include comments in plain Maple source code files. Even within worksheets, you may find # comments more convenient than text regions for annotating program code, although text regions are more convenient for longer comments between segments of code. (This book was written as a collection of Maple worksheets containing text regions interspersed with Maple execution groups to illustrate the text.)

Vertical bar (|)

The vertical bar symbol (|) is used doubled to *concatenate* or join identifiers or strings. More precisely, the left operand of the || operator must be either an identifier or a string, which determines the type of the result. The right operand can be any (scalar) expression, the value of which is appended to the left operand if possible, e.g.,

```
> a||"b";
```

$$ab$$

```
> "a"||b;
```

"ab"

Beware that the left operand is never evaluated. (However, the function `cat` concatenates the values of all its arguments, and generally its use is preferred over ||.) The || operator can be used when it is necessary to assign to a variable whose name must be constructed dynamically, since it can appear on the left of an assignment, e.g.,

```
> i := 3:  a||i := foo;
```

$$a3 := foo$$

This can save a lot of typing in interactive use of Maple. For example, if you have three expressions assigned to variables called `expr0`, `expr2`, and `expr5` (which is not an uncommon situation after some interactive experimentation) and you want to replace the values of these expressions by their numerators, you could use a loop such as this:

```
> for i in 0,2,5 do
      expr||i := numer(expr||i)
  end do:
```

If the right operand of || is a sequence or range then effectively the concatenation is mapped over the explicit or implicit sequence, like this:

```
> x||(0,2,5);
```

$$x0,\ x2,\ x5$$

```
> x||(0..3);
```

$$x0,\ x1,\ x2,\ x3$$

Note that the parentheses are necessary in the above two examples for Maple to interpret the expression correctly. This provides a convenient shorthand for (say) solving the above set of expressions for a set of related variables. In the following example I use the ditto operator to save typing the sequence of indices twice. There is no solution because I have not set up any actual problem to solve.

```
> 0,2,5:  solve({expr||%}, {x||%});
```

However, if you feel the need to use these techniques within a procedure (or module) then you should probably consider whether some other data structure, such as a table or array, might be more appropriate. Note that there may be two vertical bar symbols on your keyboard, one of which will not work as above and is best avoided. On my (U.K.) keyboard, the vertical bar symbol that does not work is on the top-left key and requires the use of the *Alt Gr* key to access it.

Backslash (\)

The backslash symbol (\) requires careful use and is generally best avoided. It plays much the same role in Maple as it does in C and many other programming languages, namely to function as an *escape character*. It can be used within strings and identifiers, and it *always* combines with the character immediately following it. The combination *x*, where *x* represents any character, is always replaced by a single character, depending on *x*.

If *x* is another backslash then the result is a backslash; i.e., a pair of backslashes \\ is interpreted as a single backslash \, as illustrated by the following *identifier*:

```
> `E:\\Program Files\\Maple 6\\intro.mws`;
```

$$E : \backslash Program\ Files \backslash Maple\ 6 \backslash intro.mws$$

However, the escape characters are preserved when a *string* is displayed automatically (or by `print` or `lprint`), and it may appear that no interpretation has taken place because the output syntax for strings is the same as the input syntax:

```
> "E:\\Program Files\\Maple 6\\intro.mws";
```

"E:\\Program Files\\Maple 6\\intro.mws"

To see how the string is *interpreted* it is necessary to use a lower level facility such as `printf`, e.g.,

```
> printf("E:\\Program Files\\Maple 6\\intro.mws"):
  E:\Program Files\Maple 6\intro.mws
```

The combination \n is interpreted as a *newline* character (see also Chapter 14), e.g.,

```
> `This identifier is\nsplit over two lines`;
```

This identifier is
split over two lines

This can be used to split plot title strings over two or more lines (see Chapter 3), although probably a better way is explicitly to break the string by holding the *Shift* key while pressing the *Return* key to force a line break (which also works for identifiers), e.g.,

```
> plot({sin,cos}, title =
     "The graph of the sine function superimposed
  on the graph of the cosine function");
```

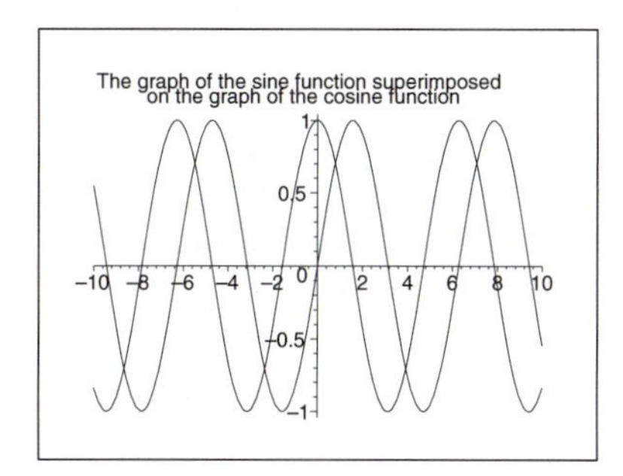

A few other characters lead to less useful special interpretations of $\backslash x$ but for all other characters $\backslash x$ is interpreted as x (i.e., effectively the backslash is ignored). This provides a way to include backward quotes within identifiers and double quotes within strings, although another way is to double the quote:

```
> `backquote\` in identifier` = `backquote`` in identifier`;
```

backquote` in identifier = backquote` in identifier

```
> "double quote\" in string" = "double quote"" in string";
```

"double quote\" in string" = "double quote\" in string"

However, other quotes within either backward or double quotes have no special significance and so do not need escaping:

```
> `Other quotes in an identifier: ' " `;
```

Other quotes in an identifier : ' "

```
> "Other quotes in a string: ' ` ";
```

"Other quotes in a string: ' ` "

Users of Microsoft Windows need to be especially aware of the interpretation of the backslash because Windows uses the backslash character as its directory separator, so the natural way for a Windows user to write a file pathname such as that above would be

```
> "E:\Program Files\Maple 6\intro.mws";
```

"E:Program FilesMaple 6intro.mws"

which does not work because the backslashes are ignored (as here) and/or interpreted, thereby corrupting the pathname. The best way to avoid this problem is to follow the Maple convention of using a forward slash as the directory separator on all platforms and let Maple interpret it appropriately for the current platform.

Comma (,)

The comma (`,`) is used primarily to build and extend expression sequences, and consequently also to build lists and sets:

```
> s := a,b,c;  # build a sequence
```

$$s := a,\ b,\ c$$

```
> s,d;  # extend the sequence
```

$$a,\ b,\ c,\ d$$

Dot (.)

The dot (`.`) not only is used to denote a floating-point constant or a non-commuting multiplication but also is doubled to denote a *range*. (Note that the range operator is formally a *pair* of dots, although Maple accepts two or more successive dots as equivalent to two dots.) When the end-points of the range are elements of a discrete ordered set, such as the integers, then a range corresponds to a sequence consisting of the end-points and all intermediate elements in *ascending* order. A range can be expanded explicitly into a sequence by applying the `$` operator, and the range is interpreted as the sequence by functions such as `seq`, `sum`, `add`, `product`, and `mul` and by the concatenation operator `||`, e.g.,

```
> $1..5;
```

$$1, 2, 3, 4, 5$$

Characters, which in Maple can only be represented as single-character strings, are also members of a discrete ordered set (where the order is essentially ASCII; see Chapter 14) and character ranges behave much like integer ranges, e.g.,

```
> $"a".."e";
```

"a", "b", "c", "d", "e"

If the end-points of a range are in decreasing order then the corresponding sequence is empty:

```
> $5..1;  $"e".."a";
```

When the end-points of a range are elements of a continuous set, such as the reals, then there is no corresponding sequence, but functions such as `plot` (see Chapter 3), `fsolve`, and `int` (see Chapter 5) interpret the continuous range as an interval, e.g.,

```
> int(sin(x), x = 0..Pi);
```

$$2$$

The *precise* interpretation of a range depends on the context in which it is used; for example, integration end-points can be integers, which are then interpreted as real numbers.

Question (?)

Finally, the question mark (?) provides "command-line" type access to the Maple help system: a question mark followed by any text causes Maple to look up the text in the help index. Space after the question mark is optional. For example:

```
> ? initialization
```

brings up the help page entitled *The Maple command and command line options*, which includes very useful information (for the more advanced user) about how to control the way Maple is initialized.

2.2 Assignment, evaluation, substitution, and simplification

These are concepts of fundamental importance in algebraic computation systems such as Maple, although most of them do not arise at all in more conventional programming languages.

2.2.1 Assignment and evaluation

Assignment can be thought of as giving a name or reference to some object so that it can be used later. For example, if we want to use a particular polynomial more than once we can assign it a name such as *poly*, like this:

```
> poly := x^2 + x + 1;
```

$$poly := x^2 + x + 1$$

The assignment operator is `:=`. It takes a reference, usually a name, on the left and an object on the right, and associates the reference with the object. As a side effect, the assignment ensures that the object is preserved in memory, so that it can continue to be referenced. Otherwise, the object may cease to exist after a while. This is necessary to prevent Maple running out of memory: any object that cannot be accessed will eventually be swept up by a process called *garbage collection*, which recovers any memory that is not being usefully employed.

This process can also be regarded as assigning a *value* to a *variable*. The variable is identified by the name on the left of the assignment and its new value is the object or expression on the right. This is the point of view normally taken in conventional programming languages. The object to which a name refers is called the value of the variable.

Whenever an assigned name is used (almost anywhere) it is automatically replaced with the object to which it refers, e.g.,

```
> poly;
```

$$x^2 + x + 1$$

```
> 2*poly - 1;
```

$$2\,x^2 + 2\,x + 1$$

One exception is when an assigned name appears on the left of an assignment, in which case the name is simply reassigned to the new object, thus:

```
> poly := y + z;
```

$$poly := y + z$$

```
> poly;
```

$$y + z$$

Another exception occurs when an assigned name appears within forward quotes (see above) and this provides the standard mechanism for undoing an assignment or *unassigning* a name:

```
> poly := 'poly';
```

$$poly := poly$$

The process of replacing an assigned name by the object to which it refers, or equivalently of replacing a variable by its value, is called *evaluation.* Variables are normally evaluated automatically, except when they appear on the left of assignments. Hence, a small modification can be made to the value of a variable by using it on both sides of an assignment: its value is used on the right but its name is used on the left. Of course, this makes sense only if the variable had previously been assigned. So, in the absence of appropriate *initialization*, this will produce an error:

```
> var := var + 1;
```

```
Error, recursive assignment
```

But after initialization, the above assignment is the standard idiom for incrementing the value stored in a variable, which is particularly useful within a loop:

```
> var := 0;
```

$$var := 0$$

```
> var := var + 1;
```

$$var := 1$$

```
> var := var + 1;
```

$$var := 2$$

There is an ambiguity in languages such as Maple about the concept of a variable. Consider the polynomial example again:

```
> poly := x^2 + x + 1;
```

$$poly := x^2 + x + 1$$

The name *poly* identifies a variable used in the programming sense as a name for some object or data structure, whereas the name x identifies a variable used in the mathematical sense as an *unknown* or *indeterminate.* Some newer computer algebra systems (such as GAP) make a distinction between these two kinds of variables, whereas Maple and older computer algebra systems do not. There are advantages and disadvantages to both approaches and what is important is to understand the ramifications of the approach taken by the system you are using. This is intimately tied up with the details of evaluation.

In Maple, a variable that has no value assigned to it evaluates to itself, e.g.,

```
> x;
```

$$x$$

This is how it is possible for Maple variables to play a double role. In conventional programming languages it is normally an error to try to use the value of a variable that has no assigned value. Every time an expression is used by Maple it is re-evaluated and all variables are replaced by their values *at that time*. This allows a complicated expression to be built up in stages, e.g.,

```
> poly := a*x^2 + b*x + c;
```

$$poly := a\,x^2 + b\,x + c$$

```
> x := y + z;
```

$$x := y + z$$

```
> poly;
```

$$a\,(y+z)^2 + b\,(y+z) + c$$

To make this replacement of the value of `x` permanent it is necessary to assign the current *value* of `poly` back to the *variable* `poly`. Otherwise, if the value of `x` is cleared (or changed) then the variable `poly` will have its old value again (or a different value), e.g.,

```
> x := 'x';
```

$$x := x$$

```
> poly;
```

$$a\,x^2 + b\,x + c$$

By default, Maple performs *full* evaluation (of global variables) and so follows a chain of evaluations to the end (i.e., until there is nothing more that can be evaluated). In mathematical parlance, the above three lines of Maple might be presented like this:

- Consider the polynomial $a\,x^2 + b\,x + c$.
- Let x have the value $y + z$.
- Then the polynomial has the value $a\,(y+z)^2 + b\,(y+z) + c$.

One disadvantage to this view of variables is that you can replace indeterminates *accidentally* by assigning to them. A language that kept the two kinds of variables distinct would prevent such errors. (There was a related problem in earlier versions of Maple that used identifiers to represent text constants!)

Evaluation can be forced by explicitly applying the standard function `eval`, which is occasionally necessary as we will see below. By default, `eval` performs full repeated evaluation to a depth of two delayed evaluations, but the depth of evaluation can be controlled by supplying a positive integer as an optional second argument.

Evaluation of local variables

The comments above apply to global variables, whereas local variables in procedures and modules (see Chapters 8 and 15) are *not* repeatedly evaluated by default. I will use a module here for purposes of illustration.

Provided `poly` is global then it is evaluated indirectly as it would be in "top-level" code:

```
> module()
      local a,b,c,x,y,z;
      global poly;
      poly := a*x^2 + b*x + c;
      x := y + z;
      print(poly)
  end module:
```

$$a\,(y+z)^2 + b\,(y+z) + c$$

But if `poly` is local then it is not evaluated indirectly:

```
> module()
      local poly,a,b,c,x,y,z;
      poly := a*x^2 + b*x + c;
      x := y + z;
      print(poly)
  end module:
```

$$a\,x^2 + b\,x + c$$

Full evaluation of any variable can be forced by explicitly applying the function `eval`:

```
> module()
      local poly,a,b,c,x,y,z;
      poly := a*x^2 + b*x + c;
      x := y + z;
      print(eval(poly))
  end module:
```

$$a\,(y+z)^2 + b\,(y+z) + c$$

But usually procedure and module bodies are written so that only direct evaluation is required. In the present example, this just means exchanging the two assignments:

```
> module()
     local poly,a,b,c,x,y,z;
     x := y + z;
     poly := a*x^2 + b*x + c;
     print(poly)
  end module:
```

$$a\,(y+z)^2 + b\,(y+z) + c$$

2.2.2 Evaluation and substitution

In mathematical usage there is often little distinction between *evaluate* and *substitute*: "evaluate the expression at $x = a$" and "substitute $x = a$ into the expression" mean essentially the same. As illustrated above, we can do this in Maple by assigning the value a to the variable x and then re-evaluating the expression (whatever it might be). But this affects *all* expressions involving x that are evaluated after the assignment to x has been made. Such an effect is called *global.* If we just want to evaluate this *one* expression at $x = a$ then this effect is called *local* (to that expression). One way is to *unassign* x immediately after its value has been used in the expression, but often a much better way to perform such local evaluation is to use the Maple function `eval` like this:

```
> poly := a*x^2 + b*x + c;
```

$$poly := a\,x^2 + b\,x + c$$

```
> eval(poly, x = y + z);
```

$$a\,(y+z)^2 + b\,(y+z) + c$$

Note that this returns a changed *copy* of the value of `poly`; it does not change the value of `poly` itself:

```
> poly;
```

$$a\,x^2 + b\,x + c$$

nor the value of `x`:

```
> x;
```

$$x$$

To change the value of the variable `poly` permanently it is necessary to *assign* a new value to it thus:

```
> poly := eval(poly, x = y + z);
```

$$poly := a\,(y+z)^2 + b\,(y+z) + c$$

```
> poly;
```

$$a\,(y+z)^2 + b\,(y+z) + c$$

There are other ways to perform local substitution in Maple by using the function `subs` and its relatives. They have two main differences from `eval`: the syntax is reversed so that the substitution equation(s) come first and the expression into which the substitution is to be made comes last, and the result is not evaluated. For example, the above substitution into the value of `poly` could also be made like this:

```
> subs(x = y + z, poly);
```

$$a\,(y+z)^2 + b\,(y+z) + c$$

This does not illustrate the lack of evaluation, since there is nothing to evaluate in this example. But the following example does illustrate the fact that the result of `eval` has been evaluated:

```
> eval(1 - sin(theta), theta = Pi/2);
```

$$0$$

whereas the result of `subs` has not:

```
> subs(theta = Pi/2, 1 - sin(theta));
```

$$1 - \sin(\frac{1}{2}\,\pi)$$

and subsequent (implicit or explicit) evaluation may be necessary, e.g.,

```
> %;
```

$$0$$

Evaluation can always be forced in Maple by using the function `eval` with only one argument. It is not often necessary, but it is more elegant to explicitly `eval` the result of the `subs` above than to implicitly evaluate it by recalling its value. Hence, `eval(subs(...))` is a common idiom, e.g.,

```
> eval(subs(theta = Pi/2, 1 - sin(theta)));
```

$$0$$

It is better to use eval with a second argument for this particular example, but here is an example where evaluation of the result returned by a close relative of subs is necessary:

```
> subsop([2,1]=Pi, [sin(a),sin(b),sin(c)]);
```

$$[\sin(a), \sin(\pi), \sin(c)]$$

```
> eval(subsop([2,1]=Pi, [sin(a),sin(b),sin(c)]));
```

$$[\sin(a), 0, \sin(c)]$$

The function subsop ***subs***titutes ***op***erands according to their location within an expression, expressed numerically as an index (exactly as used by the function op). This invocation of subsop picks operand 2 of the list, namely $\sin(b)$, and then picks operand 1 of that, namely b, and substitutes for it the value on the right of the equation, namely π, but the result is not evaluated.

Procedure bodies are not evaluated when they are defined, only when they are executed, as illustrated by the difference between the values of the following two inputs:

```
> sin(Pi);  # normal evaluation
```

$$0$$

```
> x -> sin(Pi);  # procedure definition
```

$$x \to \sin(\pi)$$

```
> %();  # and execution by application
```

$$0$$

2.2.3 Multiple substitutions

We have seen that the function eval accepts an equation as an optional second argument and the function subs requires an equation as its first argument; both functions also accept a list or set of equations, in which case they perform all the substitutions specified simultaneously, e.g.,

```
> eval(a+b+c, {a=A, b=B, c=C});
```

$$A + B + C$$

```
> subs({a=A, b=B, c=C}, a+b+c);
```

$$A + B + C$$

In fact, the function `subs` accepts an arbitrary number of arguments and substitutes all but the last into the last *sequentially*. Occasionally, it is important to specify either sequential or simultaneous substitution; the difference is illustrated by the following trivial example:

```
> subs(a=b, b=c, a);  # sequential
```

$$c$$

```
> subs({a=b, b=c}, a);  # simultaneous
```

$$b$$

However, `eval` does not support multiple *sequential* substitutions:

```
> eval(a, a=b, b=c);

   Error, wrong number (or type) of parameters in function eval
```

2.2.4 Simplification

Simplification is the process of rewriting an expression into an alternative form that is in some (technical) sense "simpler". It is easy to confuse it with evaluation, but the two are quite distinct. If a change is not prevented by forward (unevaluation) quotes then it is simplification rather than evaluation. Maple performs some "low-level" simplification automatically and this *cannot* be prevented. (Note that the forward quotes in the following examples are there solely to confirm that the changes shown are due to simplification rather than evaluation; the quotes would not be required in normal usage.) For example, Maple always simplifies purely numerical algebraic expressions (and sub-expressions of more general expressions):

```
> '1 + 2*3/4';
```

$$\frac{5}{2}$$

It always collects like terms in polynomials:

```
> 'x + x + 2*x^2 - x^2';
```

$$2\,x + x^2$$

It always simplifies sets into a canonical order with no duplicates:

```
> '{b,a,b,a}';
```

$$\{a, b\}$$

It always simplifies Boolean expressions:

```
> 'a > b and true';
```

$$b < a$$

It always simplifies procedure bodies (which are never evaluated at the time they are defined even without the quotes):

```
> 'x -> 2/4*x + x';
```

$$x \to \frac{3}{2}x$$

```
> 'proc(x) 2/4*x + x end proc';
```

$$\textbf{proc}(x)\, 3/2 * x\, \textbf{end proc}$$

Sometimes, procedures defined using arrow syntax are simplified to the point where they are no longer *manifestly* procedures, although the following example is perfectly correct and illustrates the fact that it is pointless to wrap another procedure around what is already a procedure:

```
> 'x -> sin(x)';
```

$$\sin$$

```
> type(%, procedure);
```

$$\mathit{true}$$

```
> 'proc(x) sin(x) end proc';
```

$$\textbf{proc}(x) \sin(x)\, \textbf{end proc}$$

Maple does not automatically perform "higher level" simplifications, such as expanding polynomials or putting rational *expressions* (as opposed to numbers) over a common denominator, which must be effected by explicitly calling an appropriate simplification function. Often, the general purpose simplifier function called `simplify` will suffice, but occasionally it will simplify too much and it is better to apply a more specific simplifier function such as `expand` or `normal`, e.g.,

```
> x*(x+1) - (x^2+x);
```

$$x\,(x+1) - x^2 - x$$

```
> simplify(%), expand(%);
```

$$0,\ 0$$

```
> 1/(x*(x+1)) - 1/(x^2+x);
```

$$\frac{1}{x\,(x+1)} - \frac{1}{x^2+x}$$

```
> simplify(%), normal(%);
```

$$0,\ 0$$

2.3 Common errors

All programming languages have their specific pitfalls, and Maple is no exception. Here are a few to beware of.

2.3.1 Juxtaposition

This means "placing next to each other". Juxtaposition is commonly used in mathematics to indicate either multiplication or application of an elementary transcendental function. However, this is clearly ambiguous and the precise meaning is context dependent; e.g., "$x\ y$" *probably* means "$x \times y$" and "$\sin\theta$" *probably* means "$\sin(\theta)$". Multiplication and function application are entirely different operations, and you must ensure that you understand which you mean. A programming language must either have well-defined rules about what juxtaposition means or must disallow it. Maple disallows it. If you juxtapose two identifiers with no space between them, then Maple will interpret this as a single identifier; if you leave space between them, then Maple will regard it as a syntax error, which means that you have broken Maple's grammatical rules.

```
> 2x;   # meant 2*x?

    Error, missing operator or `;`

> x y;   # meant x*y?

    Error, missing operator or `;`
```

```
> sin theta;  # meant sin(theta)?
```

```
Error, missing operator or `;`
```

Neither multiplication nor function application can be implied by juxtaposition in Maple; it is necessary to use an explicit multiplication symbol such as * for every multiplication and explicit parentheses () for every function application. Thus, it is necessary to write:

```
> 2*x, x*y, sin(theta);
```

$$2\,x,\ x\,y,\ \sin(\theta)$$

Note that automatic syntax correction (the button on the context bar labelled with a tick in parentheses) interprets juxtaposition as multiplication, which may not be the logically correct choice even though it is syntactically correct! Moreover, Maple does not show the multiplication operator in its typeset-style (pretty-printed) output but indicates multiplication only by a space, as we see above, which is misleading because it does not accept this for input!

2.3.2 Upper and lower case letters

The Maple language is case sensitive, which means that spelling in Maple includes the correct choice of upper case (capital) or lower case (small) letters. There is no fixed convention, but it is frequently the case that a Maple function spelt with an initial capital letter is an *inert* version of the same function spelt with all lower case letters. Very few predefined Maple identifiers use all upper case. Inert functions perform no computation, but are often recognized by Maple and displayed specially, e.g.,

```
> Int(x,x) = int(x,x);
```

$$\int x\,dx = \frac{1}{2}\,x^2$$

To avoid some potential confusion, Maple displays an inert integral slightly differently from an unevaluated active integral:

```
> Int(f(x),x) = int(f(x),x);
```

$$\int \mathrm{f}(x)\,dx = \int \mathrm{f}(x)\,dx$$

The inert integral on the left above is input as `Int` and output with the integral sign and the differential symbol d displayed in black, whereas the active integral on the right above is input as `int` and output with the integral sign and the differential symbol displayed in blue. Of course, these colour distinctions are lost in this book as they would be in any black-and-white printout of a Maple worksheet, so you need to type some integrals into a real

Maple worksheet to see the difference! A similar distinction applies to the display of inert and active repeated sums and products.

The purpose of the standard function `value` is to convert various standard inert functions into their active counterparts — see the online help for a full list of the functions that it converts. For example, another (and generally better) way of producing the first equation above is this:

```
> Int(x,x):  % = value(%);
```

$$\int x\,dx = \frac{1}{2}\,x^2$$

But note that `value` is not interchangeable with `eval`. Applying `eval` to an inert function will have no effect at all and similarly using `value` where `eval` is required works only indirectly because `value` is equivalent to a special case of `eval` with an appropriate substitution equation as its second argument.

Maple symbolic constants must also be spelt correctly; in particular, the circular constant π must be spelt `Pi`, with an initial capital. Unfortunately, the worksheet interface displays the identifier `pi` in the same way as `Pi`, but Maple does not interpret `pi` to mean the circular constant:

```
> pi <> Pi;
```

$$\pi \neq \pi$$

```
> evalf(pi) <> evalf(Pi), sin(pi) <> sin(Pi);
```

$$\pi \neq 3.141592654, \sin(\pi) \neq 0$$

2.3.3 The exponential function and the letter e

The letter e has no special significance in Maple and the only way to obtain e^x, where e is the base of natural logarithms, is to use the exponential *function* `exp`:

```
> exp(x) <> e^x;
```

$$\mathbf{e}^x \neq e^x$$

Hence, the only way to input the "exponential number" (the base of natural logarithms) is as `exp(1)`.

```
> exp(1):  % = evalf(%);
```

$$\mathbf{e} = 2.718281828$$

To avoid confusion between this symbolic constant and the identifier e, Maple displays the exponential number in bold face, which is distinguishable from the italic face used to display normal identifiers:

```
> exp(1) <> e;
```

$$\mathbf{e} \neq e$$

If you really want to be perverse and obtain e^x by using the ^ operator you can do it as follows

```
> simplify(exp(1)^x);
```

$$\mathbf{e}^x$$

but I strongly recommend using `exp(x)` directly.

2.3.4 Assignments and equations

It is *essential* not to confuse these two structures.

The assignment operator `:=` (colon equal) associates a *value* with a *variable*, like this:

```
> variable := value;
```

$$variable := value$$

After such an assignment, `variable` evaluates to *value*, whereas *value* is unaffected:

```
> variable;
```

$$value$$

```
> value;
```

$$value$$

Hence, an assignment is not symmetrical: it associates the value on the right with the variable on the left.

The equal operator `=` just constructs an equation, which is an expression or data structure, like this:

```
> expression_a = expression_b;
```

$$expression\_a = expression\_b$$

Constructing an equation has no effect on the expressions on either side of the equation, and it does *not* perform any assignment:

```
> expression_a;
```

$$expression\_a$$

```
> expression_b;
```

$$expression\_b$$

The mathematical significance of an equation is symmetrical; it is a relation that asserts the logical proposition that the objects on either side have the same value. An equation has the same status as any other expression, such as a list or a sum.

The main purpose of an equation is to build a data structure to be interpreted by some Maple function. The function `evalb` determines the truth of the equation as a logical proposition; the function `solve` and its relatives attempt to solve for the values of variables that make the equation true; the function `assign` assigns the value on the right to the value on the left, thereby essentially converting the equation into an assignment (except that the value on the left is evaluated).

2.4 Sequences, lists, and sets

A data type of fundamental importance in Maple is the *expression sequence*, which I will usually abbreviate to just *sequence*. It is represented as a collection of values separated by commas, like this:

$$a, b, c$$

Sequences frequently arise as parts of other structures. A *list* is represented as a sequence enclosed in *square brackets*, like this:

$$[a, b, c]$$

and a *set* is represented as a sequence enclosed in *curly braces*, like this:

$$\{a, b, c\}$$

The arguments of a function also constitute a sequence, e.g.,

$$\mathrm{f}(a, b, c)$$

Lists, sets, and (unevaluated) functions in Maple can all be regarded as sequences that have been *wrapped* in different ways, but the underlying naked sequence behaves the same way in all cases. This is why sequences are of such central importance in Maple, although there is nothing quite the same in most other languages.

2.4.1 Parallel assignment

An expression sequence is essentially a way of combining several values into a single object in a way that is unambiguous and can be used in any way a single value could be used *except* as an element of another sequence, in which case the two sequences get immediately merged. This has advantages and disadvantages, but it is essential to remember that it happens. As an additional convenience, Maple allows sequences to appear on both sides of an assignment operator, which means that the elements of the two sequences are assigned in parallel. Hence,

```
> a,b,c := d,e,f:
```

is a shorthand for:

```
> a := d:
  b := e:
  c := f:
```

Here is a more useful example:

```
> x1, x2, x3 := x, x^2, x^3:
  x1, x2, x3;
```

$$x,\, x^2,\, x^3$$

Unfortunately, it does not appear to be possible to construct dynamically the sequence of variables that appears on the left:

```
> x||(1..3) := x, x^2, x^3;
```

```
Error, invalid left hand side in assignment
```

This sequence assignment facility is probably most useful with a function that returns a short sequence of values on the right. It is also useful for exchanging or "rotating" values, which would otherwise require the use of temporary variables, e.g.,

```
> a,b,c := 1,2,3;
```

$$a,\, b,\, c := 1,\, 2,\, 3$$

```
> a,b,c := b,c,a;
```

$$a,\, b,\, c := 2,\, 3,\, 1$$

```
> unassign('a','b','c');
```

2.4.2 Functions and their argument sequences

Mathematically, a *function* operates on an element of its *domain* and returns an element of its *range*. The domain element is called the *argument* of the function, and the range element is called the *value* returned by the function. More generally, a function can operate on several domains at once, in which case it takes more than one argument. The arguments of a function are just a sequence enclosed in *parentheses* (*round brackets*), like this:

$$\mathrm{f}(a, b, c)$$

It is a moot point whether this is regarded as a function that operates on three domains in parallel or one three-dimensional domain, the elements of which are represented by three-element sequences. Maple does not care about your point of view; if you care then it is up to you to apply your point of view consistently.

Maple is a highly dynamic language, which is illustrated by the fact that the argument sequence of a function can be constructed dynamically like this:

```
> s := a,b,c:  f(s);
```

$$\mathrm{f}(a, b, c)$$

Just occasionally this technique is useful, although it would not be possible (or not so straightforward) in most languages.

2.4.3 Automatic simplification; inert functions

Expression sequences, lists, and function argument sequences behave very similarly. Their elements are *never automatically* changed (except in the case of a `NULL` element). By contrast, the elements of a set *are* automatically changed to put them into a *standard* (*canonical*) order and eliminate any duplicate elements. Compare the following:

```
> b,c,b,a;
```

$$b, c, b, a$$

```
> [b,c,b,a];
```

$$[b, c, b, a]$$

```
> f(b,c,b,a);
```

$$\mathrm{f}(b, c, b, a)$$

```
> {b,c,b,a};
```

$$\{a, b, c\}$$

These changes (called *simplifications*) in the case of a set are necessary so that the Maple definition of a set matches the standard mathematical definition and Maple recognizes two sets as identical if they are mathematically identical: Maple ignores repeated elements by explicitly deleting them and ignores the order of elements by forcing them into a standard order. The standard order is fixed only within one Maple session and will generally differ from one session to another. This is a general feature of Maple. It was designed this way for speed of execution, and the ordering of objects is based on their physical locations (addresses) in memory. Hence, no Maple computation should ever rely on the ordering of elements within a set. (Some other computer algebra systems use orderings that are reliably fixed and do not vary between sessions.)

An *inert* function is one that does not evaluate or simplify to something else. Standard Maple functions with capitalized names are often (but not always) inert; e.g., `Sum` and `Product` are inert:

```
> Sum(r, r = 1..5), Product(r, r = 1..5);
```

$$\sum_{r=1}^{5} r, \prod_{r=1}^{5} r$$

but `Re` and `Im` are not:

```
> Re(2 + 3*I), Im(2 + 3*I);
```

$$2, 3$$

A function is inert unless it is a built-in active function or you have defined an evaluation rule (a procedure body) for it, thus functions are inert by default, as illustrated earlier.

A *list* is equivalent to an *inert* function denoted by a special square bracket syntax, so most operations on the elements of a list apply also to the arguments of a function. By contrast, a *set* is equivalent to an *active* function denoted by a special curly brace syntax that simplifies its arguments by sorting them and discarding duplicates and returns a modified version of the input form.

2.4.4 Functions and operators

The only difference between a *function* and an *operator* is the syntax and nomenclature used: the *arguments of an operator* are called *operands* and no parentheses are required to delimit them. In some languages the same symbol can be used as either a function or an operator but, in Maple, backquotes are required to use an operator symbol as a function identifier (unless it begins with `&`; see later); e.g., the easiest way to add or multiply all the elements of a sequence is to apply the `+` or `*` operator as a function to the sequence, like this:

```
> `+`(a,b,c);
```

$$a + b + c$$

```
> `*`(a,b,c);
```

$$a\, b\, c$$

Hence, many Maple functions, such as op and nops, can be applied uniformly to expressions, lists, sets, and functions (but not directly to sequences) to extract or count their operands (hence, the names op and nops), elements, or arguments.

2.4.5 Operations on sequences, lists, and sets: op

Most operations on lists and sets proceed via the sequence form, which is extracted by applying the standard Maple function op:

```
> op([b,c,b,a]);  # list
```

$$b,\ c,\ b,\ a$$

```
> op(f(b,c,b,a));  # function
```

$$b,\ c,\ b,\ a$$

```
> op({a,b,c});  # set
```

$$a,\ b,\ c$$

The only difference between a sequence and a list is that a list is treated as a single object, and it is probably best to think of sequences and lists as alternative forms of the same basic object. A sequence of sequences is just a longer sequence, in which the sub-sequences lose their identity, but in a sequence (or list) of lists the lists (or sub-lists) retain their identity:

```
> (a,b,c), (a,b,c);
```

$$a,\ b,\ c,\ a,\ b,\ c$$

```
> [a,b,c], [a,b,c];
```

$$[a,\ b,\ c],\ [a,\ b,\ c]$$

This shows how to *append* or join two sequences, or equivalently how to attach elements to either end of a sequence (append to the end or prepend to the beginning):

```
> s := a,b,c:  t := d,e,f:
  s, t;  # appending or joining sequences
```

$$a, b, c, d, e, f$$

```
> s, z;  # appending an element
```

$$a, b, c, z$$

```
> z, t;  # prepending an element
```

$$z, d, e, f$$

Hence, the comma is not only the symbol used to separate the elements of sequences but also the operator used to construct sequences. A single element is equivalent to a sequence of one element in almost all circumstances.

The same operation on a *list* (or set or function argument sequence) must be performed via the *sequence form*, obtained using the op function:

```
> L1 := [a,b,c]:  L2 := [d,e,f]:
  [op(L1), op(L2)];  # appending or joining lists
```

$$[a, b, c, d, e, f]$$

```
> [op(L1), z];  # appending an element
```

$$[a, b, c, z]$$

```
> [z, op(L2)];  # prepending an element
```

$$[z, d, e, f]$$

2.4.6 Operations on elements of sequences, lists, and sets: op, [ ]

When a sequence needs to be passed as a *single argument* to a function, it must be *packaged* or *wrapped* as a list. For example, the op function can be applied to a sequence only in list form:

```
> S := a,b,c:  op(2,S);
```

```
Error, wrong number (or type) of parameters in function op
```

```
> op(2,[S]);
```

$$b$$

Note, however, that the selection or indexing operation can be applied directly to a variable that evaluates to a sequence:

```
> S[2];
```

$$b$$

More generally, any operation except passing it as a single argument to a function can be applied to a sequence without wrapping it as a list, and naked sequences can be used within loop control structures, as we will see later. The above selection operation is equivalent to this more explicit (but rather pointless) selection:

```
> (a,b,c)[2];
```

$$b$$

The general way to *change* an operand of an expression or an element of a data structure specified by its *index* or position number within the expression or structure is to use the `subsop` function, which returns a *copy* of the original data structure with the specified operand or element changed. The function `subsop` is like `subs` except that the index number of the operand or element to be substituted goes on the left of the equation, e.g.,

```
> L := [a,b,c];  subsop(2=B, L);
```

$$L := [a, b, c]$$
$$[a, B, c]$$

Alternatively, in the case of a list (or an array or table) only, direct assignment can be used to change an element of the original data structure, e.g.,

```
> L[2] := B;  L;
```

$$L_2 := B$$
$$[a, B, c]$$

(However, direct assignment to list elements is generally best avoided and is only allowed on small lists anyway for efficiency reasons.) Note that `subsop` returns a *copy* of the original data structure with the element changed, whereas direct assignment changes the element in the *original* data structure. Neither operation can be performed directly on a sequence, which must first be packaged as a list and then unpackaged again using `op`, e.g.,

```
> S := a,b,c;  op(subsop(2=B, [S]));
```

$$S := a, b, c$$
$$a, B, c$$

```
> S[2] := B;
```

```
Error, cannot assign to an expression sequence
```

A sub-sequence of elements or a sub-list can be selected using a *range*:

```
> L := [a,b,c]:
  op(2..3, L);
```

$$b, c$$

```
> L[2..3];
```

$$[b, c]$$

Note the important different that `op` *always* returns a sequence, whereas selection from a sequence, list or set using a range in square brackets returns a structure of the same data type as that to which it was applied, e.g.,

```
> (a,b,c)[2..3];
```

$$b, c$$

```
> {a,b,c}[2..3];
```

$$\{b, c\}$$

Hence, `op` can be used to *insert* elements into a sequence (or list, etc.) by using ranges to split the structure apart:

```
> S := a,b,d,e:
  op(1..2,[S]), c, op(3..4,[S]);
```

$$a, b, c, d, e$$

```
> L := [a,b,d,e]:
  [op(1..2,L), c, op(3..4,L)];
```

$$[a, b, c, d, e]$$

It can similarly be used to remove or *delete* elements by appending the subsequences before and after the element(s) to be removed, e.g.,

```
> L := [a,b,c,d,e]:
  [op(1..2,L), op(4..5,L)];
```

$$[a, b, d, e]$$

Incidentally, applying the selection operation (i.e., indexing) with no index to a sequence, list, or set is equivalent to applying the function op with a single argument and (slightly inconsistently) extracts the *sequence* of elements:

```
> L[] = op(L);
```

$$(a, b, c, d, e) = (a, b, c, d, e)$$

However, I recommend op as the clearer syntax.

Negative indices count from the end of a structure in selection operations on all data structures except arrays. This provides an easy way to reverse the elements of an ordered structure, such as a sequence or list, and -1 provides a very convenient alternative to using nops to access the last element in a structure, e.g.,

```
> L[-1] = L[nops(L)];
```

$$e = e$$

The functions op and subsop allow indexing by a list of indices of the form $[i_1, i_2, \ldots, i_n]$, which is equivalent to accessing operand i_1, then accessing operand i_2 of operand i_1, etc., thus:

$$\mathrm{op}([i_1, i_2, \ldots, i_n], \mathit{expression}) = \mathrm{op}(i_n, \ldots, \mathrm{op}(i_2, \mathrm{op}(i_1, \mathit{expression})), \ldots)$$

but the list form is both more succinct and more efficient, e.g.,

```
> op([2,2], [x=a, y=b, z=c]);
```

$$b$$

For some data types, such as functions and indexed symbols, the 0th operand can be accessed and corresponds to the function or symbol name, e.g.,

```
> subsop(0=g, f(a,b));
```

$$\mathrm{g}(a, b)$$

2.4.7 The empty sequence: NULL

The empty sequence has the special name NULL (all capitals), and is a name for "nothing". Hence, NULL is only ever used for input and never appears in output:

```
> NULL;
```

The only situation in which the elements of a sequence (or list, etc.) are automatically changed is when any of them evaluate to NULL, in which case they disappear. Hence, another way to delete an element from a sequence (or list, etc.) is to replace it by NULL, e.g.,

```
> L := [a,b,c,d,e]:
  subsop(2=NULL, L);
```

$$[a, c, d, e]$$

NULL is also useful to initialize a variable to an empty sequence when a sequence is to be constructed iteratively, as we will see in the next section.

Beware that it would be dangerous to try to use subs to change a specific element of a structure such as a list, because subs would change *all* occurrences of the given sub-expression. For example, we might intend the following code to delete the third element of the list, but in fact it replaces *all* occurrences of c which leads to an error (because arithmetic operations cannot be performed on NULL):

```
> subs(c=NULL, [0,1,c,c^2,c^3]);

   Error, invalid terms in product
```

2.4.8 Constructing sequences, lists, and sets

There are essentially three ways to construct a sequence. It is then trivial to construct a list or set from the sequence, or apply a function to it. In order of decreasing efficiency but increasing flexibility, these are the ways:

- Write it explicitly:

  ```
  > S := 1, 4, 9, 16, 25;
  ```

  $$S := 1, 4, 9, 16, 25$$

- Use the seq function:

  ```
  > S := seq(i^2, i = 1..5);
  ```

  $$S := 1, 4, 9, 16, 25$$

  or the $ operator:

  ```
  > S := j^2 $ j = 1..5;
  ```

  $$S := 1, 4, 9, 16, 25$$

- Use a do-loop:

```
> S := NULL:
  for i to 5 do
     S := S, i^2
  end do:
  S;
```

$$1, 4, 9, 16, 25$$

If possible, use `seq` to construct a non-trivial sequence, because it constructs the whole sequence in one step. Using a do-loop to construct a sequence is inefficient because it constructs all the intermediate sub-sequences. However, it is the most flexible method because it allows arbitrarily complicated code to be executed in order to generate each element. When elements of a sequence depend on previous elements, a do-loop is the only way to construct the sequence. The control variable, `i` in the above examples, is treated differently by `seq` and `for`: in `seq` it is completely local; in `for` it is not evaluated initially but is left set to the value that terminates the loop, 6 in the above example.

The sequence *operator* `$` is an older alternative to the `seq` function. In general, the `$` operator is less efficient and less convenient than `seq` because it uses different evaluation rules and so is best avoided for general sequence construction. The `seq` function does not evaluate its arguments before using them to construct the sequence, which is usually what is both required and expected, whereas the `$` operator does evaluate its operands before using them. Hence, it is frequently necessary and generally advisable to use unevaluation quotes liberally when using `$`. If I had wished to use `i` as the control variable in the above example of using `$`, then, because `i` has an assigned value left over from its previous use in the do-loop, `$` would see the nonsense operands:

```
> i^2, i = 1..5;
```

$$36, 6 = 1..5$$

and complain:

```
> i^2 $ i = 1..5;
```

```
Error, wrong number (or type) of parameters in function $
```

It works correctly only if i is suitably unevaluated:

```
> 'i'^2 $ 'i' = 1..5;
```

$$1, 4, 9, 16, 25$$

However, `$` is a great shorthand for expanding a range into a sequence or for generating several copies of the same object:

```
> $1..10;
```

$$1, 2, 3, 4, 5, 6, 7, 8, 9, 10$$

```
> x$5;
```

$$x, x, x, x, x$$

The latter syntax is particularly useful for constructing multiple derivatives, e.g.,

```
> diff(f(x), x$5) = diff(f(x), x,x,x,x,x);
```

$$\frac{\partial^5}{\partial x^5} \mathrm{f}(x) = \frac{\partial^5}{\partial x^5} \mathrm{f}(x)$$

Moreover, $ can remain symbolic, whereas `seq` cannot:

```
> f(j) $ j=1..n;
```

$$\mathrm{f}(j)\,\$\,(j = 1..n)$$

```
> seq(f(j), j=1..n);
   Error, unable to execute seq
```

2.4.9 Relatives of sequences: sums and products

Sums and products are closely related to sequences and can be easily generated from them by applying the appropriate operator (`+` or `*`) as a function, e.g.,

```
> `+`(seq(x^r, r = 1..5));
```

$$x + x^2 + x^3 + x^4 + x^5$$

There are also functions that can be used to construct sums and products directly: `sum` and `add` construct sums, and `product` and `mul` construct products, e.g.,

```
> add(x^r, r = 1..5);
```

$$x + x^2 + x^3 + x^4 + x^5$$

The main difference is that `sum` and `product` are intended primarily to construct and evaluate *symbolic* repeated sums and products, whereas `add` and `mul` are intended to simply add or multiply a *known* number of terms. Hence, `add` and `mul` require limits that must be explicit integers, whereas `sum` and `product` do not require limits at all, and if limits are supplied they can be symbolic:

```
> sum(x^r, r = 1..n);  # add would fail!
```

$$\frac{x^{(n+1)}}{x-1} - \frac{x}{x-1}$$

```
> expand(normal(eval(%, n=5)));  # as above
```

$$x + x^2 + x^3 + x^4 + x^5$$

```
> sum(x^r, r);  # add would fail!
```

$$\frac{x^r}{x-1}$$

The functions sum and product are equivalent to add and mul if the limits are explicit integers. However, add and mul are more efficient if simple explicit repeated addition or multiplication is required. There are inert versions of sum and product, but not of add and mul:

```
> Sum(x^r, r = 1..n);  # inert version
```

$$\sum_{r=1}^{n} x^r$$

```
> % = value(%);  # a common use of inert forms
```

$$\sum_{r=1}^{n} x^r = \frac{x^{(n+1)}}{x-1} - \frac{x}{x-1}$$

The functions seq, add, and mul were all added to Maple fairly recently to provide convenient and efficient ways to generate sequences and to perform explicit repeated addition and multiplication. They all support iteration over a data structure whereas $, sum, and product allow iteration only over explicit ranges, e.g.,

```
> a+b+c;  mul(t, t=%);
```

$$a + b + c$$
$$a\,b\,c$$

The function int is a continuous analogue of sum, whereas add, product, and mul do not have any continuous analogues. The index or control variable in $, sum, product, and int is global and so is fully evaluated, hence it should normally be declared local within a procedure, although it is left unbound. However, that in seq, add, and mul is automatically local and not evaluated, so it need not be declared local within a procedure. The control variable in a for-loop is not evaluated and retains its final value after the loop terminates; it should also normally be declared local within a procedure.

2.5 Characters and strings

The only way to represent a character in Maple is as a unit-length string, hence a string can be considered as a concatenation of characters. In all programming languages, text characters are encoded internally as non-negative integers. The simplest character set that is commonly used nowadays is the ASCII character set, which is encoded using the integers from 0 to 127 inclusive. Non-English and other special characters, such as mathematical symbols, require a much larger character set, for which Unicode is becoming widely adopted. (This topic is discussed in more detail in Chapter 14.) Whatever the precise details, characters are members of a discrete ordered set and therefore behave much like small non-negative integers. This relationship between characters and integers is transparent in some languages, such as C and C++, but less so in other languages, such as Maple, although it is nevertheless there. A string is essentially a sequence of characters, and strings in Maple are to some extent analogous to sequences or lists of characters.

Whatever the underlying character set is, you can almost certainly assume that it contains the ASCII character set, which is all that I will use explicitly in this book. (The mathematical symbols provided by Maple are not directly accessible, and the symbol font is only accessible within plots and only useful in the `textplot` commands.) Your computer should provide some facility for viewing fonts and selecting specific characters, many of which are not directly or easily accessible via the keyboard. For example, Microsoft Windows provides an accessory called "Character Map", which shows (among a lot of other useful information) that all decimal digits come before all upper case English letters which come before all lower case English letters.

Hence, Maple can iterate over a range or sequence of characters, characters and strings can be compared, and selection operations can be performed on strings. Characters that appear earlier in the collation sequence (e.g., ASCII or Unicode) compare less than characters that appear later. Having ascertained that the space character is the first graphic (i.e., printable) character in ASCII and tilde (~) is the last, then you can easily generate a string of all the ASCII graphic characters in order by concatenating the sequence of characters obtained by expanding the range. However, for purposes of display it is more convenient to split it up like this:

```
> cat($" ".."@");
```

" !\"#$%&'()*+,-./0123456789:;<=>?@"

```
> cat($"A".."`");
```

"ABCDEFGHIJKLMNOPQRSTUVWXYZ[\\] ^_`"

```
> cat($"a".."~");
```

"abcdefghijklmnopqrstuvwxyz{|}~"

Hence, as you might expect

```
> evalb("1" < "2"), evalb("a" < "b");
```

true, *true*

although you might not expect

```
> evalb("A" < "a");
```

true

The ASCII characters before space are non-graphic control characters, such as carriage return, line feed, and tab. (The names come from the era of teletypewriters.) They can normally be generated from the keyboard by holding the Control (*Ctrl*) key and pressing the appropriate character key, which is the only way to generate those that do not have dedicated keys. (Beware that this can easily happen by accident. If Maple complains about a program that looks correct, try turning on display of invisible characters [by clicking the toolbar button marked with a paragraph sign or selecting the item *Show Invisible Characters* in the View menu] and look for empty square boxes, which indicate uninterpreted control characters; delete any that should not be there!)

The characters within a string can be accessed by using the selection or subscript operation, e.g.,

```
> "abcde"[3];
```

"c"

```
> "abcde"[2..4];
```

"bcd"

However, the `op` function does not provide access to the characters within a string and empty selection does not convert a string into a sequence; the only way to do that is to use `seq` like this:

```
> seq(i, i="abcde");
```

"a", "b", "c", "d", "e"

Some control characters can be input using backslash escape sequences, as described above, and arbitrary strings can be generated by converting from a list of positive integers that represent the underlying character encodings. In fact, this conversion can be used both ways and is effected by the `convert` function with second argument `bytes`, e.g.,

```
> convert("ab\ncd", bytes);
```

$$[97, 98, 10, 99, 100]$$

```
> convert(%, bytes);
```

"ab\ncd"

Experimentation suggests that Maple uses only the integers 1 to 255 inclusive to encode text characters, where 1 to 127 encode the ASCII character set and 128 to 255 encode accented and other European letters and a few special symbols. To check this on your version of Maple, try evaluating the expression `convert([$1..255], bytes)` and variations on it. I will not try to display the result! The only character that cannot be represented in Maple appears to be the ASCII NUL character, which is encoded as the integer zero, presumably because it is used to denote the end of a string, as in C, and characters are represented as unit-length strings in Maple.

For further details, see Chapter 14 and also the Maple online help reference `examples,string`, which is an example worksheet entitled *Using Strings in Maple.*

2.6 Exercises

Expressions

Enter the following expressions *so that they are displayed essentially as shown.* (You may need to prevent Maple's automatic evaluation.) *Then*, by accessing each expression using `%`, simplify it to 0 *without* using the general-purpose Maple simplification function `simplify`. If explicit simplification is necessary then use the *most appropriate* specific simplification function, such as `expand` or `normal`. (Use the online help to check the precise purpose of these simplifiers.)

$$(x+y)^5 - x^5 - 5\,x^4\,y - 10\,x^3\,y^2 - 10\,x^2\,y^3 - 5\,x\,y^4 - y^5$$

$$\frac{x+y}{x-y} + \frac{\frac{1}{x}+\frac{1}{y}}{\frac{1}{x}-\frac{1}{y}}$$

$$\log(\tan(\frac{\pi}{4}))$$

Sequences

Construct the Maple expression sequence:

$$1,\ x,\ \frac{x^2}{2!},\ \frac{x^3}{3!},\ \frac{x^4}{4!},\ \frac{x^5}{5!}$$

(but with the factorials evaluated) in two ways: using the `seq` function, and using a `do`-loop. Convert it into a sum using the `convert` function. (Look up how to do this in the online help if necessary.) Do the same by making the sequence the argument sequence of the operator `+` used as a function.

Sums and products

Either omit the plotting later in this exercise until after you have read the first section of the next chapter or refer to the online help for guidance.

Use `Sum` and `value` to construct the equation,

$$\sum_{r=0}^{5} \frac{x^r}{r!} = 1 + x + \frac{1}{2}\,x^2 + \frac{1}{6}\,x^3 + \frac{1}{24}\,x^4 + \frac{1}{120}\,x^5$$

and assign the expression on the right-hand side to the variable `e1`. Note that this is the 5th-degree Taylor polynomial of e^x about 0, and that $e^x = (e^{x/n})^n$, which for $x \ll n$ is approximately equal to $(1 + x/n)^n$. Assign the expanded form of the binomial $(1 + x/n)^n$ for $n = 5$ to the variable `e2`. (Observe the differences between the values of `e1` and `e2`, which are two different approximations to e^x.) Plot the exponential function e^x (using `exp(x)`) and the two approximations `e1` and `e2` together for $-2 \le x \le 2$, with a suitable title. (To plot several expressions together make them the elements of a set or list and plot that.)

Observe that $\sin(x) \sim x$ for small x and that $\sin(n\pi) = 0$ for integer n. Hence, based on its zeros, $\sin(x)$ is proportional to

$$\prod_{r=-\infty}^{\infty} (x - r\pi) = x \prod_{r=1}^{\infty} (x^2 - (r\pi)^2)\,,$$

which is proportional to

$$x \prod_{r=1}^{\infty} \left(1 - \left(\frac{x}{r\pi}\right)^2\right).$$

This expression clearly tends to x as x tends to 0. Hence,

$$\sin(x) = x \prod_{r=1}^{\infty} \left(1 - \left(\frac{x}{r\pi}\right)^2\right).$$

Define the Maple mapping S such that

$$S(n,\, x) = x \prod_{r=1}^{n} \left(1 - \left(\frac{x}{r\pi}\right)^2\right).$$

On one plot, display the graph of $\sin(x)$ together with graphs of $S(n,\, x)$ $n = 1,\, 2,\, 3$; use $0 \le x \le 4\pi$ and vertical axis values in the range -10 to 1 and give the plot a suitable title.

Compare $S(3,\, x)$ with the Taylor polynomial of the same degree, namely $T(3,\, x)$, as follows. Define the Maple mapping T such that

$$T(n,\, x) = \sum_{r=0}^{n} \frac{(-1)^r\, x^{2r+1}}{(2r+1)!}.$$

Assign the *expanded* value of $S(3,\, x)$ to the variable `e1`, and the value of $T(3,\, x)$ to the variable `e2`. Evaluate `e1` and `e2` using floating-point approximation (which will evaluate the coefficients of the polynomials, leaving the indeterminate x symbolic) and observe the difference. (Use the default precision.) On one plot, display the graph of $\sin(x)$ together with the graphs of $S(3,\, x)$ and $T(3,\, x)$, for $0 \le x \le 4\pi$ and vertical axis values in the range -10 to 10, as before, with a suitable title.

Go back and use the `legend` option to annotate your plots.

Chapter 3

Two-Dimensional Plotting

This chapter is about graphics that logically exist in two dimensions. The next chapter is about graphics that logically exist in three dimensions, even though in practice they must be projected into two dimensions. Two-dimensional graphics arise as representations, called *graphs*, of real-valued functions of a single real variable, and being able to plot such graphs easily is crucially important in many mathematical investigations, in which the underlying data being plotted may be either algebraic or numeric. But two-dimensional graphics also arise from problems in two-dimensional geometry, which can be much easier to understand visually than in the abstract. These two aspects of graphics merge to some extent in algebraic or coordinate geometry.

After a review of simple two-dimensional plotting in the first section, the second section looks at more sophisticated methods for constructing two-dimensional plots, in particular by combining plots constructed in different ways, and at two-dimensional animation. Animation can be very helpful in visualizing how things work and time essentially provides another axis. For this we need to go beyond the basic plotting facilities in the main Maple library and use the `plots` package. The third section is essentially a case study that uses geometrical animation to illustrate the facilities introduced in the previous two sections. Maple's plotting facilities can be used to generate graphics for use outside Maple, and the last section discusses how to do this.

3.1 Simple plotting

This section describes the basic facilities for two-dimensional plotting that are available using the `plot` function in the main Maple library.

3.1.1 Plotting graphs of functions

The `plot` function in the main Maple library accepts various arguments and requires *at least one argument*; all other arguments are *optional*. The first argument must always be the data to plot, which in the simplest cases must be either an *expression containing one free variable* or a simple mapping (an abstract function of one variable not applied to any argument). If the first argument is an expression then the second argument must be either the free variable or an equation with the free variable on the left; if the first argument is a mapping then the second argument must *not* specify a variable. (Minimalists will appreciate that the latter case, namely plotting a simple mapping, allows the `plot` function to be called with only a single argument.)

Hence, the following two examples are almost equivalent, the only difference being that in the first case the horizontal axis is labelled with the name of the variable:

```
> plot(sin(x), x);
```

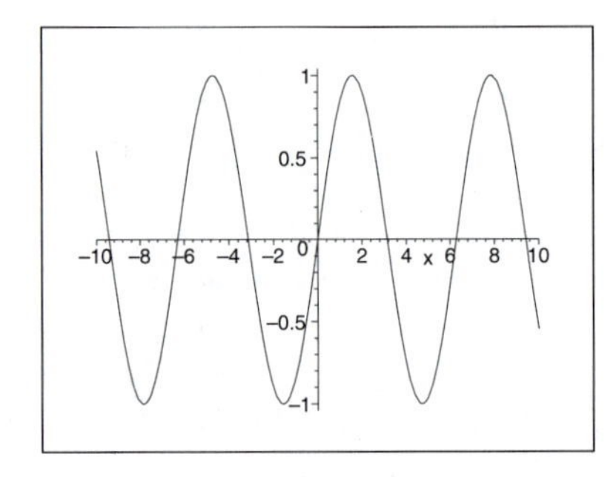


```
> plot(sin);
```

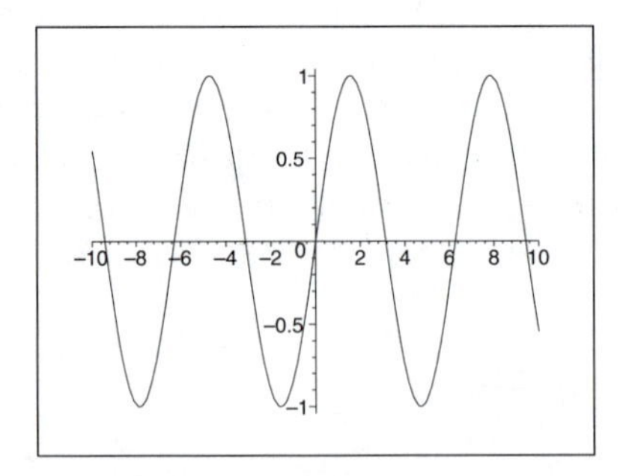

But the following two examples both fail (in different ways):

```
> plot(sin(x));

    Plotting error, empty plot
```

```
> plot(sin, x);
```

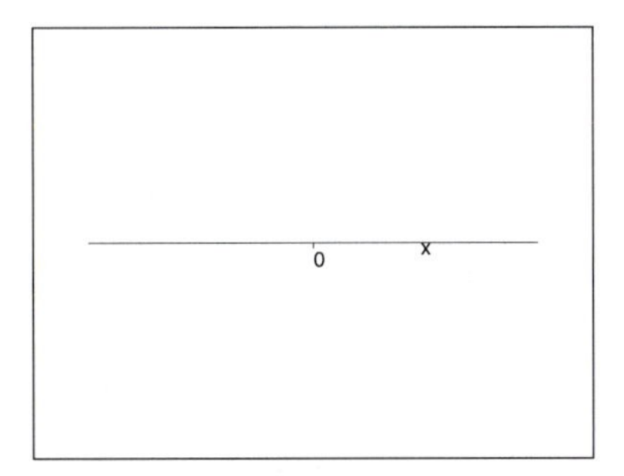

If you get strange plotting errors then check that you have specified the independent variable if you are plotting an expression and that you have not specified an independent variable if you are plotting a mapping.

Plot arguments are evaluated by Maple before being processed by the plot function. (This applies to the arguments of almost all Maple functions.) Hence, any variables in plot arguments that have been assigned values will be replaced by those values. A free variable is one that evaluates to itself. Hence, the following input lines will all produce the same plot (except for axis labelling). (Multiple identical plots are not shown. To convince yourself that they really are all the same, I suggest that you type the examples into Maple yourself.)

```
> plot(sin(Pi*x), x);
```

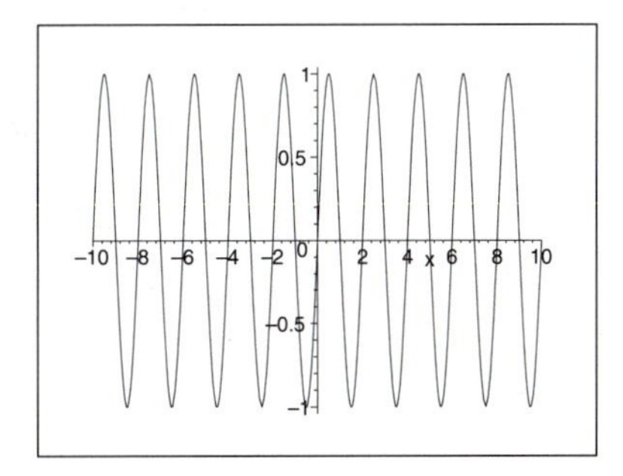


```
> k := Pi:  plot(sin(k*x), x):

> f := x -> sin(Pi*x):  plot(f):

> plot(x -> sin(Pi*x)):

> plot(x -> sin(k*x)):
```

If necessary, evaluation of assigned variables can be prevented by using forward quotes:

```
> x := 23:  plot(sin(Pi*'x'), 'x'):  # not recommended!
```

But this is not entirely reliable in more complicated situations, and it is safest explicitly to unassign (clear) variables that need to be free before using them, e.g.,

```
> x := 'x':
```

Note that plotting *mappings* rather than *expressions* avoids this problem!

3.1.2 Plotting graphs of "strictly numerical" functions

A problem can arise when trying to plot graphs of functions that are defined only for strictly numerical argument values and cause an error when applied to a symbolic argument. This can happen with functions that you have written yourself or functions that have been generated by Maple, and the cause of the problem is usually that the function tries to make an inequality test on its argument. While such functions can be evaluated internally within the numerical context of the `plot` function, they cannot be evaluated as arguments to the `plot` function in the form of *expressions with a symbolic argument.*

We will see some more realistic examples of this problem later, but for now let me illustrate the problem with the Maple function `evalhf`, which *requires* a numerical argument (and evaluates it using hardware floating point). You might expect that this would plot the straight-line graph $y = x$, but ...

```
> plot(evalhf(x), x);
```

```
Error, cannot handle unevaluated name `x` in evalhf
```

This error can be avoided either by *unevaluating* the first argument to `plot`, so that it is not evaluated until it is within a valid numerical context, or by plotting the related *mapping* rather than the expression:

```
> plot('evalhf(x)', x);
```

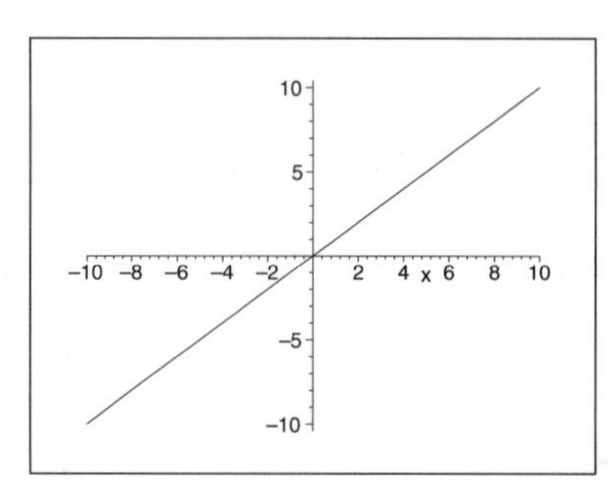

```
> plot(evalhf);
```

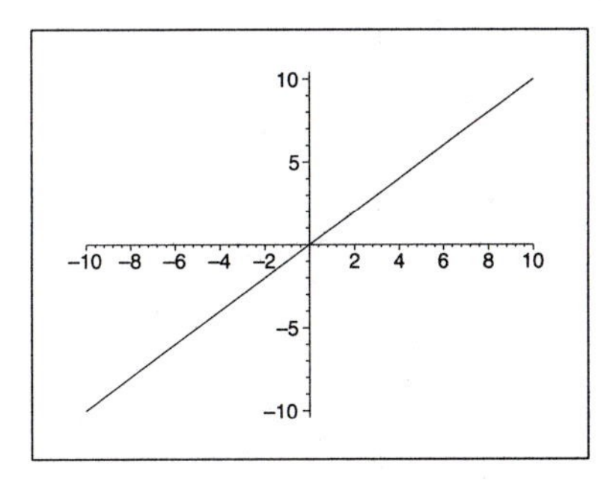

3.1.3 Plot ranges

Usually, it is desirable to specify the horizontal plot range, which can be done by giving `plot` a second argument in the form of a range `a..b`. The default horizontal range is `-10..10`, as illustrated above. If the plot variable is also to be specified then it and the range *must be combined* into an equation (e.g., `x = a..b`), in which case the axis is labelled with the variable, e.g.,

```
> plot(sin(x), x = 0..2*Pi);
```

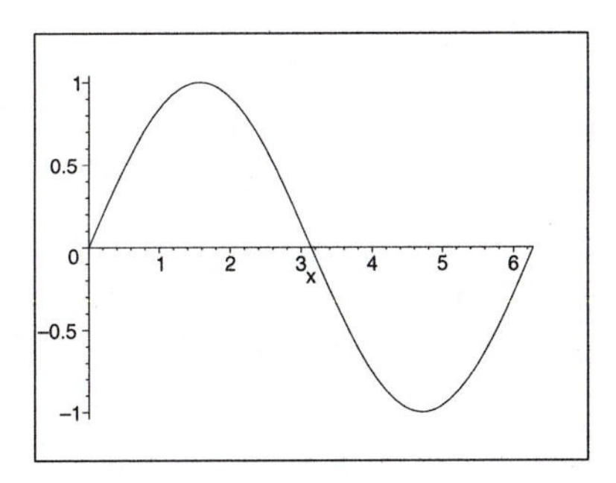

The default vertical range is computed from the values of the expression or mapping being plotted, but it can be specified explicitly as the third argument. This can be useful to select some particular region of the plot, which may be necessary if the values being plotted become very large. If the third argument is specified as an equation with the vertical range on the right and a name on the left, then the name is used to label the vertical axis. (A name on its own just labels the axis without specifying a range.) For example,

```
> plot(1/x^2, x = -4..4, y = 0..4);
```

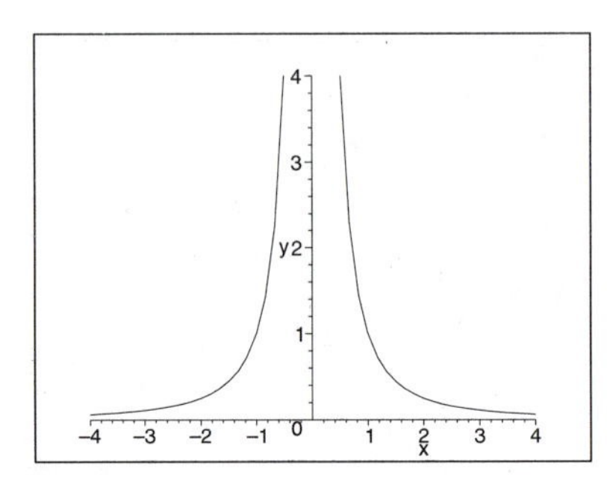

is probably more useful than

```
> plot(1/x^2, x = -4..4, y, axes = BOXED);
```

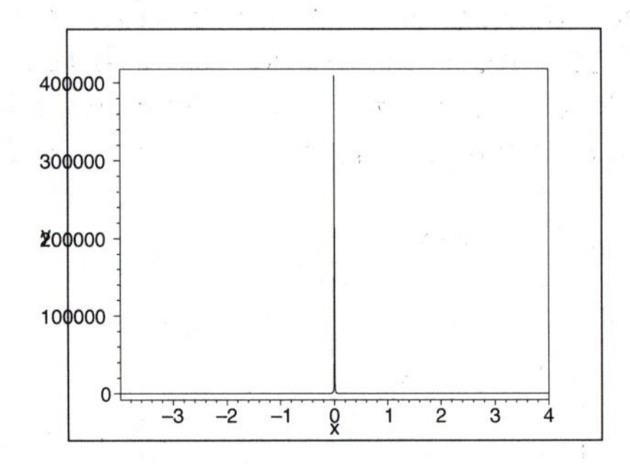

(Without colour it may not be clear that the graph appears as a spike with the shape of an inverted T that almost coincides with the axes.)

3.1.4 Multiple plots

Several plots can be superimposed by using a *set* or *list* of expressions (all with the same free variable) or mappings as the first argument of a plotting command. (This works for most plotting commands, not only `plot`, although in some cases it matters whether you use a set or list as we will see later. But note that expressions and mappings cannot be mixed in a multiple plot.) By default, the plot parameters will be the same for all the plots except that different colours will be used for different plots if possible, e.g.,

```
> plot({sin, cos}, -Pi..Pi);
```

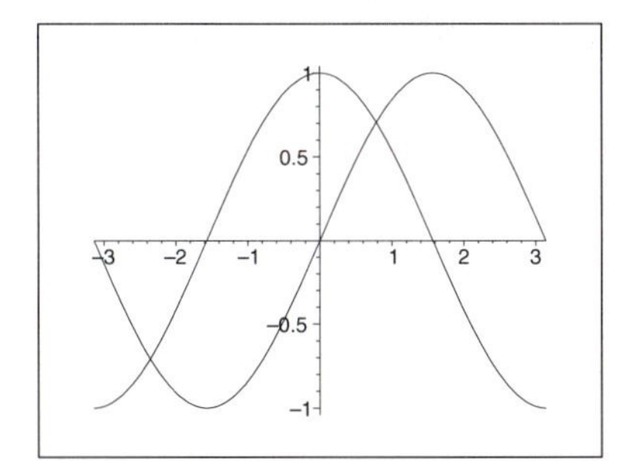

With multiple plots it can be important to distinguish and label them clearly. This can be done using plot options, some of which are described below, in which case it is necessary to use a list of plot data rather than a set, because it is essential to preserve the order of the different plots.

3.1.5 Parametric plotting

A plot can be defined *parametrically*, which means that both the horizontal and vertical coordinates of each point of the plot are specified as functions of a *parameter*, which is a variable that is not in general identical to either the horizontal or vertical coordinate, but varies monotonically along the curve. The Maple syntax for a parametric plot is

```
plot([x(t), y(t), t = a..b], options)
```

where `t` represents the parameter and `x(t)` and `y(t)` represent functions of the parameter that define the horizontal and vertical coordinates of a general point on the curve to be plotted. The first argument to `plot` is the whole parametric representation of the curve *contained in a list.*

A parametric plot is a generalization of a plot of a single expression, which is equivalent to a parametric plot in which the horizontal coordinate function is the identity; in other words, the following two plots would be identical:

```
plot(y(x), x = a..b)
plot([x, y(x), x = a..b], x)
```

(The optional argument `x` in the second plot is to label the horizontal axis to match the first plot.)

The advantage of parametric plots is that they can be used to plot curves that are not graphs of (single-valued) functions. This is the case for any curve that "folds over" itself. Mathematically, such a curve can be represented by an equation of the form $f(x, y) = 0$ that does not have a unique solution for y in terms of a single-valued function of x (or vice versa). For example, this function-call plots a circle of unit radius parametrically:

```
> plot([cos(theta), sin(theta), theta = 0..2*Pi],
    scaling = CONSTRAINED);
```

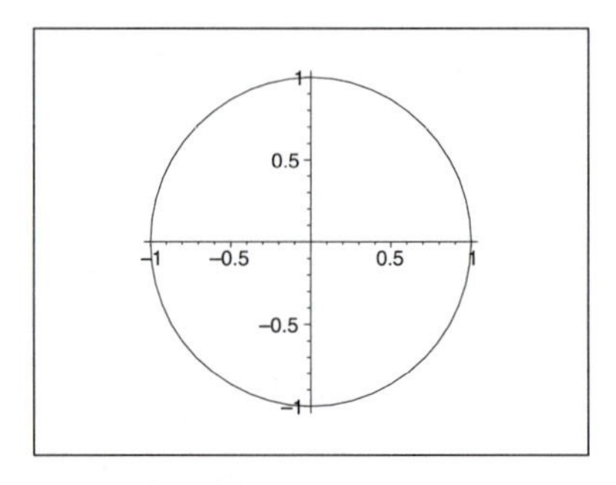

The option `scaling = CONSTRAINED` is necessary to constrain the scaling of the vertical and horizontal axes to be the same, so as to preserve the geometry of the plot, i.e., in this case to ensure that the circle looks circular. The parametrization of a particular curve is not unique; the above parametrization of a circle corresponds to its natural representation in circular polar coordinates and is hence its natural and simplest parametric representation.

The equation representing this circle in Cartesian coordinates is $x^2 + y^2 = 1$, which has the following *two* solutions for y as a function of x:

```
> solve(x^2+y^2=1, y);
```

$$\sqrt{-x^2+1}, -\sqrt{-x^2+1}$$

Hence, the circle is the *union* of the graphs of *two* single-valued functions, which requires a *multiple* plot like this:

```
> plot({%}, x = -1..1, scaling = CONSTRAINED);
```

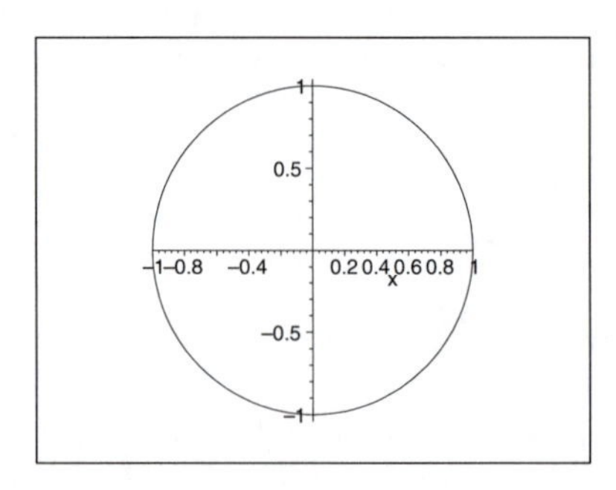

(By default, Maple plots the top and bottom semi-circles in different colours, which confirms that they are different graphs.) It is best to plot curves parametrically if they cannot conveniently be represented in terms of a vertical coordinate that is a (single-valued) function of the horizontal coordinate; parametric plotting is much more flexible than plotting graphs of expressions or functions.

An ellipse is a generalization of a circle, obtained by stretching (or squashing) it in some direction. The longest and shortest distances from the centre to the perimeter of an ellipse are called its *semi-major* and *semi-minor* axes, often denoted as a and b, respectively. They generalize the radius of a circle and their directions define natural Cartesian axes. An ellipse can be plotted using its natural parametrization like this:

```
> a := 2:  b := 1:  # semi-major and semi-minor axes
  plot([a*cos(theta), b*sin(theta), theta = 0..2*Pi],
     scaling = CONSTRAINED);
```

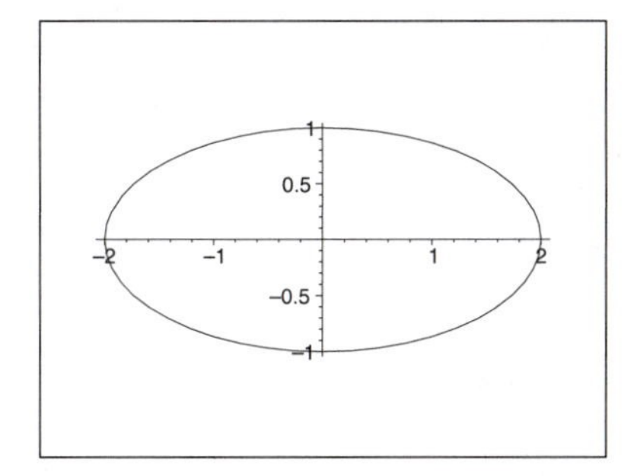

3.1.6 Plotting points

Sometimes it is more appropriate to define a plot in terms of specific points on it rather than a formula for the general point. The former would be the case when plotting a polygon (any figure composed of straight-line segments) or when plotting numerical data such as might be obtained from an experiment or a numerical computation. Each *point* is specified as a *list* containing its horizontal and vertical coordinates in that order. The points to be plotted are put into a list in the order that they are to be plotted, and this *list of lists* is supplied as the first argument to `plot`:

`plot([[`x_1`, `y_1`], [`x_2`, `y_2`], [`x_3`, `y_3`], ...],` *`options`*`)`

For example:

```
> plot([[0,0],[1,2],[4,3]]);
```

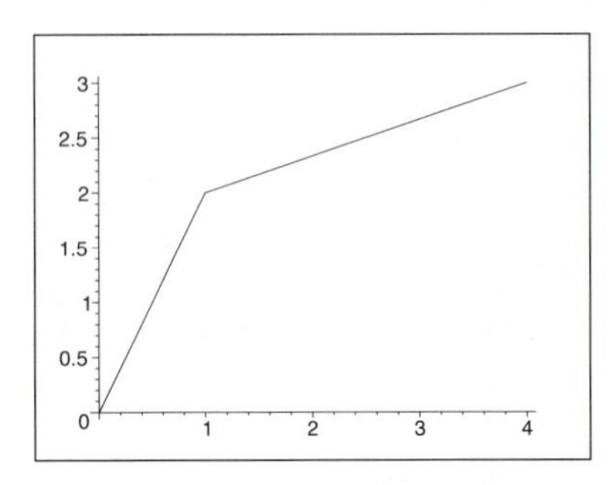

This form of input is similar to that of the internal `PLOT` data structure. The list of lists can be generated in any way that is convenient.

By default, the points plotted are joined by straight lines, but it is possible — and occasionally useful — to plot just one or more points without any lines by specifying the appropriate plot option (see below). Either way, this is called a *point plot*, which refers to the form of input data rather than the resulting output.

3.1.7 Plot options

All arguments to `plot` after the first are in general optional, except that the free variable in expressions to be plotted must be specified in the second argument. If a vertical plot range is to be specified then it must be the third argument. Other optional arguments can be specified in any order and take the form of *equations* with a keyword on the left and its value on the right. Option keywords should be in lower case and option value keywords should mostly be in upper case, thus:

`option = VALUE`

(The case and spelling of option keywords is actually more flexible than this. The documentation, which I will follow, uses all upper case, although all lower case also seems to work. However, mixed case seems not to work.)

Here are some important options (for the other options, see the online help for `plot[options]` or other documentation):

`title =` *`string`*
: For example, `title = "A very interesting plot"`. This option specifies a title for the plot. The default is no title. The value can be either a string or an identifier. To split the displayed plot title over two or more lines either simply split the title string (or identifier) over two or more lines by typing *Shift-Return* at the end of each line or include newline (`\n`) characters. (The `titlefont` option can be used to change the

font of the entire plot title, but there is currently no way to change the font within the title and hence no way to include special symbols.)

`axes = BOXED, FRAME, NONE, NORMAL`
This option specifies the location and style of axes. The default is `NORMAL`, meaning through the origin. Changing this can be useful when the axes obscure some important detail. `FRAME` means at the left and bottom; `BOXED` is a closed version of `FRAME`. `NONE` may be appropriate when plotting geometrical figures for which the axes are not important or are distracting.

`scaling = CONSTRAINED, UNCONSTRAINED`
The default is `UNCONSTRAINED`, meaning that Maple changes the relative scales of the axes to fit the plot to the shape of the frame. `UNCONSTRAINED` means that Maple must preserve the relative scales of the axes to be 1:1, which is necessary to preserve the geometry of a plot (e.g., to make circles appear circular and avoid them being squashed).

`style = LINE, POINT`
The default is `LINE`, meaning join the plotting points (with straight lines). `POINT` style plots points only, which is most likely to be useful when plotting specific points using the "list of lists" input format.

`symbol = BOX, CIRCLE, CROSS, DIAMOND, POINT`
This option specifies the symbol for plotting points when using `style = POINT`. The default depends on the output device. (The option `symbol = POINT` may produce a dot that is too small to see on some output devices.)

The following options are most useful for distinguishing plots within a multiple plot:

`color` **or** `colour = black, red, green, blue, ...`
This option specifies the colour of the actual plot; the axes, labelling, etc. are always black. There are many more colours available. It is sometimes useful to force all plots to be black so that they print visibly. Alternatively, it can be useful to link specific colours with specific plots in a multiple plot to distinguish them, although that is perhaps better achieved by using the `legend` option.

`linestyle = 1, 2, 3, 4`
This option specifies the style of line used in the actual plot. The numbers have the following meanings: 1, continuous (the default); 2, dots; 3, dashes; 4, alternating dots and dashes.

`legend = `*`string`*
: This option provides annotation that is most useful to distinguish the plots in a multiple plot.

If a multiple plot is specified using a list as its first argument then, where it makes sense, option values may be given as lists of values corresponding to the list of objects to plot. (See the online help topic `plot[multiple]` for further details.) The following example illustrates a multi-line title and the use of different line styles and a legend to annotate an all-black plot:

```
> plot([sin, cos], -Pi..Pi, colour = black,
     linestyle = [1, 3], legend = ["Sine", "Cosine"],
     title = "The graph of the sine function superimposed
  on the graph of the cosine function");
```

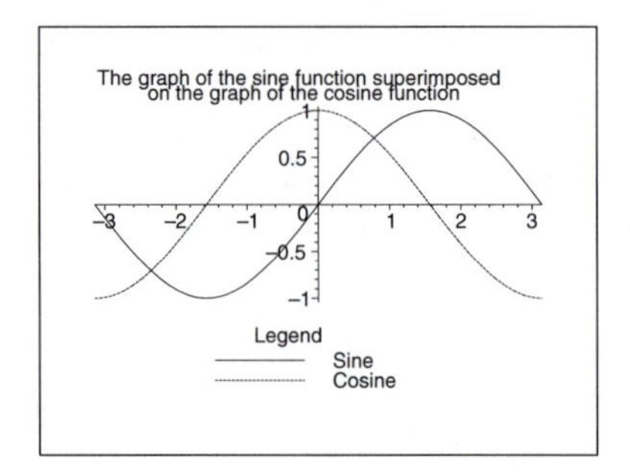

The most important options, which include all those listed above except the title, can be changed interactively once a plot is displayed. This provides a useful way to choose the best options, which can then be edited into the plot command to preserve the choice if desired.

3.1.8 Plotting discontinuous functions

Maple assumes that functions to be plotted are continuous; if they are not, then it will probably plot a vertical line at a discontinuity, e.g.,

```
> plot(1/x, x=-5..5, -5..5, axes = BOXED);
```

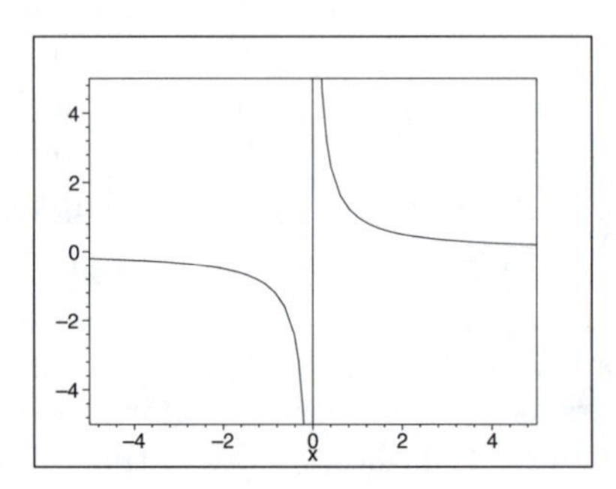

If the plot option `discont = true` is given then Maple tries to detect any discontinuities and to draw continuous graphs only between the points of discontinuity, e.g.,

```
> plot(1/x, x=-5..5, -5..5, discont = true, axes = BOXED);
```

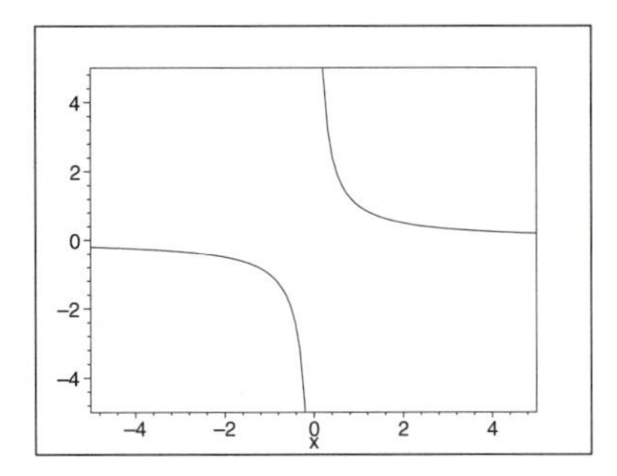

This option also works when plotting mappings. Note that if `true` and `false` are used as option values then they must be specified in lower case because they are standard Boolean constants rather than values that are specific to plotting.

3.1.9 Displaying and manipulating plots

Maple consists of two main components: the Maple *kernel* that performs all the computation and is essentially the same on all platforms; the Maple *user interface* (called Iris) that performs all the user input and output and may be different on different platforms. The interface between the two components is controlled by the `interface` function, which is detailed in the online help. The (read-only) interface variable `version` provides information about the version of the user interface (Iris). For example, at the time of writing I am running this interface version:

```
> interface(version);
```

Maple Worksheet Interface, Maple 6.01, IBM INTEL NT, Jun 9 2000 Build ID 79514

The `interface` function provides a lot of useful control over plotting and allows plots to be directed to different plot devices or files using different output formats, instead of simply incorporating them into the worksheet, which is the default behaviour.

To produce graphics, the kernel constructs a *plot data structure* and Iris displays it, in much the same way that it displays formulae. A plot data structure is much like any other Maple data structure — in fact, it is an inert function called either `PLOT` or `PLOT3D` (for a two-dimensional or three-dimensional plot, respectively), e.g.,

```
> lprint(plot([[0,0]])):

    PLOT(CURVES([[0.,0.]],COLOUR(RGB,1.0,0.,0.)),AXESLABELS("",""),
    VIEW(DEFAULT,DEFAULT))
```

Maple has two main output formats. The normal format displays data in the most natural way possible: it displays mathematical expressions using essentially normal typeset style and it displays plots graphically. The other main format is that produced by the command `lprint` (which stands for "linear print"), which simply displays the internal form of data line-by-line without any special formatting, essentially in the one-dimensional format used for input. Maple default display normally uses typeset style (although this can be changed using the *Output Display* option in the Options menu or the `interface` command).

Plot data structures can be manipulated exactly like any other Maple data, and in particular they can be *assigned* to variables. A plot is displayed graphically *only* if it is the *sole data structure* and not if it is part of another data structure. Hence, a plot structure in an assignment (or any other structure such as a sequence or list) is displayed in "linear printed" format, and *generally automatic display of assignments or other structures containing plots should be inhibited by terminating the statement with a colon.* But, of course, the value of the assigned variable can immediately be displayed.

Hence, the following are equivalent:

```
> plot(sin);
```

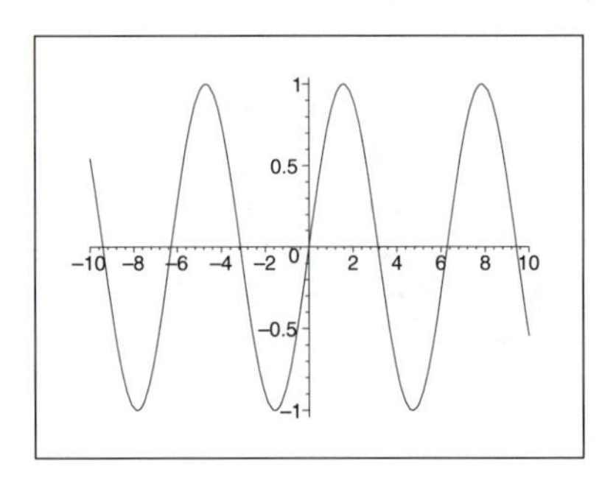


```
> p := plot(sin):  p;
```

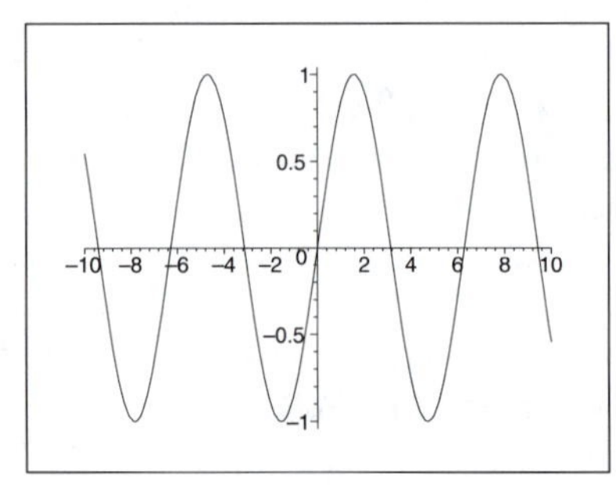

To construct complicated graphics it is necessary to combine several simple plot data structures, as we will see later, and to do this it is often necessary to assign them to variables and/or incorporate them into data structures such as sequences, lists, or arrays and to display only the final composite plot structure.

Plots can be displayed either directly in the worksheet (inline, the default) or in separate windows, depending on the setting of *Plot Display* in the Options menu. In either case, they can be manipulated interactively, once they have been displayed, by first clicking the mouse over them. In the inline case, this activates a frame with "handles" that can be dragged with the mouse to change the size and shape of the plot (but not its position), and it activates menus and a toolbar that can be used to change plot options. *Right*-clicking on a plot activates a context menu that provides even faster access to the plot and export options. Plots behave essentially like large and complicated single characters (hieroglyphs) and so the positions of inline plots can be controlled in the same way that the positions of characters can be controlled, via the paragraph format options. To do this, click just to the right of the plot to de-select it but put the cursor on the relevant paragraph marker (the style should be shown as "Maple Plot" in the box at the left of the context bar). Plots can be cut, copied, and pasted, and generally manipulated with a considerable degree of flexibility (but only in ways that could also be applied to single characters).

Clicking on a two-dimensional plot also shows in the box at the left of the context bar the coordinates of the mouse pointer when the click happened with respect to the axes and scale of the plot. This provides a very useful way of reading off approximate position information, such as the locations of intersections or extrema.

It is worth stressing that for a plot to be displayed it must be "printed", which usually happens implicitly as a result of evaluating some plotting function but can be forced by using the `print` command to "print" the plot. It is important to remember that plot display is not a side effect of executing a plotting command. Hence, when plotting functions appear within Maple control structures, procedures, or modules their values will *normally* not be printed and so the plots will not be displayed, which can be disconcerting. The best solution to this problem is to ensure that the required plots are printed, either by explicitly *applying* the `print` command or possibly by *returning* a plot structure from a procedure. If you want to use a module (see later) to localize the code necessary to construct a plot then the only way to ensure that the plot appears is explicitly to print it. Here is a trivial example, just to reinforce this important point:

```
> module() plot(sin) end module;
```

module() end module

```
> module() print(plot(sin)) end module:
```

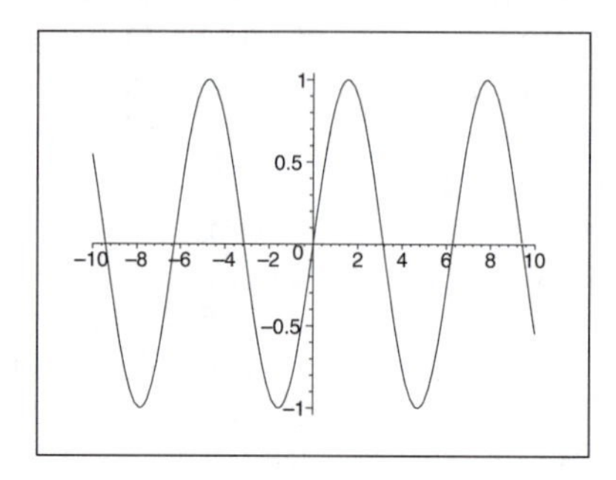

3.2 The `plots` package and animation

Many additional plotting facilities are provided in the standard Maple *package* called `plots` and this section will introduce a few of the more fundamental ones. The easiest way to use the package is to execute the function-call:

```
> with(plots):
```

```
Warning, the name changecoords has been redefined
```

Alternatively, to make just a few functions from the `plots` package easily available use, e.g.,

```
> with(plots, display, animate):
```

If the `with` command is not used then functions from the `plots` package must be called using their full names, e.g., `plots[display]` instead of simply `display`. I will assume for the rest of this chapter that `with(plots)` has been executed.

3.2.1 Aside: packages and the `with` command

There are two implementations of packages in Maple based on tables and modules. Modules are a new facility in Maple 6 and new packages should be implemented as modules, whereas older packages are still implemented as tables. The `plots` package is such an older package. For simple purposes, it makes no difference which implementation is used.

The functions in a table-based package are assigned to elements of a Maple table having the same name as the package, so that they can always be accessed as elements of the table by using the name of the function as the table index. This is the reason for the square-bracket syntax. You can see this easily by printing (or evaluating) the package name, e.g., by executing `print(plots)`. The effect of the `with` command is to make assignments of the form:

```
display := plots[display]
```

after which the functions that have been reassigned can be accessed without specifically including the name of the package.

By contrast, you can see what a package implemented as a module looks like by executing `print(LinearAlgebra)`, for example. However, functions within module-based packages can be accessed using the same selection syntax as for table-based packages:

```
> print(plots[display]):
```

$$\textbf{proc}(F)\ \ldots\ \textbf{end proc}$$

```
> print(LinearAlgebra[Trace]):
```

$$\textbf{proc}(M{::}Matrix)\ \ldots\ \textbf{end proc}$$

and the `with` command still has essentially the same effect for both package implementations.

3.2.2 Simple animation: animate

A plot can be *animated* (like a cartoon film) by displaying several related plots one after the other. A family of functions of one variable can be animated easily using `plots[animate]`. If $f(x)$ is a function of one variable x then $f_i(x)$ represents a *family of functions* of one variable x, where i specifies the member of the family. To animate this family means to plot $f_i(x)$ for each i in some discrete index set; e.g., $i = 1..20$. In Maple, $f_i(x)$ is any expression with *two* free variables or mapping taking two arguments, and the simple animation syntax is

```
animate(f(i,x), x = a..b, i = j..k)
```

or

```
animate(f, a..b, j..k)
```

where x is the plot variable and i is the animation variable; e.g., this animation produces a travelling sine wave (moving to the right):

```
> animate(sin(x-i), x = -Pi..Pi, i = 0..2*Pi);
```

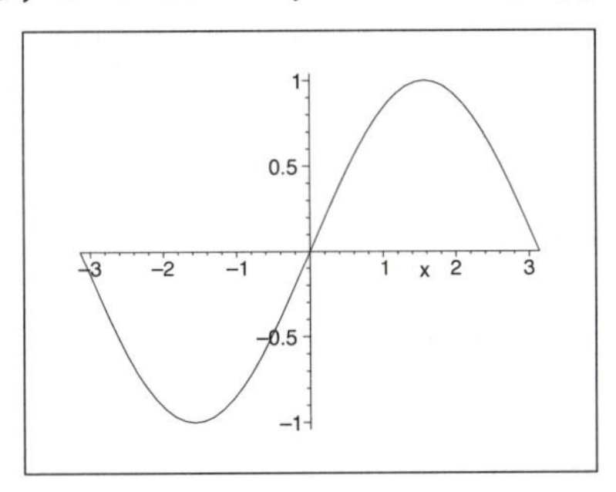

Note that all the frames (separate plots) of an animation are computed before any is displayed. This may take a long time, and Maple may run out of memory before it finishes. ***So save your worksheet before attempting any animation (or complicated plotting).*** If Maple does run out of memory, try reducing the number of frames by specifying the option `frames = n`; the default is 16 frames so try using a value of n less than 16. Alternatively, for a smoother animation you could try increasing the number of frames.

When an animation has been generated, the first frame is displayed like a plot but it is not animated. When the animation is activated by selecting it with the mouse, the context bar shows a set of buttons similar to those on a VCR. These can be used to start and stop the animation, select continuous play, etc. The menus available when an animation is active are exactly the same as those for a static plot, plus an animation menu that provides the same operations as the animation context bar. Note that the context bar has a pair of up-down arrow buttons at the right-hand side that can be used to switch it between the static plot context bar and the animation context bar.

In fact, the first argument of the `animate` function can be anything that is acceptable as a first argument to the `plot` function but with an additional free variable (or argument). Here is a more complex trigonometric animation based on a parametric plot, using t as the parameter and k as the animation variable:

```
> animate([sin(2*t), cos(3*t+k), t = 0..2*Pi],
      k = 0..2*Pi, frames = 32, axes = NONE);
```

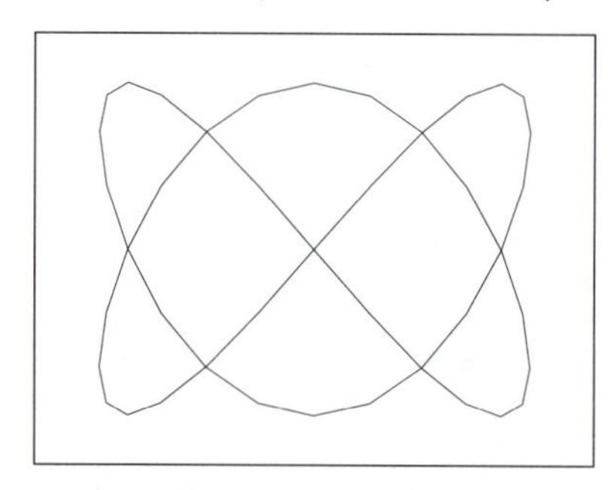

This is actually a "Lissajous figure" with the phase shift k varying between frames, which makes it look like a sinusoid on a rotating cylinder.

The `animate` command is like `plot` except that the second range equation specifies the frame sequence. However, `animate` is not very flexible and is useful only for trivial animation. More general animation can be produced using the `display` function (see later).

3.2.3 Combining and animating plots: `display`

Probably the most useful tool in the `plots` package is the `display` function, which can combine plots in *space by superimposing or grouping* them or in *time by animating* them. The plots to be combined should be plot structures, produced by calls to plotting functions such as `plot`, that have been assigned to variables or collected into sets or lists. The `display` function can also be used to change some of the options of a plot.

- To superimpose a number of plots (combine them in space) the first argument of `display` should be a *set or list* of plot structures.

- To animate a sequence of plots (combine them in time) the first argument of `display` *must be a list* of plot structures (to preserve their order), and the display option `insequence = true` must also be given.

- To display a number of plots together but keep them *distinct* the first argument of `display` should be an *array* of plot structures. (However, the individual plots may not appear quite the same as they would if plotted separately.)

Other options can be specified as for `plot` and can be used to set or change some options in the resulting plot. However, options specified in `display` cannot change properties of individual curves such as their colour or line style but only properties of the entire plot such as the axes and title. In fact, `display` can change only those options that can be changed interactively. If options differ in plots that are combined then there seems to be no guarantee which will be used although appropriate options specified in `display` take precedence. It is probably best not to specify options in individual plots that will be specified later in `display`. The best way to set options (such as the colour) for *all* the curves that will be combined by `display` is to change the default using `setoptions` (see below) *before* constructing the plot structures to be combined.

A set or list of plot structures to be combined by `display` can be produced in any convenient way, e.g., by `seq` with a call of `plot` as its first argument. It is probably best to use a variable to hold the number of frames to be plotted (i.e., the number of plots in the set or list) so that this value can be easily and reliably changed if necessary, either to decrease the number of plots to save memory or to increase it to give a smoother animation. For example:

```
> n_plots := 16:
  plot_seq :=
     seq(plot(sin(x-2*Pi*i/(n_plots-1)), x = -Pi..Pi),
        i = 0..n_plots-1):
```

Superimpose in space:

```
> display({plot_seq});  # could also use a list
```

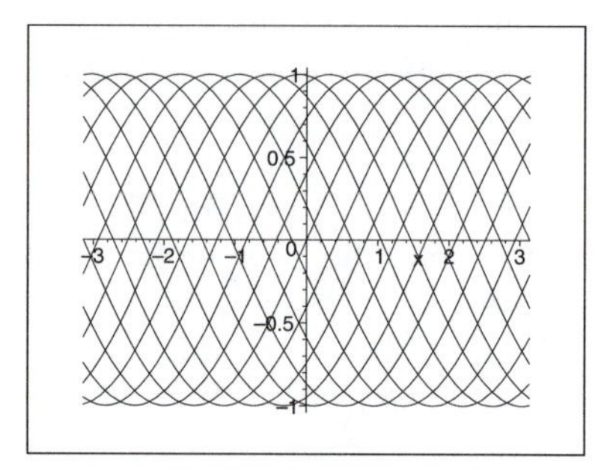

Animate in time:

```
> display([plot_seq], insequence = true);
```

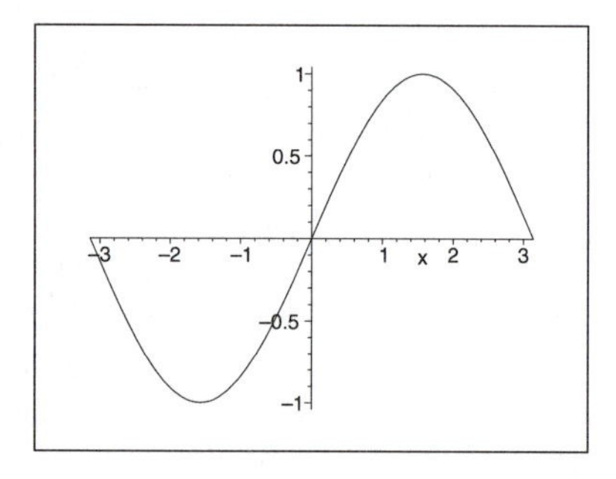

This animation is the same as that produced earlier using the `animate` function. However, `display` provides much more flexible animation, as we will see.

An animation can be displayed over a static background plot by including the animation and static plot in a set or list and leaving `insequence` set to false (the default), as we will see later. This is more elegant (and more efficient) than including the static background in each frame of the animation. In fact, it appears to be possible to combine an arbitrary number of animations and static plots in this way.

Displaying an array of plot structures is similar to displaying each plot separately and then moving them around in the worksheet. However, an array plot is a single plot structure that can be exported as a single file and some of the details of the plots may be changed slightly when they are incorporated into the array plot, so it is less flexible than keeping the plots entirely separate. On the other hand, it is easier to recompute an array plot, and the disposition of the plots within the array plot is fixed and corresponds to the disposition of the elements of the array when regarded as a matrix. Here is a trivial example of a 2×2 array of plots:

```
> display(array([[plot(sin),  plot(cos)],
                 [plot(sinh), plot(cosh)]]));
```

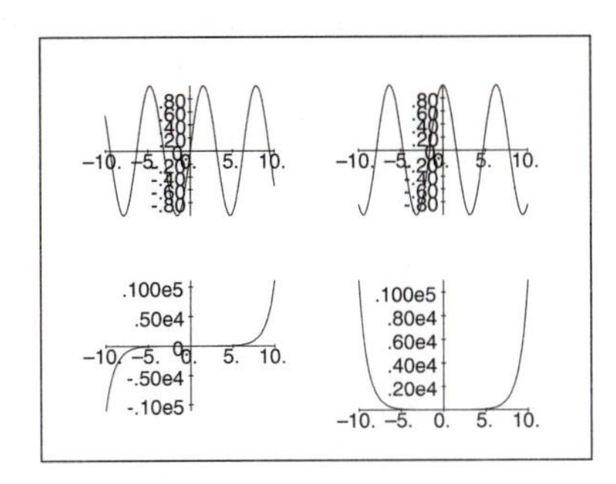

Plotting a one-dimensional array of plot data is equivalent to plotting a two-dimensional array containing a single row.

3.2.4 Changing default plotting options: `setoptions`

The *default* options are the values that plotting commands use for options that are not specified. They can be changed by giving them as arguments to the command `plots[setoptions]`, which can save a lot of typing. For example, when plotting pictures of geometrical objects it might be convenient to specify at the start:

```
> setoptions(axes = NONE, scaling = CONSTRAINED,
     colour = black);
```

3.2.5 Annotating plots: `textplot`

A plot containing text at specified coordinates can be produced by using the function `plots[textplot]` and text plots can be combined with other plots by using `display`. The syntax is

textplot([*x*, *y*, *text*], *options*)

where x and y are the horizontal and vertical coordinates and *text* is a string of text. The `textplot` function also accepts a set or list of such triples as its first argument, and it accepts alignment options (see the online help for details). The other options available are as for plot.

One particularly useful option has the form `font = [SYMBOL]` and causes the symbol font to be used instead of the default text font. This allows plots to be annotated with Greek letters and (some) mathematical symbols. An easy way to find the mapping between the desired symbol and the keyboard character required in `textplot` is to use the accessory provided by your operating system that gives access to characters other than those available on the keyboard. (For example, on Windows it is called *Character Map* and is available in the *Programs/Accessories* item on the Start menu). Select the symbol font and insert the appropriate symbol into Maple via the clipboard. What will appear in Maple is not the selected symbol but the ASCII character with the same encoding. (See Chapter 14 for further background.) But if the symbol font is selected in `textplot` then the correct symbol will appear in the plot. Alternatively, the online help topic `[worksheet,reference,symbolfont]` shows the mapping between keyboard characters and Greek letters in the symbol font.

3.2.6 A simple geometrical example

This simple geometrical plot illustrates most of the plotting techniques introduced so far.

Plot two radial lines joined by an arc of a circle:

```
> theta := Pi/4:  r := 0.75:  t := 't':
```

```
> P := plot({[[0,0], [1,0]],
      [[0,0], [cos(theta),sin(theta)]],
      [r*cos(t),r*sin(t), t = 0..theta]}):
```

Label the angle:

```
> t := theta/2:  r := 0.5:
  Q := textplot([r*cos(t),r*sin(t), "q"], font = [SYMBOL]):
```

Display the combined plot:

```
> display({P, Q});
```

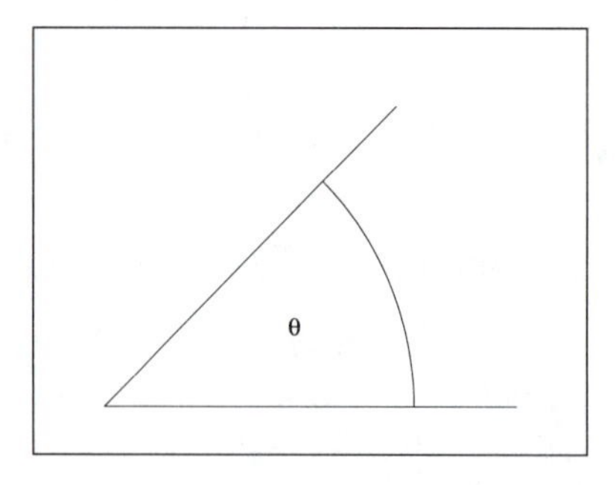

3.3 A graphical investigation of the cardioid

This section illustrates in some detail two-dimensional parametric plotting and animation of geometrical figures. (It is based on a pedagogical paper by

P. Drijvers [5].) We will investigate a locus called a *cardioid* or epicycloid of one loop.

First, we allow the plots package functions to be called with short names:

```
> restart;  with(plots):

  Warning, the name changecoords has been redefined
```

and set both axes to have the same scale by default:

```
> setoptions(scaling = CONSTRAINED);
```

Suppose we take a fixed circle F of unit radius, centred on the origin, and roll a moving circle M of unit radius (and hence diameter 2) around it, so that the two remain in contact without slipping. Then we need a total plotting area of 3 units around the origin. This allows us to fix appropriate horizontal and vertical plotting ranges as follows:

```
> setoptions(view = [-3..3, -3..3]);
```

Circles are most conveniently plotted parametrically, based on a polar coordinate representation of a point on a circle and using the angle as parameter. Remember that trigonometric functions in Maple require their arguments to be expressed in *radians* and not in degrees.

3.3.1 The basic geometry

First we store the plot structure for the fixed circle F so that we can use it again later, and then display it (with a title) to check that it is correct:

```
> F := plot([cos(t), sin(t), t = 0..2*Pi]):

> display(F, title = "The Fixed Circle F");
```

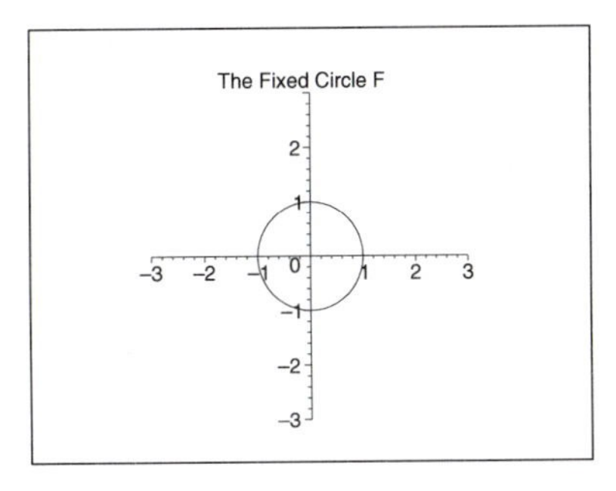

The centre of the moving circle M must always be a distance of 2 units from the origin, because it is required always to touch the fixed circle F.

Its position can be specified by the polar angle s of its centre. Hence, it is convenient to define a simple mapping or function of one variable (s) that plots the moving circle M when its centre has polar coordinates $(2, s)$; i.e., we use the same technique as above but with the centre of the circle *shifted* to polar coordinates $(2, s)$, which in Cartesian coordinates is $(2\ \cos(s),\ 2\ \sin(s))$. We use Cartesian coordinates for the actual plotting to make this shifting easier. (Try using polar coordinates if you don't believe me!)

```
> M := s -> plot([2*cos(s)+cos(t), 2*sin(s)+sin(t), t=0..2*Pi]):
```

As a test, let us plot the moving circle M with its centre on the horizontal axis (i.e., at polar angle $s = 0$), together with the fixed circle F:

```
> display({F, M(0)}, title = "The Moving Circle M");
```

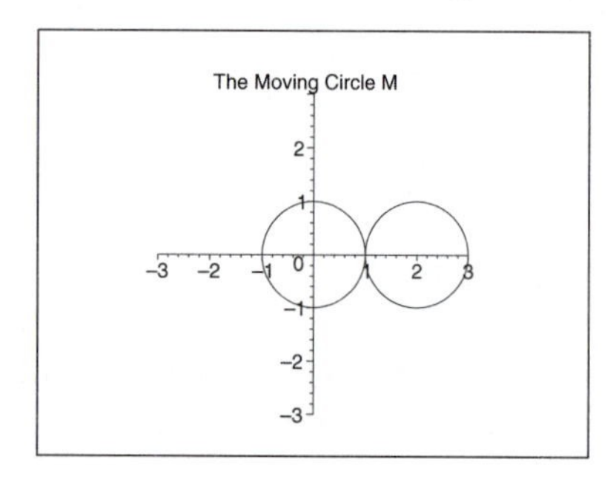

Suppose we mark the point P on the circumference of the moving circle M where it touches the fixed circle F. Then, when the centre of M is at polar coordinates $(2,\ s)$ the point P can be plotted as a small circle (using a point plot) by the following function. (The position of P depends solely on the position of M, which is specified by the angle s. The moving circle M must rotate through twice the angle of its centre [i.e., $2s$], and P starts on the left side of M as in the plot above. The coordinates of P can be easily determined relative to the centre of M to give the coordinates relative to the origin shown below, as explained in more detail later.)

```
> P := s -> plot([[2*cos(s) - cos(2*s), 2*sin(s) - sin(2*s)]],
     style = POINT, symbol = CIRCLE):
```

Note the subtle difference between the syntax of the *parametric* plot used for F and M and the *point* plot used for P. The data used to generate a point plot must always be a list of lists, even if there is only one point to plot, in which case the plot data is a list containing only one list.

When the moving point P is on the horizontal axis the combined plot looks like this, where I have used boxed axes so that P can be seen more clearly:

```
> display({F, M(0), P(0)}, title = "The Moving Point P",
     axes = BOXED);
```

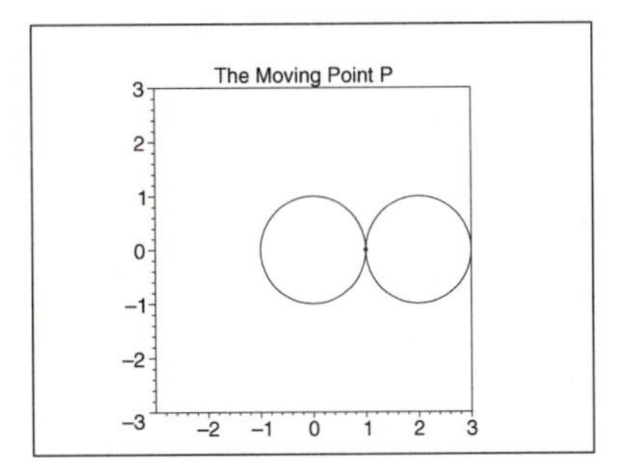

3.3.2 The geometry in more detail

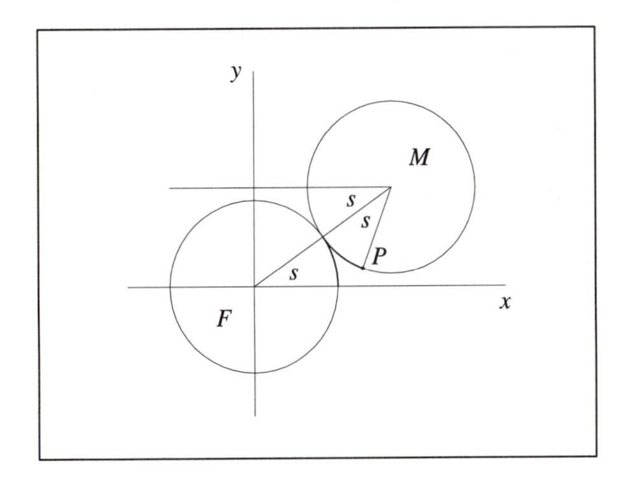

Figure 3.1: Geometry of rolling.

Suppose that, as shown in Figure 3.1, the centre of the moving circle M has moved from its initial location on the x axis at $x = 2$, $y = 0$ to Cartesian coordinates:

$$x = 2\,\cos(s),\;\; y = 2\,\sin(s).$$

Rolling means touching and not slipping, so the contact point between M and the fixed circle F moves the same distance around the circumference of both circles. Hence, the two arcs traced out, which are shown bold in Figure 3.1, subtend the same angle s at the centre of each circle. The radius of each circle is 1. Hence, relative to the centre of M, the moving point P is at an angle $2s$ below the horizontal and so at Cartesian coordinates:

$$x = -\cos(2\,s),\;\; y = -\sin(2\,s).$$

The coordinates of P relative to the origin can be found by going first to the centre of M and then to P (i.e., by adding the above coordinates) to give:

$$x = 2\cos(s) - \cos(2\,s), \; y = 2\sin(s) - \sin(2\,s).$$

3.3.3 Animating the geometry

Now we prepare to animate the rolling of the moving circle M around the fixed circle F using a number (`n_frames`) of "frames". To do this, we construct a sequence of plot structures (with no axes) where the centre of M is at a number of different angles uniformly spaced through the range $0..2\,\pi$. It looks best if this range includes both end-points so that the locus is closed. The following computation does all the work and so takes a few moments to complete:

```
> n_frames := 8:  step := 2*Pi/n_frames:
  frame_seq := seq(
     display([M(i*step), P(i*step)], axes = NONE),
        i = 0..n_frames):
```

The frames can now be displayed in sequence to animate the rolling of the circle, as follows. The first display animates the moving circle M and moving point P (using `insequence = true`) and the second displays the animation with the fixed circle F as fixed background (using `insequence = false`, the default).

```
> display([frame_seq], insequence = true):
  display({F, %});
```

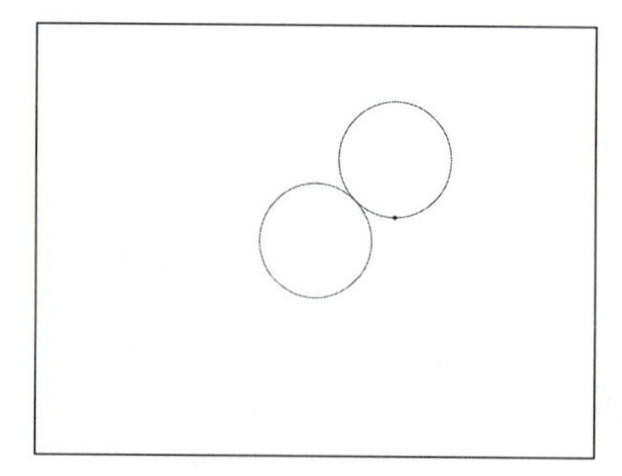

Shown above is a single frame. Within Maple you can select the plot and use the "VCR buttons" on the animation context bar to run the animation continuously, forwards, backwards, in single steps, etc. If the computer you are using is sufficiently powerful then try increasing the number of frames (`n_frames`) to 16 or 32.

The *locus* or trajectory of the moving point P is called a "cardioid" because of its heart shape. We can plot it by plotting the moving point P alone *parametrically* as the angle s runs continuously from 0 to $2\,\pi$:

```
> cardioid := plot(
      [2*cos(s)-cos(2*s), 2*sin(s)-sin(2*s), s=0..2*Pi]):
  display(cardioid, title = "The Cardioid");
```

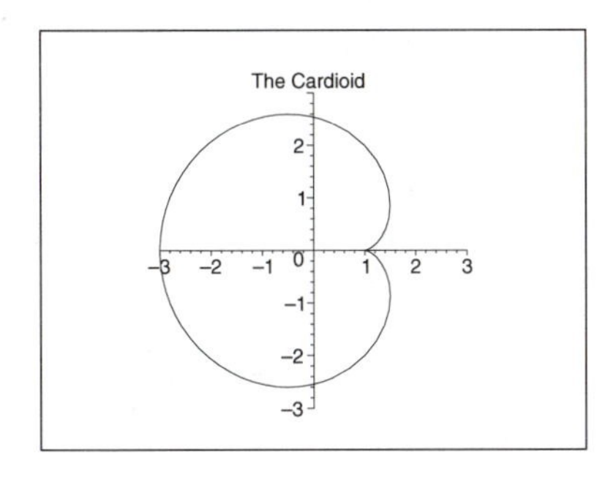

When plotted together with the fixed circle F and the moving circle M in one position (with no axes) it looks like this:

```
> display([F, cardioid, frame_seq[3]],
      title = "The Cardioid and Moving Point", axes = NONE);
```

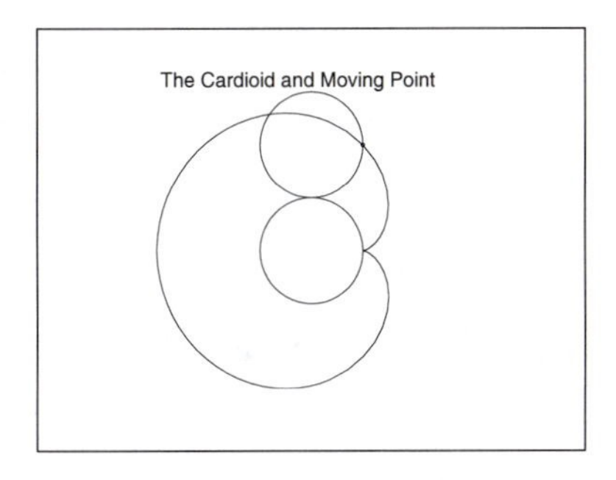

When plotted together with the moving circle M in *several* positions (with no axes) it looks like this:

```
> display([F, cardioid, frame_seq],
      title = "The Cardioid and Moving Point", axes = NONE);
```

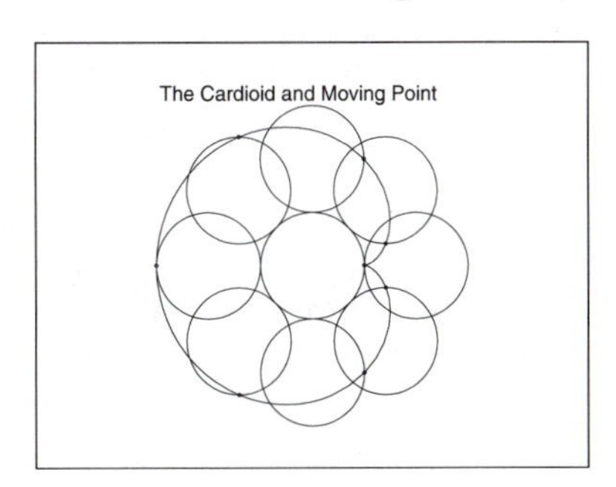

Finally, it is easy to modify this code so that the moving point P really appears to trace out the cardioid as the moving circle M rolls:

```
> Cardioid := proc(S) local s;
     # cardioid between angles 0 and S
     plot([2*cos(s)-cos(2*s), 2*sin(s)-sin(2*s), s = 0..S])
  end:
  Frame_seq :=  # including the cardioid traced so far
     seq(display([M(s*step), P(s*step), Cardioid(s*step)],
           axes = NONE),
        s = 0..n_frames):
  display([Frame_seq], insequence = true):
  display({F, %}, title = "Tracing the Cardioid");
```

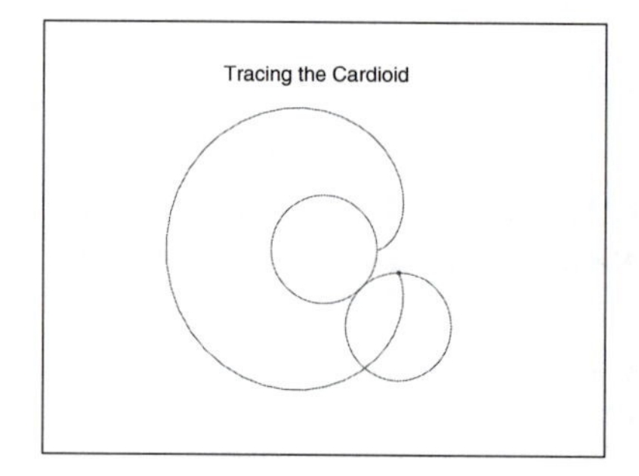

3.4 Exporting plots (and worksheets)

It may be convenient to save a complete Maple worksheet or just the plots in a different format for use with some other software (as I did to produce this book). The following comments apply equally to two- and three-dimensional plots, but they apply only to interactive export via menu options; other formats and much more control over the output of plots is available via the `plotsetup` function.

If a Maple worksheet is exported in a different format then any plots in it are also automatically exported in appropriate formats. If the worksheet is exported as HTML (HyperText Markup Language) for use with a web browser then the plots are exported a GIF (Graphics Interchange Format) files. (In fact, each piece of mathematical notation is also exported as a separate GIF file, which can lead to a huge number of small GIF files. Future versions of Maple will avoid this proliferation by exporting the mathematics as MathML embedded in the HTML file.) If the worksheet is exported as LaTeX for typesetting then the plots are exported as EPS (Encapsulated PostScript) files. The LaTeX file is constructed so that a suitable DVI (DeVice Independent) file driver [9, §1.2.2], such as `dvips`, will automatically include the plots in the PostScript that it generates, and a suitable previewer will also display the plots (although it may need the support of a PostScript interpreter such as the free GhostScript interpreter). Exporting to Maple Explorer also save the plots as GIF files and may be a convenient way to save all the plots as GIF files without also saving the mathematics as GIF files. Exporting to RTF (Rich Text Format) saves the plots internally in the RTF file and exporting as text format does not save the plots at all.

Alternatively, an individual plot can be saved in various formats. After selecting a plot by clicking the mouse on it, the menu bar changes to show an Export menu, which can also be accessed as a submenu of the context menu called "Export As" by *right*-clicking (or on a Macintosh platform *Option*-clicking) on the plot. This menu offers various export formats including EPS and GIF, some of which may be proprietary and depend on the platform on which you are running Maple (e.g., Windows Bitmap BMP and Windows Metafile WMF formats on Windows platforms).

If an animation (either two- or three-dimensional) is exported to a GIF file then it automatically produces an animated GIF, which should automatically display as an animation in a suitable viewer, such as a web browser. Other formats do not support animation and will normally just show the first frame.

3.5 Exercises: the cycloid

The aim of this exercise is to investigate graphically the motion of a point P on a circle C of unit radius rolling along a straight line L through one complete revolution from left to right. The locus traced out is called a *cycloid*. Use the cardioid example discussed in the text as a model.

The basic geometry

Allow functions in the `plots` package to be called with short names and set the default scaling to be constrained (to 1:1).

Construct a plot representing a circle *C1* of unit radius with its centre at the Cartesian coordinates $(-\pi, 1)$ and assign it to a variable for later use. (Use parametric plotting.) Construct a plot representing a circle *C2* of unit radius with its centre at the Cartesian coordinates $(+\pi, 1)$ and assign it to a variable for later use. Construct a plot representing the straight line L from $(-\pi - 1, 0)$ to $(+\pi + 1, 0)$ and assign it to a variable for later use. (Specify the end points of L as points and let Maple plot the line between them.) Display *C1*, *C2*, and L together on the same plot, without axes (which would obscure L).

The two circles *C1* and *C2* represent the limits of the motion of the circle C as it rolls. Choose a horizontal range that fully displays both circles, and keep it fixed for the rest of this exercise.

The rolling circle

Consider the circle C with its centre at $(0, 1)$. Suppose a point P is marked on the highest point of the circle. Now roll the circle to the *right* away from this position along the line L, keeping P fixed relative to C. Let θ denote the angle from the vertical to the radius to P. The geometry is shown in Figure 3.2. By using elementary circle geometry to determine the distance that the

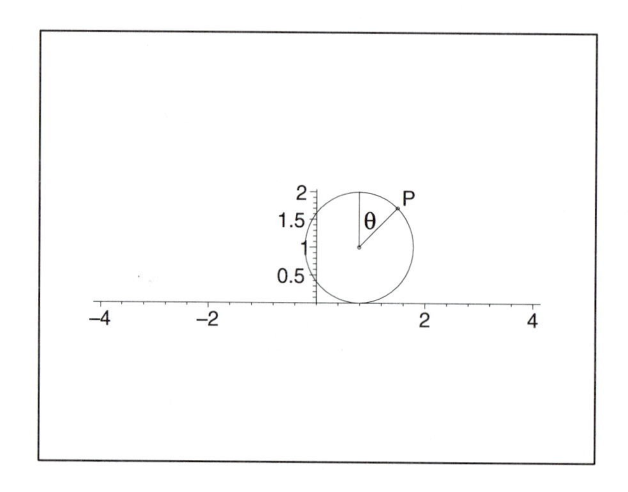

Figure 3.2: The rolling circle.

circle has rolled along L after it has rotated through an angle θ, convince yourself that the coordinates of the centre of C as a function of θ are $(\theta, 1)$. Hence, write and *test* a Maple function `C(theta)` that returns a plot structure representing the circle C after it has rolled along L through an angle θ, i.e., the circle (only) shown in Figure 3.2.

By using elementary geometry, convince yourself that the coordinates of the point P *relative* to the centre of C are $(\sin(\theta), \cos(\theta))$ and hence that the *absolute* coordinates of P (relative to the origin) as a function of θ are $(\theta+\sin(\theta), 1+\cos(\theta))$. Hence, write and *test* a Maple function `P(theta)` that returns a plot structure representing the point P after the circle C has rolled along L through an angle θ, i.e., the point P (only) shown in Figure 3.2.

Write and *test* a Maple function `CP(theta)` that combines $C(\theta)$ and $P(\theta)$ into one plot. Use function `CP(theta)` to construct a sequence of 17 plot structures, each combining $C(\theta)$ and $P(\theta)$, for uniformly spaced values of θ in the range $[-\pi, \pi]$ inclusive (i.e., in steps of $\pi/8$). Display the sequence as an *animation* of the circle C rolling along the line L as a fixed background, without axes (which would obscure L).

The cycloid

Now use the formula specifying the coordinates of P to plot its *locus* (i.e., the path that it traces as the circle rolls) for $-\pi \leq \theta \leq \pi$, together with a suitable title (and axes). Finally, combine the locus with the sequence of rolling circles and points used for the animation and display the combined plot (as a static superposition, not as an animation) together with the line L (with no axes).

Modify the animation so that the point on the moving circle actually draws the cycloid as the circle rolls.

3.6 Appendix: code to plot the figures

Modules (see Chapter 15) are used here to provide a local context within which to isolate this code from the rest of the worksheet.

Geometry of rolling (Figure 3.1)

This code calls plotting functions written in Section 3.3.

```
> module()
     local s, t, L, A, T, F;
     plots[setoptions](scaling=CONSTRAINED, colour=black);
     # Choose a convenient location for M:
     s := 35*Pi/180:  # 35 degrees
     # Construct the three straight lines:
     L := plot({[[0,0], [2*cos(s),2*sin(s)]],
        [[-1,2*sin(s)], [2*cos(s),2*sin(s)]],
        [[2*cos(s)-cos(2*s),2*sin(s)-sin(2*s)],
           [2*cos(s),2*sin(s)]]}):
     # Construct the two thick arcs:
     A := plot({[cos(t), sin(t), t = 0..s],
        [2*cos(s)-cos(t), 2*sin(s)-sin(t), t = s..2*s]},
        thickness=3):
```

```
    # Plot the text:
    T := plots[textplot]({[0.5*cos(s/2),0.5*sin(s/2), "s"],
       [2*cos(s)-0.5*cos(1/2*s),
           2*sin(s)-0.5*sin(1/2*s), "s"],
       [2*cos(s)-0.5*cos(3/2*s),
           2*sin(s)-0.5*sin(3/2*s), "s"],
       [2*cos(s)-cos(2*s)+0.2,2*sin(s)-sin(2*s)+0.15, "P"],
       [-0.5*cos(Pi/4),-0.5*sin(Pi/4), "F"],
       [2*cos(s)+0.5*cos(Pi/4),
          2*sin(s)+0.5*sin(Pi/4), "M"],
       [3, -0.2, "x"], [-0.2, 2.5, "y"]},
       font=[TIMES,ITALIC,12]):
    # Combine the plots:
    F := plot([cos(t), sin(t), t = 0..2*Pi]):
    plots[display]([F, M(s), P(s), L, A, T],
       view=[-1.5 .. 3, -1.5 .. 2.5],
       tickmarks=[0,0]);
    # Print the plot (modules do not return values):
    print(%)
 end module:
```

The rolling circle (Figure 3.2)

```
> restart;
```

```
> module()
    local theta, xrange, t;
    theta := Pi/4;  xrange := -Pi-1..Pi+1;
    plots[setoptions](scaling=CONSTRAINED, colour=black);
    print(plots[display]({
       plot([theta+cos(t), 1+sin(t), t = 0..2*Pi], xrange),
       plot([[theta+sin(theta), 1+cos(theta)]], xrange,
          style=POINT, symbol=CIRCLE),
       plot([[theta, 1]], xrange, style=POINT, symbol=CIRCLE),
       plot([[theta, 1],
          [theta, 2]], xrange),
       plot([[theta, 1],
          [theta+sin(theta), 1+cos(theta)]], xrange),
       plots[textplot](
          [theta+sin(theta)+0.2, 1.2+cos(theta), "P"]),
       plots[textplot](
          [theta+0.2, 1.5, "q"], font=[SYMBOL])
    }))
 end module:
```

Chapter 4

Three-Dimensional Plotting

This chapter is about graphics that exist logically in three dimensions, even though, with current readily available technology, they can exist physically only in two dimensions. However, Maple preserves much of their three-dimensional nature by allowing real-time rotation, apparently in three dimensions. Just as in two dimensions, three-dimensional graphics arise as graphs of functions and from problems in three-dimensional geometry, and three-dimensional plots can be animated.

The first section describes how to use the `plot3d` function in the main Maple library to produce simple three-dimensional plots. The next two sections are concerned with controlling plots interactively and programmatically. The `plot3d` function can plot only surfaces; another important class of graphics is curves, which in three dimensions can in general be twisted space curves, although contours or level curves are by definition plane curves. To plot curves in three dimensions or manipulate three-dimensional plots it is necessary, as in two dimensions, to use the `plots` package, as described in Section 4.4. Section 4.5 is a case study that illustrates a simple three-dimensional geometrical animation and Section 4.6 explains the export formats that preserve three-dimensional information. Finally, an appendix contains the Maple code used to construct the figures.

Three-dimensional (3D) plotting is largely an extension of two-dimensional (2D) plotting, but beware that there are a number of undesirable syntactic inconsistencies! The result of any 3D plotting function is a 3D plot structure represented by the inert function `PLOT3D`. These structures can be manipulated like 2D plot structures (represented by the inert function `PLOT`). But 2D and 3D plot structures can never be combined. Beware that both constructing and displaying 3D plots take more time and memory than 2D plots. When working with 3D plots, I recommend that you save your worksheet frequently and delete any unnecessary plots.

4.1 The `plot3d` function

Many 3D plotting functions have names that are the same as their 2D analogues but with "3d" appended. This is the case with the main 3D plotting function, which is the 3D analogue of `plot` and is called `plot3d`. It is not quite as flexible (i.e., overloaded) as its 2D analogue. It cannot be used to plot lines, curves, or points, although other functions are available for these purposes.

One of the main problems with 3D plotting is how to represent three dimensions on a two-dimensional display. It is resolved by making a projection, which is intended to represent three-dimensional space in essentially the way that we see it (but using only one eye). There is also the problem of how to represent a surface. To illustrate the solution, let us begin with a brief look at surfaces from a mathematical point of view.

4.1.1 Graphs of real-valued functions of two real variables

What is the *graph* of a real-valued function of two real variables and how does one plot it? A function of two variables,

$$f : \mathbb{R}^2 \to \mathbb{R}, \quad (x, y) \mapsto f(x, y),$$

can be regarded as a *family* of functions of one variable:

$$f_y : \mathbb{R} \to \mathbb{R}, \quad x \mapsto f_y(x).$$

The graph of f_y is a curve in the $(x,\ f)$ plane $\mathbb{R} \times \mathbb{R}$. For each y in (some interval of) $\mathbb{R}$, we can stack these curves together to define a *surface* in the $(x,\ y,\ f)$ space $\mathbb{R}^2 \times \mathbb{R}$, as developed from left to right in Figure 4.1. This

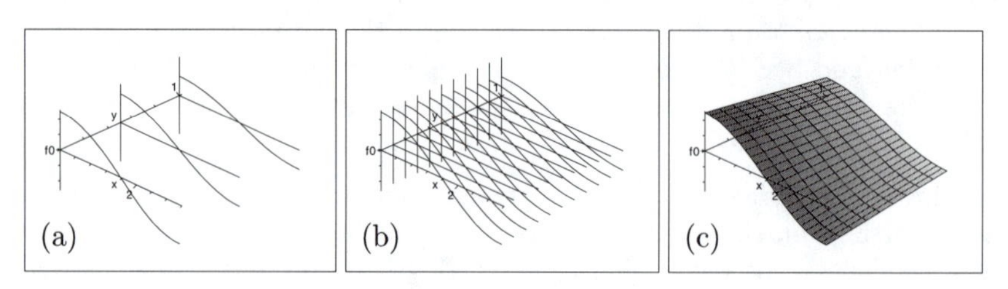

Figure 4.1: Constructing a surface by stacking curves.

surface S is the *graph* of f. If f is a (single-valued) function then there can be at most one point of S above each point (x, y) in the domain $\mathbb{R}^2$ (i.e., the graph cannot "fold over"). If the graphs of f_y and f_x for discrete sets of values of y and x are superimposed then they form a "rectangular" grid within the graph of f, as in Figure 4.1(c). This grid is the basis for a graphical representation of the surface. It can be displayed as an open "wire frame", or the cells of the grid can be shaded to give a solid appearance, or both representations can be used together.

In Maple, the main function for plotting graphs of functions of two variables is `plot3d`, and its syntax is similar to that of `plot`. If f is a mapping that takes two arguments then the main syntax for plotting f as an expression is

```
plot3d(f(x,y), x = a..b, y = c..d)
```

and for plotting f as a mapping is

```
plot3d(f, a..b, c..d).
```

For example:

```
> plot3d(x*sin(y), x = -1..1, y = -Pi..Pi);
```

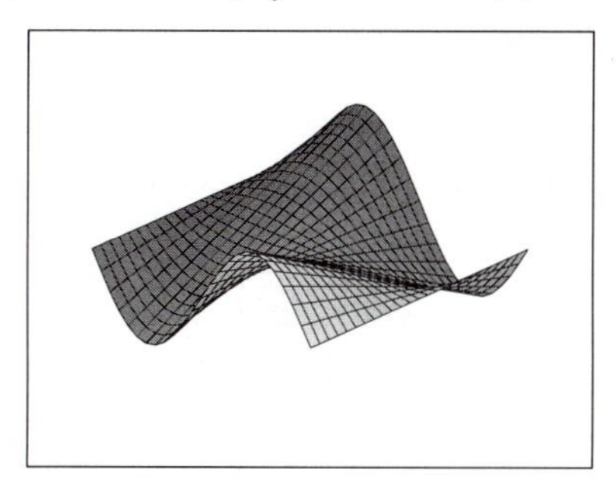

The following mapping version

```
> plot3d((x,y) -> x*sin(y), -1..1, -Pi..Pi):
```

produces an identical plot. Notice that the default for 3D plotting is to have no axes. The first three arguments to `plot3d` as used above are always required and there are no default plot ranges.

4.1.2 Parametric plotting of surfaces

Surfaces that are not graphs of (single-valued) functions can be specified parametrically using *two* parameters — the coordinates of each point of the surface are specified as functions of two variables, e.g.,

$$(x(s,t),\ y(s,t),\ z(s,t)).$$

In Maple, these three expressions or functions are specified as the elements of a list, thus:

$$\texttt{plot3d([}x(s,t),y(s,t),z(s,t)\texttt{], } s = a..b,\ t = c..d\texttt{)}$$

Note that the parameter ranges are not included in the list for a 3D plot, whereas they are for a 2D plot! The coordinates can also be specified as mappings, in which case the parameter names are omitted from the range specifications also, thus:

$$\texttt{plot3d([}x,y,z\texttt{], } a..b,\ c..d\texttt{)}$$

where x, y, z must be Maple mappings defined with two arguments. Hence, the following are equivalent:

$$\texttt{plot3d(}f(x,y)\texttt{, } x = a..b,\ y = c..d\texttt{)}$$

$$\texttt{plot3d([}x,y,f(x,y)\texttt{], } x = a..b,\ y = c..d\texttt{)}$$

For example, the above modulated sine surface could have been plotted parametrically like this:

```
> plot3d([x, y, x*sin(y)], x = -1..1, y = -Pi..Pi):
```

Spheres and spheroids

A point on a *sphere* of radius r can be expressed parametrically (using spherical polar coordinates as parameters) as:

$$x = r\sin(\phi)\cos(\theta),\ y = r\sin(\phi)\sin(\theta),\ z = r\cos(\phi),$$

$$0 \le \theta \le 2\pi,\ 0 \le \phi \le \pi.$$

The significance of the parameters is shown in Figure 4.2. Note that the x and y axes play symmetrical roles, but the z axis plays a distinguished role in this parametrization.

Hence, a sphere of unit radius centred at the origin can be plotted in Maple like this:

```
> plot3d([sin(phi)*cos(theta), sin(phi)*sin(theta), cos(phi)],
      theta = 0..2*Pi, phi = 0..Pi, scaling = CONSTRAINED);
```

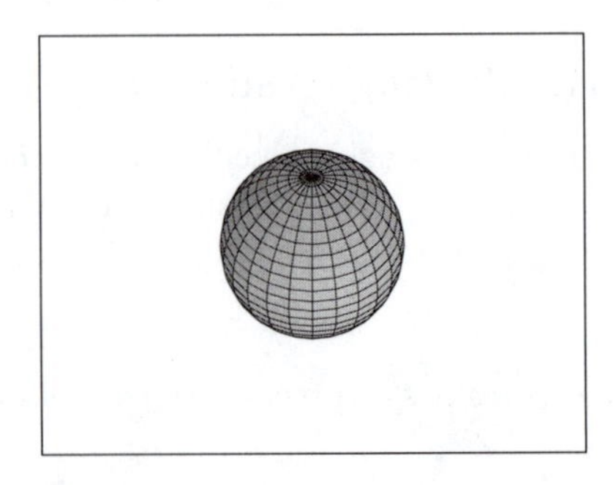

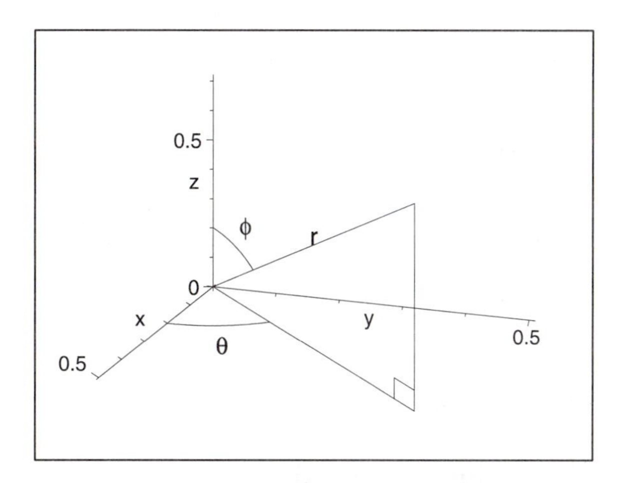

Figure 4.2: Spherical polar coordinates.

A *spheroid* is a sphere that has been squashed or stretched in one direction and is intermediate between a sphere and a general ellipsoid. A squashed sphere is called an *oblate* spheroid and a stretched sphere is called a *prolate* spheroid. The direction of squashing or stretching is the only axis of rotation symmetry of a spheroid, and when a spheroid is described using spherical polar coordinates it is simplest to make this symmetry axis the z axis. Thus, a point on a *spheroid* can be expressed parametrically (again using spherical polar coordinates as parameters) as:

$$x = a\sin(\phi)\cos(\theta),\ y = a\sin(\phi)\sin(\theta),\ z = b\cos(\phi),$$

$$0 \le \theta \le 2\pi,\ 0 \le \phi \le \pi.$$

Relative to a sphere, it has been stretched along the z axis if $b > a$ and squashed if $b < a$, e.g.,

```
> plot3d([sin(phi)*cos(theta),sin(phi)*sin(theta),2*cos(phi)],
     theta = 0 .. 2*Pi, phi = 0 .. Pi,
     title = "Stretched along z",
     scaling = CONSTRAINED);
```

```
> plot3d([sin(phi)*cos(theta),sin(phi)*sin(theta),1/2*cos(phi)],
     theta = 0 .. 2*Pi, phi = 0 .. Pi,
     title = "Squashed along z",
     view = [-1..1,-1..1,-2..2], # only for alignment
     scaling = CONSTRAINED);
```

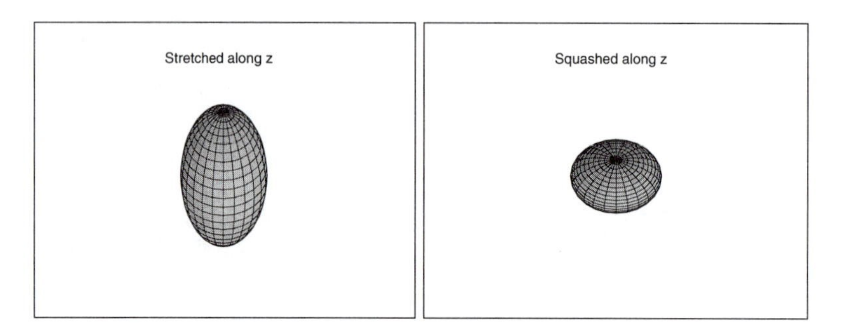

We will use these ideas again in Section 4.5 to animate a bouncing ball.

4.1.3 Multiple three-dimensional plots

The first argument of `plot3d` may be a *set* of any of the above data forms to superimpose or merge a set of surfaces, e.g.,

```
> plot3d({sin(x),sin(y)}, x = -Pi..Pi, y = -Pi..Pi);
```

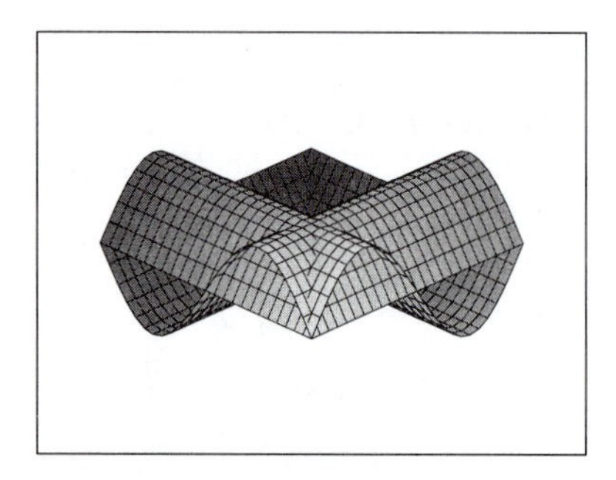

However, unlike `plot`, `plot3d` does not accept a list of plot data forms:

```
> plot3d([sin(x),sin(y)], x = -Pi..Pi, y = -Pi..Pi);
```

```
Error, (in plot3d) first argument must be either in standard or
parametric form
```

4.2 Three-dimensional plotting options

These are specified, as for 2D plotting, as equations of the form `option = VALUE` *after* the first three compulsory arguments. Many options are the same as for 2D plotting. The full set of 3D plotting options is available in the online help page `plot3d,option`. The following are some particularly important options:

`title` — as for 2D.

`axes` — as for 2D except that `NONE` is the default.

`scaling` — as for 2D.

`style = WIREFRAME, HIDDEN, PATCH, PATCHNOGRID`
This specifies how the surface is to be drawn. Some of the options may not be available on some devices. The default style is `HIDDEN` (for hidden line rendering).

`WIREFRAME` means take a rectangular grid in the domain and project it onto the surface.

`HIDDEN` is the same as `WIREFRAME`, but suppresses any lines that are behind another part of the surface.

`PATCH` means shade the surface to make it look solid, plus `HIDDEN`.

`PATCHNOGRID` means shade only.

`style = CONTOUR, PATCHCONTOUR`
I will discuss the contouring options later.

`orientation` = $[\theta, \phi]$
This specifies the orientation of the plot in terms of the *viewing angle* using spherical polar coordinates as shown in Figure 4.3. The orientation is specified in *degrees* (whereas normally Maple and most other programming languages use radians) and the default is $[45°, 45°]$.

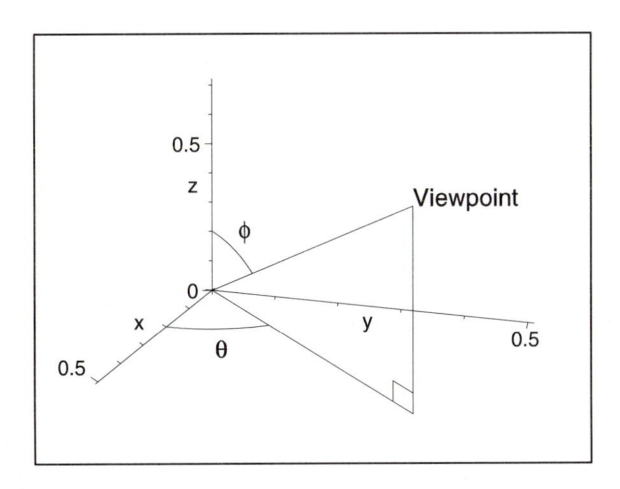

Figure 4.3: The orientation parameters $[\theta, \phi]$.

4.3 Interacting with three-dimensional plots

To interact with a 3D plot in a worksheet, click the mouse somewhere on the plot. This displays a frame around the plot with "handles" that can be dragged with the mouse to resize the plot. It also changes the context bar and menus to display options for controlling 3D plots. The plot orientation is specified in terms of the viewing angle (see Figure 4.3) at the left of the context bar. This can be changed by simply editing it, by using the "spinner" buttons, or by dragging the mouse within the plot frame, which causes the plot to rotate in space. This is "real-time" rotation, which means that it happens with no delay and it provides some of the facilities of a simple virtual reality browser (but see also Section 4.6). Clicking the right mouse button on a 3D plot (*Option*-clicking on a Macintosh) brings up a context menu offering as sub-menus the same plotting menus that appear on the main menu bar.

4.4 Using the `plots` package

The `plots` package provides additional support for 3D as well as 2D plotting. The functions described in the rest of this chapter are all in the `plots` package, which must first be activated using the `with` command before the functions in it can be called using only their simple short names:

```
> with(plots):

  Warning, the name changecoords has been redefined
```

4.4.1 Setting default options and plotting text: `setoptions3d`, `textplot3d`

These are 3D analogues of `setoptions` and `textplot`.

4.4.2 Combining and animating three-dimensional plots: `animate3d`, `display3d`

The functions `animate3d` and `display3d` are analogous to the 2D functions `animate` and `display`. We will consider a simple 3D bouncing ball animation in Section 4.5 that uses `display3d`. Remember that 2D and 3D plot structures cannot be combined, so the elements in a set or list of plot structures to be combined by `display` or `display3d` must be either *all 2D* or *all 3D* plot structures. As for `display`, if plot structures have conflicting options then `display3d` picks those that it sees first. Hence, it is usually best to set any important options explicitly in the plotting functions call or in `setoptions3d`. (The identifier `display3d` is actually just a synonym for `display`, so it does not matter which is used, although it is good practice to use the correct name.)

4.4.3 Plotting curves in space: spacecurve

A *curve* is a one-dimensional continuous set of points — an image of a line under some mapping. A curve can be embedded in a space of any higher dimension, such as $\mathbb{R}^2$ or $\mathbb{R}^3$. The natural way to describe a curve is parametrically in terms of a parameter that varies monotonically along the curve. (One important such parameter is the arc length or distance along the curve.)

Space curves in $\mathbb{R}^3$ can be plotted in Maple *only* by using the function `plots[spacecurve]`. Its use is similar to that of the `plot` function for parametric plotting of curves in $\mathbb{R}^2$. A space curve C can be defined as a list of three expressions or simple mappings, or as a list of points each represented as a list of three coordinates, i.e., as any one of the following, where *fx*, etc. evaluate to simple mappings and *x1*, etc. evaluate to numbers:

$$C \;\texttt{:=}\; \texttt{[}fx\texttt{(}s\texttt{)},\; fy\texttt{(}s\texttt{)},\; fz\texttt{(}s\texttt{)},\; s \;\texttt{=}\; a\texttt{..}b\texttt{]}$$

$$C \;\texttt{:=}\; \texttt{[}fx,\; fy,\; fz,\; a\texttt{..}b\texttt{]}$$

$$C \;\texttt{:=}\; \texttt{[[}x1,\; y1,\; z1\texttt{]},\; \texttt{[}x2,\; y2,\; z2\texttt{]}\; \texttt{...]}$$

A single space curve is plotted by a call of the form:

$$\texttt{spacecurve(}C,\; options\texttt{)}$$

and several space curves can be plotted together by a call of the form:

$$\texttt{spacecurve(}\{C1, C2, \ldots\}\texttt{,}\; options\texttt{)}$$

If all the space curves are represented parametrically then the first *option* may be a range in either of the forms `s = a..b` or `a..b`, in which case it sets a default for any range omitted from a spacecurve definition in the first argument. For example, three orthogonal circles could be plotted as follows. (I have turned on display of axes, without which it is difficult to appreciate the 3D structure.)

```
> c := cos(theta):  s := sin(theta):
  spacecurve({[0,c,s], [s,0,c], [c,s,0]},
     theta = 0..2*Pi, scaling = CONSTRAINED,
     axes = NORMAL, orientation = [30,60]);
```

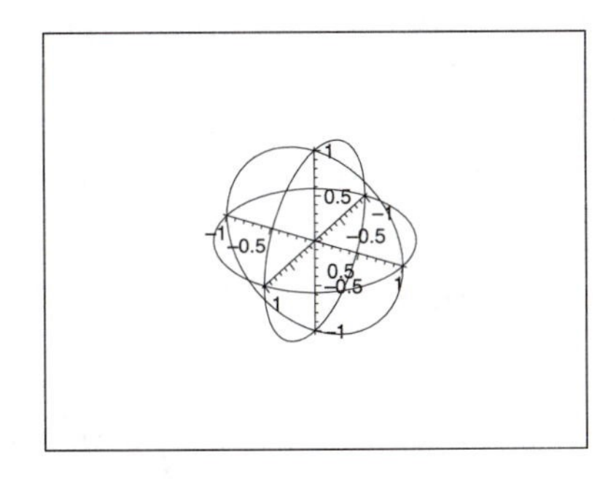

The `spacecurve` function also allows *points* to be plotted in three-space; they are joined together by straight lines unless the option `style = POINT` is specified. As an example of plotting several straight spacecurve segments joining specific points, the coordinate axes themselves could be plotted together with frame axes like this:

```
> spacecurve({[[-1,0,0],[+1,0,0]],
              [[0,-1,0],[0,+1,0]],
              [[0,0,-1],[0,0,+1]]}, axes = FRAME);
```

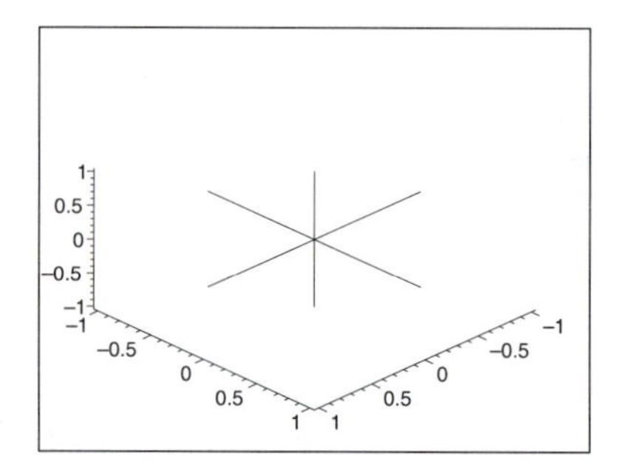

4.4.4 Plotting contours: `contourplot3d`, `contourplot`

The *contours* of a surface are its intersections with a set of parallel planes at different fixed heights z, as shown in Figure 4.4. If a surface is the graph of a function $f(x, y)$ then its contours are the curves defined (implicitly) by:

$$f(x, y) = z_i$$

for some discrete set of heights $z = z_i$. It is common to project the contour curves down onto the domain of the function, i.e., the (x, y) plane, which gives the same contours as on a geographical map, as shown for the above example in Figure 4.4(d).

However, contours can also be plotted as space curves in (x, y, z). The option `style = CONTOUR` causes the function `plot3d` to plot *only* contours as space curves in (x, y, z); the option `style = PATCHCONTOUR` causes it to plot contours as space curves in (x, y, z) and *also* shade the surface to make it look solid, as in the 3D contour plot in Figure 4.4(c). The function `contourplot3d` is generally the most convenient function for plotting contours. It is effectively the same as `plot3d` except that the default style is `CONTOUR`. The contours can be projected onto the (x, y) plane by setting the orientation so that the viewpoint is on the positive z axis and the (x, y) axes are oriented conventionally. The appropriate option is `orientation = [-90,0]`. Although the result appears to be two dimensional, it is still represented by a 3D plot structure and so *cannot* be combined with 2D plot structures. The function `contourplot` produces essentially the same appearance, except that it returns a 2D plot

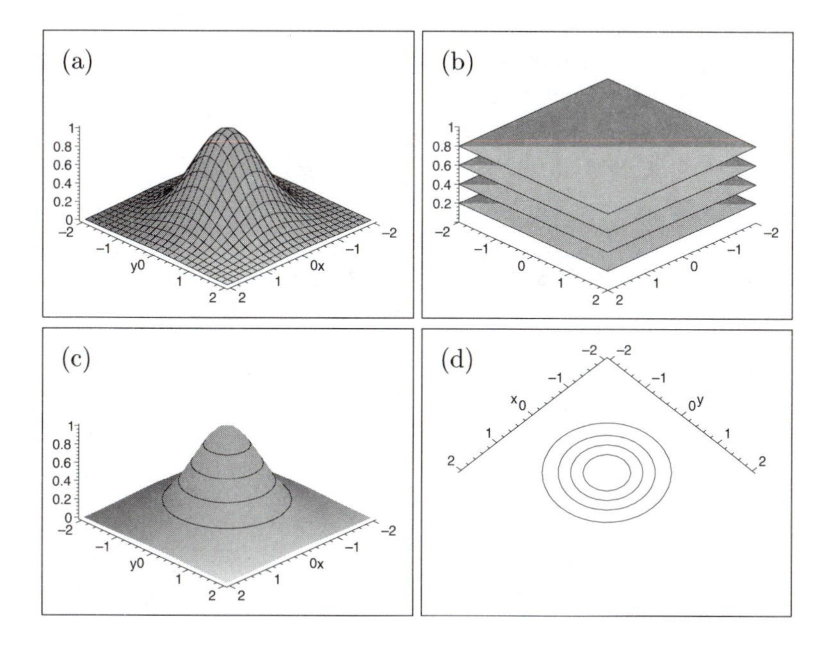

Figure 4.4: Contouring by slicing a graph and projecting.

structure. However, it is very slow! Note that `contourplot` neither accepts nor requires an `orientation` option.

Beware that if the 3D plotting option `orientation = [-90,0]` is used together with `axes = NORMAL` then the axes may appear to be mislabelled, e.g.,

```
> contourplot3d(exp(-x^2-y^2), x = -3..3, y = -2..2,
    orientation = [-90,0], axes = NORMAL);
```

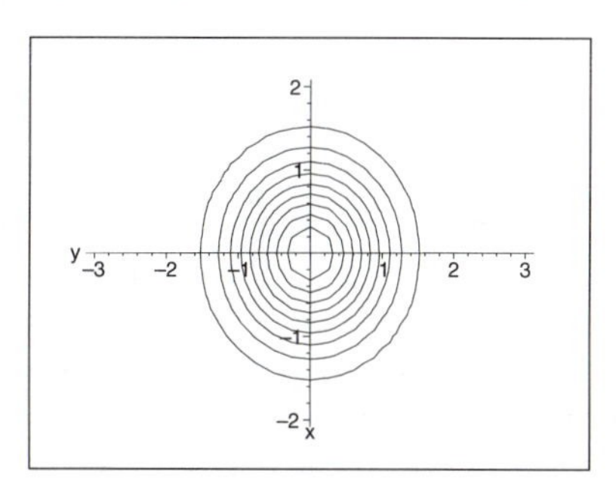

I believe this is caused by an error in the code that tries to optimize the position of the axis labels. If you interactively make very small changes to the orientation of the above plot you will see that the positions of the axis labels

are unstable and jump from one end of the axes to the other. I think that the code is trying to label each axis near the origin, as it does with `FRAME` or `BOXED` axes, but with `NORMAL` axes the other axis is in the way, which pushes the label to the end of the other axis. By choosing the axes to be of different lengths, as above, one sees that the axes themselves are correct.

The difference between `contourplot3d` and `contourplot`

The function `contourplot3d` returns a 3D plot structure representing the entire surface and the actual contour computation is done by Iris, the Maple GUI that is essentially written in C and so is fast. By contrast, `contourplot` is a procedure written in the Maple language that computes the locations of the projected contour curves and returns a 2D plot structure representing a set of plane curves that is then just plotted by Iris. Because interpreted Maple code is much slower than compiled C code, `contourplot` is much slower than `contourplot3d`. Using `contourplot` is equivalent to using `implicitplot`, which is described below.

Beware that the online help states that both functions take the same arguments, which is not strictly true because `contourplot3d` accepts 3D plotting options whereas `contourplot` accepts 2D plotting options, and the 2D and 3D plotting options are not identical. Moreover, the defaults are slightly different.

Specifying contour levels

The plotting functions `plot3d`, `contourplot3d`, and `contourplot` all accept an option having one of the following forms

- `contours =` *`number of contours`*, where the levels are chosen automatically
- `contours =` *`list of specific contour levels`*

The online help is inconsistent, but the default appears to be 8 uniformly spaced contour levels for `contourplot` and 10 for `contourplot3d` and `plot3d`.

Critical points of functions

There are a number of important local features of functions that have characteristic graphs and contours. Some of the simplest and most important are the simple *critical points* of functions of two variables, which are *maxima*, *minima*, and *saddle points*. A maximum was illustrated above and a minimum is the same but the other way up. Surface and contour plots of a saddle point are shown in Figure 4.5. Both the surface and the contour plot have the same (x, y) orientation to facilitate comparison. Surface and contour plots together can provide a lot of information about the nature of a function of two variables. (Notice that the contour plot is missing part of the horizontal contour line. This is a common problem with contouring routines. In general,

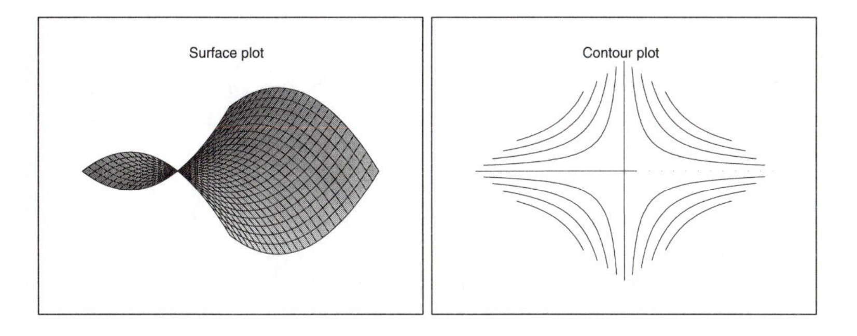

Figure 4.5: Surface and contour plots of a saddle point.

they are bad at handling contour lines that intersect, because the intersection is a singularity of the family of curves.)

4.4.5 Plotting gradient fields: `gradplot`

The *gradient* of a real-valued multivariate function is a vector field in the domain of the function. At any point, the direction of the gradient vector is the direction in which the function increases fastest and the magnitude of the gradient vector is the rate of fastest increase (i.e., it is the direction and magnitude of steepest slope of the graph of the function). A vector field can be depicted as a field of small arrows, where the direction and length of an arrow at any point represent the direction and magnitude of the corresponding vector at that point, e.g.,

```
> plot3d(y^2, x = 0..1, y = 0..1, axes = NORMAL,
     style = PATCH, orientation = [-65,75],
     title = "Surface plot");
```

```
> gradplot(y^2, x = 0..1, y = 0..1, grid = [10,10],
     arrows = THICK, title = "Gradient field");
```

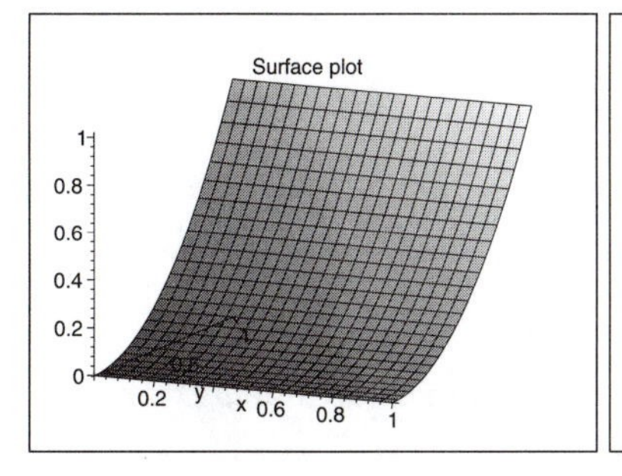

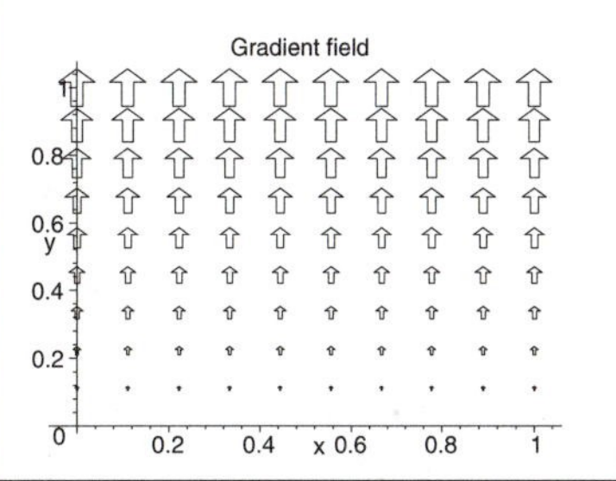

The function `plots[gradplot]` plots the gradient field of a function on a 2D domain and its arguments are the same as for `plot3d`. This function returns a 2D plot structure, which can therefore be combined with other 2D plot structures, such as that returned by `contourplot`. It is an easy exercise in vector calculus to prove that the gradient vector of a real-valued function at any point is perpendicular to the tangent to the contour through that point (in any number of dimensions), which can be demonstrated in 2D by using Maple to superimpose a `gradplot` and a `contourplot`. Note that we cannot use the faster `contourplot3d` routine because it returns a 3D plot structure, and remember to set `scaling = CONSTRAINED` to ensure that the contour lines and gradient vectors are manifestly perpendicular!

4.4.6 Implicit plotting: `implicitplot`

A curve in a two-dimensional space or a surface in a three-dimensional space can be defined *implicitly* by an equation of the form $f(x, y) = 0$ or $f(x, y, z) = 0$, respectively, e.g.,

$$\frac{x^2}{a^2} + \frac{y^2}{b^2} = 1 \quad \text{and} \quad \frac{x^2}{a^2} + \frac{y^2}{b^2} + \frac{z^2}{c^2} = 1$$

define the general ellipse in two dimensions and ellipsoid in three dimensions, respectively. In order to plot figures defined implicitly, the equations that describe them essentially have to be solved in some way, which generally is *slow*. (Of course, the particular equations for the ellipse and ellipsoid shown above can be solved parametrically, which provides a fast way to plot them.) The general 2D curve defined by $f(x, y) = 0$ can be plotted in Maple by the function `plots[implicitplot]` which takes the same arguments as `plot3d` except that the first argument can be an equation as well as an expression or mapping, e.g.,

```
> implicitplot(x^4 + y^4 = 1, x = -1..1, y = -1..1,
    scaling = CONSTRAINED);
```

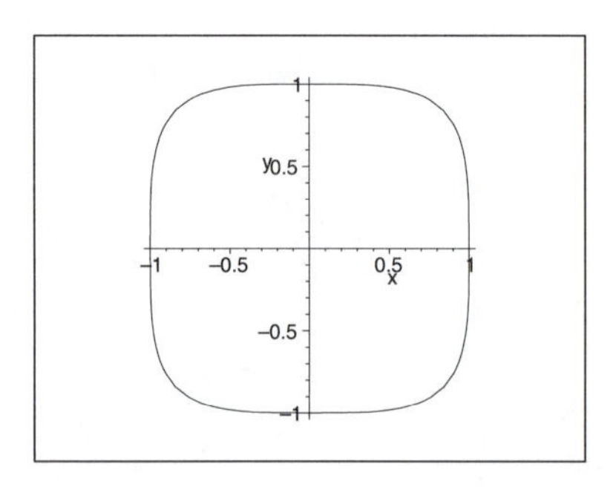

The analogous function `implicitplot3d` plots surfaces defined implicitly in 3D by equations of the form $f(x, y, z) = 0$.

There is a close link between implicit plotting and contour plotting: the curve defined implicitly by $f(x, y) = 0$ is the zero contour of $f(x, y)$, and the contours of $f(x, y)$ are the curves defined implicitly by $f(x, y) = z_i$ for some discrete set of height values z_i. Hence, the above implicit plot is identical to this contour plot:

```
> contourplot(x^4 + y^4 - 1, x = -1..1, y = -1..1,
     contours = [0], scaling = CONSTRAINED);
```

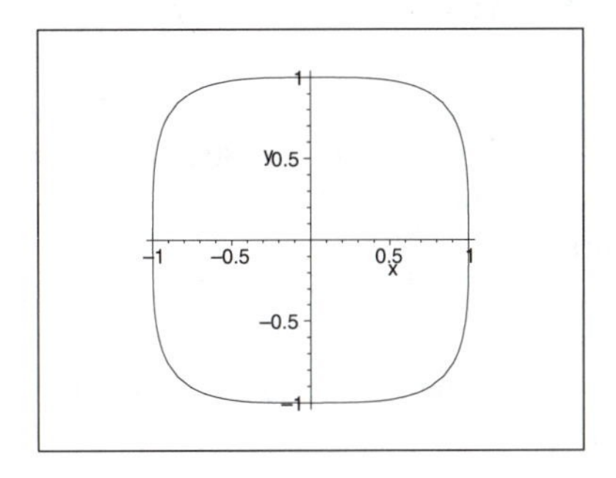

and `contourplot3d` can be used to produce essentially the same plot much faster like this:

```
> contourplot3d(x^4 + y^4 - 1, x = -1..1, y = -1..1,
     contours = [0], orientation = [-90,0], axes = NORMAL,
     scaling = CONSTRAINED);
```

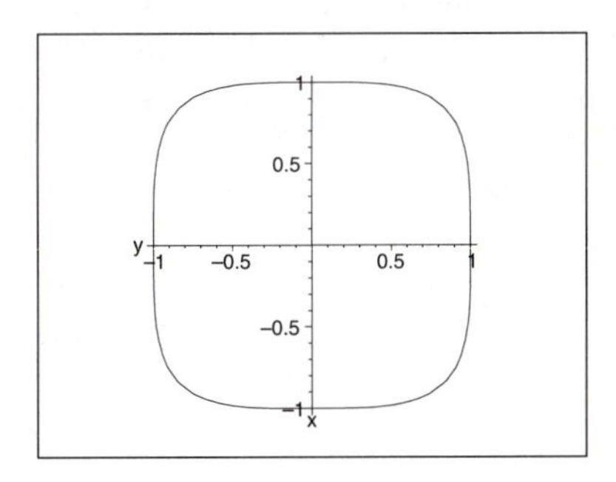

4.5 Three-dimensional bouncing ball animation

Animation in three dimensions is essentially the same as in two dimensions, but 3D animations involve one more coordinate and take more computer time and memory to generate. Here is a short case study of a simple bouncing ball animation, which we can develop in easy stages.

```
> setoptions3d(scaling = CONSTRAINED);
```

While the ball is in the air we assume its shape is a sphere. Let us make it bounce up and down the z axis. Then it is convenient to write a function, which I will call `Ball`, that plots a sphere with its centre at (0, 0, z), using a shifted version of the code presented earlier to plot a sphere:

```
> Ball := z -> plot3d([cos(theta)*sin(phi),
        sin(theta)*sin(phi), z + cos(phi)],
     theta = 0..2*Pi, phi = 0..Pi):
```

To test this, we could call the function `Ball` with various arguments and display it with axes turned on.

When the ball hits whatever it is bouncing off, it will squash down the z axis and then expand again before actually bouncing off in the opposite direction. We can represent the squashed ball as an oblate spheroid, and in this coarse animation sequence the squashed ball frame represented by the following single plot seems to suffice:

```
> Squash := plot3d([cos(theta)*sin(phi),
        sin(theta)*sin(phi), -10.5+0.5*cos(phi)],
     theta = 0..2*Pi, phi = 0..Pi):
```

Now we can build the animation sequence. I have used 11 frames with the centre of the ball falling from $z = 0$ to $z = -10$, then one squashed ball frame, then 11 frames with the centre of the ball rising from $z = -10$ back to $z = 0$. The object the ball is bouncing off is assumed to be at $z = -11$, and the lower surface of the ball touches it in the last falling frame, the squashed ball frame, and the first rising frame. The location of the origin was chosen for convenience; no axes are displayed so the observer does not know what origin we have chosen. The argument of `Ball` is a quadratic function of the frame parameter k, which represents time, in order to capture the dynamics of the vertical motion of a body under gravity. (There is no resistance, so this ball will bounce forever!) Here is the animated frame sequence:

```
> Bounce := display3d(
     [seq(Ball(-k^2/10), k = 0..10), Squash,
      seq(Ball(-k^2/10), k = -10..0)],
     insequence = true):
```

To complete the animation, let us provide the ball with a small square plane to bounce off, which we will use as a static background to the animation, as follows

```
> Plane := plot3d(-11, -2..2, -2..2):
```

```
> display3d({Plane, Bounce}, title = "Bouncing Ball");
```

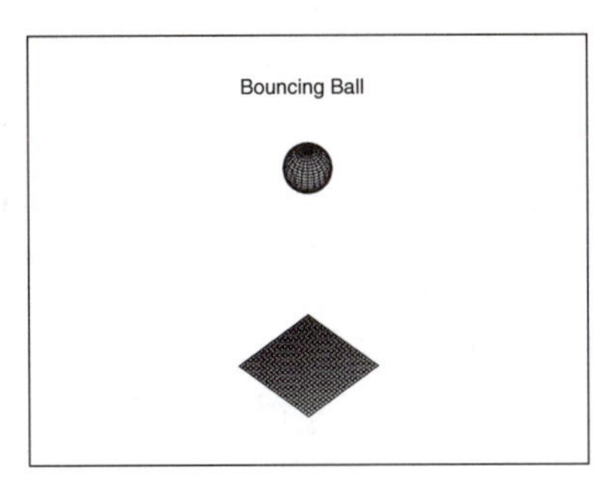

I can only illustrate one frame of the animation in this book, but if you type the code into Maple then you can run the animation for yourself. The animation probably looks best if you make it as large as possible by selecting the plot and dragging a corner until it fills the screen before you click on the play button. (You could also try turning off the tool, context, and status bars to maximize screen area.) If you select continuous animation then you can just sit back and watch the ball bounce forever!

Alternatively, if you select the animation and then select *Graphics Interchange Format (GIF)...* from the Export menu (or just *right*-click on the animation and use the context menu), you can put the resulting animated GIF file on your web site, or open it directly from your filestore in a web browser (e.g., just *double*-click it in Windows Explorer), to watch the ball bouncing with no further help from Maple! (Or visit the book web site advertised in the Preface, where I have already done this for you.)

4.6 Exporting three-dimensional plots

Three-dimensional plots can be exported in exactly the same way as two-dimensional plots, but the result is a fixed projection into two dimensions and the ability to rotate the plot in three dimensions is lost. However, Maple can save three-dimensional plots to the following file formats that preserve their three-dimensional structure:

DXF — AutoCAD's Drawing Exchange Format

POV — Persistence of Vision Raytracer (POVRay) scene format

VRML — Virtual Reality Modelling Language

Here is a little more detail about each of these formats and software for viewing them (in each case extracted from the web site quoted).

AutoCAD® software is a *commercial* customizable 2D and 3D design and drafting environment and toolset. Architects, engineers, drafters, and design-related professionals use it to create, view, manage, plot and output, share, and reuse accurate, information-rich drawings. For further details see the web site http://www.autodesk.com/autocad.

Persistence of Vision Raytracer is a high-quality and *totally free* tool for creating stunning three-dimensional graphics. It is available in official versions for Windows 95/98/NT, DOS, the Macintosh, i86 Linux, SunOS, and Amiga. For further details see the web site http://www.povray.org/.

Virtual Reality Modelling Language is the International Standard (ISO/IEC 14772) file format for describing interactive 3D multimedia on the Internet. For further details see the web site http://www.vrml.org/.

Cosmo Player is a VRML browser from Cosmo Software that runs as a web browser plug-in. It is available *free* for Windows and Macintosh platforms. I have tested Cosmo Player 2.1 on Windows NT 4.0 with both Netscape Navigator 4.7 and Microsoft Internet Explorer 5.5 and it appears to work fine. But beware that it is no longer supported! For further details see the web site http://www.cosmosoftware.com, and for information about other VRML browsers see the web site http://www.web3d.org/vrml/browpi.htm.

The first two formats, DXF and POV, are available as items on the Export menu, as described at the end of Chapter 3. VRML can currently be generated only by the function `vrml` in the `plottools` package, which requires two arguments, a plot structure and the name of a file. The standard file extension for VRML data files is "`.wrl`" (short for world). Here is an explicit example of how to generate a VRML file. One needs a shape without too much symmetry, so I will construct a "mushroom", which has only rotation symmetry about its vertical axis. (It's a magic mushroom with no visible connection between its stem and cap!)

```
> plot3d([sin(phi)*cos(theta), sin(phi)*sin(theta), cos(phi)],
     theta = 0..2*Pi, phi = 0..Pi/2):
  plot3d([cos(theta)/2, sin(theta)/2, z],
     theta = 0..2*Pi, z = -1..0):
  display3d({%,%%}, scaling = CONSTRAINED);
```

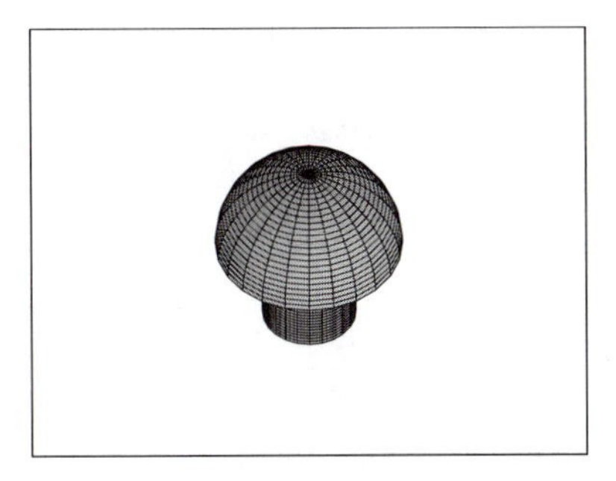

```
> plottools[vrml](%, "/tmp/mushroom.wrl");
```

Beware that the default view in a VRML browser may not be the same as in Maple. Also, it takes a little effort to learn to control a VRML browser, but it's worth it! For viewing Maple three-dimensional plots with Cosmo Player, the "Examine" controls seem to be more useful than the "Movement" controls.

4.7 Exercises

Beware that three-dimensional plotting is very memory intensive. ***Save your worksheet frequently.*** If necessary, close and then re-open Maple between exercises — just executing a `restart` command may not suffice.

Surface and contour plotting of simple functions

For each of the following expressions z, plot the graph of z, i.e., the surface such that z represents the height above the point with Cartesian coordinates (x, y) in the horizontal plane, using frame axes and values of x and y in the range -1 to 1. If necessary, change the default orientation to show the shape of the surface clearly. For each expression, also plot separately the contours (in two dimensions) using normal axes. (Use `contourplot3d` together with the option `orientation = [-90,0]`, which is much faster than `contourplot`.)

$$z = x^2 - y^2, \quad z = x^2 - y, \quad z = x^2, \quad z = \sqrt{x^2 + y^2}, \quad z = x^2 + y^2$$

Produce a contour plot for the expression $z = x^2 + y^2$ by using the function `contourplot` (which is much slower than `contourplot3d` and neither requires nor accepts the `orientation` option). Plot the *gradient* of z as a field of arrows by using the `gradplot` function. Note that both these functions return 2D, not 3D, plot structures. Superimpose the two (2D) plot structures using constrained (1:1) scaling. Note that the gradient field lines are orthogonal to the contours (provided you used constrained scaling!).

Produce an alternative contour plot for the expression $z = x^2 + y^2$ by plotting the set of curves defined *implicitly* by equating z to five uniformly spaced values in the range from 1/5 to 1 *inclusive.* (This is essentially what `contourplot` does, and it is equally slow!)

Parametric surface plotting

A prolate spheroid is a sphere that has been stretched in one direction; an oblate spheroid is a sphere that has been squashed in one direction. Plot parametrically a prolate and an oblate spheroid together, with common centre and axis of rotation symmetry (the stretching or squashing direction), so that they *intersect* and both are visible. (For example, the prolate spheroid could be a sphere that has been stretched by a factor of 2 along the z axis and the oblate spheroid could be a sphere that has been stretched by a factor of 2 in the (x, y) plane.) Use constrained (1:1) scaling and choose an orientation that shows the intersected structure clearly.

Plotting of simple space curves

A "circular helix" of radius r and pitch k around the z axis can be defined (parametrically) by the following equations for some appropriate range of values of θ:

$$x = r\cos(\theta),\ y = r\sin(\theta),\ z = k\,\theta.$$

The "axial line" of the helix is the line along the z axis onto which the helix projects, i.e., all points on the z axis that have the same z coordinates as points on the helix. Choosing convenient values for r and k, plot three complete turns of a circular helix around the z axis (i.e., three complete rotations in the (x, y) plane) *together with the axial line* of the helix. Use frame-style plotting axes. (Note that the helix is a *space curve* and not a surface, and so *cannot* be plotted using the `plot3d` function.)

Plotting singular functions of two variables

The following two expressions are discontinuous (and undefined) at the origin. Experiment with surface and contour plots that use different plot ranges and surface orientations to try to display the discontinuities as clearly as possible. Singular functions are very difficult to plot; the results can be strange but interesting and should certainly alert you to the singular nature of the function!

$$z = \frac{2\,x\,y^2}{x^3 + y^3},\quad z = \frac{x^3\,y}{y^2 + 2\,x^6}$$

4.8 Appendix: code to plot the figures

```
> restart;
```

Constructing a surface by stacking curves (Figure 4.1)

```
> proc(n)
     plots[spacecurve]({
        seq([[0,i/n,-1], [0,i/n,1]], i=0..n),
        seq([[0,i/n,0], [Pi,i/n,0]], i=0..n),
        seq([x, i/n, (1-i/2/n)*cos(x), x=0..Pi], i=0..n)},
        orientation = [-45,45], axes = NORMAL,
        labels = ["x","y","f  "], tickmarks = [2,2,2])
   end:

> %(2);  %%(10);

> plot3d((1-y/2)*cos(x), x=0..Pi, y=0..1,
     orientation = [-45,45], axes = NORMAL,
     grid = [25,10], labels = ["x","y","f  "],
     tickmarks = [2,2,2]);
```

Spherical polar coordinates (Figure 4.2)

```
> proc()
     local c,s,r,x,y,z,F1,T1,c2,s2,T2;
     c := cos(Pi/4);  s := sin(Pi/4);  r := 0.2;
     x := c*s;  y := s*s;  z := c;
     F1 := plots[spacecurve]({[[0,0,0], [x,y,z]],
        [[0,0,0], [x,y,0]], [[x,y,0], [x,y,z]],
        [[0.9*x,0.9*y,0], [0.9*x,0.9*y,0.1*z]],
        [[0.9*x,0.9*y,0.1*z], [x,y,0.1*z]],
        [r*c*sin(phi), r*s*sin(phi), r*cos(phi),
           phi = 0..Pi/4],
        [r*cos(th), r*sin(th), 0, th = 0..Pi/4]});
     if nargs = 0 then
        # Figure 4.2: Spherical polar coordinates
        T1 := plots[textplot3d]([x/2,y/2,z/2, "r"],
           align = ABOVE, font = [HELVETICA, 12])
     else
        # Figure 4.3: The orientation parameters
        T1 := plots[textplot3d]([x,y,z, "Viewpoint"],
           align = {ABOVE,RIGHT}, font = [HELVETICA, 12])
```

```
    end if;
    c2 := cos(Pi/8);  s2 := sin(Pi/8);  r := 0.3;
    T2 := plots[textplot3d]([[r*c*s2, r*s*s2, r*c2, "f"],
       [r*c2, r*s2, 0, q]], font = [SYMBOL]);
    plots[display3d]({F1, T1, T2}, style = LINE,
       orientation = [20, 65], axes = NORMAL,
       tickmarks = [2,2,2], labels = ["x","y","z"],
       color = black)
  end proc:
```

```
> %();
```

The orientation parameters $[\theta, \phi]$ (Figure 4.3)

```
> %%("Viewpoint");
```

Contouring by slicing a graph and projecting (Figure 4.4)

```
> plot3d(exp(-x^2-y^2), x = -2..2, y = -2..2,
    style = PATCH, axes = FRAME);
```

```
> plane := z ->
    [ [[-2,-2,z],[+2,-2,z]], [[-2,+2,z],[+2,+2,z]] ]:
  plots[surfdata](
    {plane(0.2),plane(0.4),plane(0.6),plane(0.8)},
    view = [-2..2, -2..2, 0..1], axes = FRAME);
```

```
> plot3d(exp(-x^2-y^2), x = -2..2, y = -2..2,
    style = PATCHCONTOUR,
    contours = [0.2,0.4,0.6,0.8], axes = FRAME);
```

```
> plots[contourplot3d](exp(-x^2-y^2),
    x = -2..2, y = -2..2, orientation = [45,0],
    contours = [0.2,0.4,0.6,0.8], axes = FRAME);
```

Surface and contour plots of a saddle point (Figure 4.5)

```
> plot3d(x^2-y^2, x = -1..1, y = -1..1,
    style = PATCH, title = "Surface plot");
```

```
> plots[contourplot3d](x^2-y^2, x = -1..1, y = -1..1,
    orientation = [45,0], title = "Contour plot");
```

Chapter 5

Numerical and Semi-Numerical Computation

Numerical computation involves *explicit* representations of real numbers as used by modern electronic calculators (e.g., 1.4142, 3.1416). Semi-numerical computation involves numbers that are represented in some other way, often symbolically rather than explicitly (e.g., $\sqrt{2}$, π). These two categories cover computation with integer, rational, real and complex numbers, and finite fields. Most conventional scientific computation is explicitly numerical, so that is what I will focus on in this chapter, but I will also touch briefly on some aspects of semi-numerical computation, for which Maple is a particularly powerful tool. The next chapter is devoted to linear algebra in general and numerical linear algebra in particular.

This chapter is primarily about the facilities that are built into Maple and some background to them, rather than about numerical computation techniques in general. The first section discusses the number systems used in mathematics and how they are represented in Maple and other programming languages. The second section focuses on computation with real numbers using the approximate floating-point representation, although for comparison I also discuss the related facilities for computing exact semi-numerical or symbolic solutions. Section 5.3 is a brief digression on multiple solutions of algebraic equations that provides background for the first section of Chapter 12, and the last two sections are very brief introductions to semi-numerical computation with algebraic numbers and computation in finite fields. (The latter is of fundamental importance for many of the algorithms that underlie computer algebra systems.)

5.1 Number systems

Before we can understand the number systems available in Maple and other programming languages we must first consider the main number systems that arise in mathematics.

5.1.1 Number systems in mathematics

Number means quantity. Number systems can be built up starting from the set of *natural numbers* $\mathbb{N} = \{0, 1, 2, \ldots\}$, which are the numbers used for the fundamental operation of counting. They are also called the *cardinal numbers.* It is a matter of dispute whether 0 should be included in the natural numbers, but the set $\{0, 1, 2, \ldots\}$ can be unambiguously called the *non-negative* or *unsigned* integers.

A set is said to be *closed* under some operation on the set if the result of the operation is always in the set. In order to have a set that is closed under the operation of subtraction (the inverse of addition) it is necessary to include negative numbers (and zero) in the set of natural numbers to give the set of *integers* $\mathbb{Z} = \{0, \pm 1, \pm 2, \ldots\}$. $\mathbb{Z}$ is an *additive group.* In order to have a set that is closed under division it is necessary to include fractional numbers in the set of integers to give the set of *rational numbers* (or simply *rationals*):

$$\mathbb{Q} = \{\pm p/q \mid p, q \in \mathbb{N},\ q \neq 0\}.$$

$\mathbb{Q} \backslash \{0\}$ is a *multiplicative group* and $\mathbb{Q}$ is a *field.* (The infix operator $\backslash$ denotes set minus, which is denoted in Maple by the infix operator `minus`; see Chapter 9.)

Filling in the gaps between the rationals gives the set of *real numbers* $\mathbb{R}$, which is the number system used for *measurement* of *continuous* quantities. In order to be able to solve an arbitrary polynomial equation with real coefficients it is necessary to include the square roots of the negative real numbers to give the set of *complex numbers* $\mathbb{C}$. $\mathbb{R}$ and $\mathbb{C}$ are both fields.

These number systems satisfy the following set inclusion hierarchy:

$$\underbrace{\mathbb{N} \subset \mathbb{Z} \subset \mathbb{Q}}_{\substack{\text{discrete}\\ \text{or}\\ \text{countable}}} \subset \underbrace{\mathbb{R} \subset \mathbb{C}}_{\substack{\text{continuous}\\ \text{or}\\ \text{uncountable}}}.$$

They are all infinite sets. An infinite set is called *countable* (or *denumerable*) if it can be put into one-to-one correspondence with $\mathbb{N}$. Such sets are also called *discrete*, because each of the elements is distinct from its neighbours.

Irrational numbers

The set $\mathbb{R} \backslash \mathbb{Q}$ is the set of *irrational numbers.* There are two types: *algebraic* and *transcendental.* Algebraic irrationals (usually called *algebraic numbers*) are zeros of polynomials with integer coefficients; e.g., $\sqrt{2}$ and the imaginary unit i ($\sqrt{-1}$) satisfy $x^2 - 2 = 0$ and $x^2 + 1 = 0$, respectively. Transcendental irrationals transcend (go beyond) polynomials. They can be defined algebraically only by using infinite series; e.g., π and e satisfy $\cos(x) + 1 = 0$ and $\ln(x) - 1 = 0$, respectively.

Infinite numbers

The sets $\mathbb{N}$, $\mathbb{Z}$, and $\mathbb{Q}$ have the same *cardinality* or number of elements, which is infinitely large, meaning that counting the elements will take forever — the process will never terminate. The set $\mathbb{R}$ has a different infinite cardinality that is larger than that of $\mathbb{N}$, because the inclusion,

$$\mathbb{Z} \subset \mathbb{R}$$

is strict. The difference is that $\mathbb{R}$ is a continuous set, whereas $\mathbb{N}$, $\mathbb{Z}$, and $\mathbb{Q}$ are discrete, meaning that there are gaps between the elements. An uncountable or continuous set contains more elements (a "higher infinity") than a countable or discrete set, and so is more difficult to work with, as we will see! The symbol ∞, called *infinity*, is used to represent any infinitely large number when the precise nature of the infinity is not important. It has the common arithmetic properties of all infinities, e.g., $\infty + k = \infty$, $\infty\, k = \infty \operatorname{sgn}(k)$, and $k/\infty = 0$ for any finite k.

Extension sets

There are many other sets of numbers intermediate between $\mathbb{N}$ and $\mathbb{C}$ that can be constructed by *extension* of a smaller set. Let us begin by considering polynomials as an extension of a number system. Each of the sets $\mathbb{Z}$, $\mathbb{Q}$, $\mathbb{R}$, and $\mathbb{C}$ is a *ring* (i.e., a set in which addition, subtraction, and multiplication are defined). Multiplication includes positive integer powers, hence ring operations are enough to define polynomials. If we treat a *symbol*, say x, the same as a number then we can construct the set of all polynomials in x with coefficients in the chosen ring R (where R is any ring, not necessarily the ring $\mathbb{R}$ of reals). This set of polynomials is denoted $R[x]$, where R is called the coefficient ring and x is called the unknown or *indeterminate*. This new set $R[x]$ is an *extension* of the ring R, and is itself a ring. If we include another symbol, y say, then we can construct the set of polynomials in two indeterminates or variables, $R[x][y]$, which turns out to be isomorphic to $R[y][x]$ and is often denoted $R[x, y]$.

If we now fix the value of x used to build $R[x]$ then we have a new number system. For example, $\mathbb{Z}[\sqrt{2}]$ is a ring that can be obtained by including $\sqrt{2}$ in the ring of integers and performing arithmetic on it as if it were an integer, or equivalently by setting $x = \sqrt{2}$ in the ring of all polynomials in x with integer coefficients. This new ring is an extension of $\mathbb{Z}$ that is intermediate between

$\mathbb{Z}$ and $\mathbb{R}$ (but distinct from $\mathbb{Q}$); i.e.,

$$\mathbb{Z} \subset \mathbb{Z}[\sqrt{2}] \subset \mathbb{R}.$$

Similarly, $\mathbb{Q}[\sqrt{2}]$ is an extension of $\mathbb{Q}$, and

$$\mathbb{Z} \subset \mathbb{Q} \subset \mathbb{Q}[\sqrt{2}] \subset \mathbb{R},$$

$$\mathbb{Z} \subset \mathbb{Z}[\sqrt{2}] \subset \mathbb{Q}[\sqrt{2}] \subset \mathbb{R}.$$

A particularly important example is the extension of the real numbers by $\sqrt{-1}$, which is usually denoted by the symbol i in mathematics (but Maple uses I and engineers often use j), to give the set of complex numbers, $\mathbb{C} = \mathbb{R}[i]$. Other important related extensions are $\mathbb{Z}[i]$, called the *Gaussian integers*, and $\mathbb{Q}[i]$, called the Gaussian rationals, which clearly satisfy the inclusion hierarchy:

$$\mathbb{Z}[i] \subset \mathbb{Q}[i] \subset \mathbb{R}[i] = \mathbb{C}.$$

Representation of algebraic numbers

The number $\sqrt{2}$ can be *defined* as the positive number x such that $x^2 - 2 = 0$. Hence, $\mathbb{Z}[\sqrt{2}]$ is the same as (i.e., isomorphic to) the set of all polynomials $\mathbb{Z}[x]$ subject to the constraint that $x^2 - 2 = 0$. In other words, it is (isomorphic to) a set of *equivalence classes*, which is often called a *quotient set* and denoted $\mathbb{Z}[x]/(x^2 - 2)$. An algebraic number can be represented by the (monic) polynomial of lowest degree of which it is a zero, which is called its *minimal polynomial.*

Finite number systems

All the number systems discussed so far have been infinite sets (some, such as $\mathbb{R}$, more infinite than others, such as $\mathbb{Q}$). Infinite number systems, and in particular continuous number systems such as $\mathbb{R}$, are impossible to work with explicitly in practice and some finite subset must be used instead. Some mathematical problems naturally use finite number systems but the rest must be solved using algorithms or approximations that use finite number systems internally.

An important class of finite number systems is the *modular integers.* In the system of integers modulo some positive integer $m > 1$, all integers of the form $n + k\,m$ form an equivalence class, which is represented by one particular member of the class, often the smallest non-negative member. The number system is called "the integers mod m" and denoted $\mathbb{Z}_m$. Hence, $\mathbb{Z}_3$ can be represented by the three numbers $\{0, 1, 2\}$ and the representative of any integer n can be computed as the remainder when n is divided by 3, i.e., as `irem(n,3)` in Maple.

The modular integers have the advantage that $\mathbb{Z}_m$ is a finite ring. Moreover, if m is a *prime number* p then $\mathbb{Z}_p$ is a *finite field.* Finite fields are of major importance in algebraic computation because they provide finite models for infinite fields such as $\mathbb{Q}$ and $\mathbb{R}$, in which computation can be performed efficiently and without any need for approximation.

The main representation for $\mathbb{Z}_p$ is $\{0, 1, \ldots, p-1\}$ but another useful representation is the *balanced* or *centred* representation:

$$\{-\tfrac{1}{2}(p-1), \ldots, -1, 0, 1, \ldots, \tfrac{1}{2}(p-1)\},$$

which is well defined when p is a prime greater than 2 because then it must be odd. Any representation must contain precisely p elements.

5.1.2 Number systems in Maple

Maple can represent a large (but finite) subset of $\mathbb{Z}$ (and hence $\mathbb{N}$). The largest integer is limited primarily by the memory available, but it should be possible to represent integers with at least 500,000 decimal digits, which is more than large enough for all normal purposes. Similarly, Maple can represent a large but finite subset of $\mathbb{Q}$ as *pairs* of integers. The representation is normalized so that the denominator is positive and the numerator and denominator have no common factor. In most situations, by default Maple regards numbers as elements of $\mathbb{Q}$ and does not make any significant distinction among $\mathbb{N}$, $\mathbb{Z}$, and $\mathbb{Q}$.

By default, Maple represents complex numbers of the form $x+i\,y$ externally (by default) as $x + \mathtt{I}\,y$, where `I` is a protected global identifier that represents $\sqrt{-1}$. In fact, the Maple user interface converts between `I` externally and the internal representation of the imaginary unit, which is `Complex(1)` and not explicitly $\sqrt{-1}$, and the interface controls the external representation:

```
> interface(imaginaryunit);
```

$$I$$

The `interface` function can be used to set the external representation of the imaginary unit to any other identifier, which automatically removes any special significance from the identifier `I`. But do not do this without good reason or you will cause yourself considerable confusion! The function `Complex` is the general complex number constructor and can be used instead of the explicit syntax $x + \mathtt{I}\,y$.

The real and imaginary parts x, y can be any valid Maple expressions; in particular, they can be explicit numbers of any kind, or they can be identifiers or unassigned variables. In the latter case, some care is required in manipulating the complex number because in general Maple assumes by default that all variables take complex values, although the complex evaluation function `evalc` assumes that all variables take only real values, as is appropriate for the real and imaginary parts of complex numbers.

A major problem in computing is the representation of *irrational* numbers, which have no representation that is both *finite* and *explicit*. There are two choices:

- exact but symbolic (e.g., $\sqrt{2}$, π)
- explicit but approximate (e.g., 1.414213562, 3.141592654)

By default, Maple (and any other computer algebra system) uses the exact but symbolic representation, e.g.,

```
> solve(x^2-2);
```

$$\sqrt{2}, -\sqrt{2}$$

```
> arccos(-1);
```

$$\pi$$

An exact explicit numerical representation of an irrational number would contain infinitely many digits, so to be practicable it must be truncated. We must choose:

- the nature of the representation, and
- the number of digits.

The usual choice is a *positional* representation in some fixed base (radix), which is usually 10 externally (but it may be 2 or a power of 2, e.g., 16, internally). For example, in base 10, π is 3.1416 to 5 significant figures or 4 decimal places.

Alternative explicit representations

Other choices of representation might be an interval or range representation, e.g.,

```
> shake(Pi);
```

$$\text{INTERVAL}(3.14159265045..3.14159265673)$$

or a continued fraction representation, e.g.,

```
> numtheory[cfrac](Pi,5);
```

$$3 + \cfrac{1}{7 + \cfrac{1}{15 + \cfrac{1}{1 + \cfrac{1}{292 + \cfrac{1}{1 + \ldots}}}}}$$

While both these representations are important, I will not pursue them further in this book. But you should be aware that representations other than the familiar positional decimal representation exist, and may be preferable for some applications. For example, interval arithmetic is used to control numerical error propagation and continued fraction representation is used to produce optimal rational approximations to real numbers.

Floating- and fixed-point representations

By default, Maple uses what is called *floating-point* representation to represent real numbers *explicitly* and hence (normally) *approximately*. However, any representation that can be applied to irrationals could also be applied to rationals, in which case it would not be infinite — it would either terminate or repeat and so could be written in a way that is both finite and exact. For example, using positional decimal representation,

$$\frac{3}{5} = 0.6, \quad \frac{5}{3} = 1.\overline{6}.$$

Hence, we can consider for *all real numbers* together representations that are generally approximate, although they could in principle be exact for the special case of rational numbers.

The Maple floating-point representation consists internally of an integer (called the *mantissa*) times a power (called the *exponent*) of 10, i.e.,

$$mantissa \times 10^{exponent}.$$

If a base different from 10 were used internally (as is the case for some systems other than Maple) it makes no difference to the principles involved. The decimal point "floats" through the integer mantissa, and its position is indicated by the exponent. By contrast, in a *fixed-point* representation the exponent has a fixed (normally negative) value. The term "floating-point representation" is often abbreviated to simply "float", which is the name used in the Maple data-type system. In Maple, the mantissa is a Maple integer of unlimited size but the exponent is a "machine" integer of fixed size (probably either 32 or 64 bits).

The function-call `Float(`*mantissa*`,`*exponent*`)` corresponds directly to the internal representation, and can be used for input (and is occasionally used for output), e.g.,

```
> Float(123,-2);
```

$$1.23$$

However, normally "scientific" form is used for both input and output and corresponds to a number in decimal positional representation, optionally including a decimal point, times a power of 10. An explicit power of 10 is used in output only when the power is large in magnitude, e.g.,

```
> Float(123,3), Float(123,7);
```

$$123000., .123\,10^{10}$$

The usual way to input a float is to use the letter `e` (or `E`) in place of "`*10^`" like this:

```
> 1.23e-4 = 1.23E-4;
```

$$.000123 = .000123$$

This is more efficient than the more explicit "`*10^`" form, which causes actual computation to be performed rather than simply inputting a constant, although the resulting value is the same:

```
> 1.23*10^(-4);
```

$$.0001230000000$$

Controlling floating-point output

The default Maple output facility, which produces the same output as the function `print`, has a fixed way of displaying floating-point numbers, e.g.,

```
> v := 1.23*10^(-1);
```

$$v := .1230000000$$

A common question is how to control the output format. Probably the easiest way is to use the *formatted* print function `printf` (or one of its close relatives; see the online help). It takes a format string as its first argument and then a sequence of values to print. The following example prints using *fixed-point* notation with precisely 4 decimal places:

```
> printf("%.4f", v);
```

```
.1230
```

whereas this example prints using *scientific* notation, which always has one digit before the decimal point, with precisely 3 digits after the decimal point:

```
> printf("%.3e", v);
```

```
1.230e-01
```

Many other output formats and much more control are possible; see the online help for full details.

Overflow and underflow

Various errors can arise when using floating-point representation. If the exponent becomes too large in magnitude for the available memory then this error is called *overflow* if the exponent is positive and *underflow* if it is negative. This is unlikely to happen in Maple because the maximum exponent magnitude is very large; e.g., on a 32-bit Windows system the float range is

```
> Maple_floats(MAX_FLOAT) .. Maple_floats(MIN_FLOAT);
```

$$.9\,10^{2147483647}\,..\,.1\,10^{-2147483645}$$

This is approximately $10^{2147483647}\,..\,10^{(-2147483646)}$, which corresponds to the use of a 32-bit *signed* integer exponent with a maximum value of

```
> 2^31-1;
```

$$2147483647$$

When a number underflows it is often replaced by 0, which is what Maple does, and when a number overflows Maple replaces it by a suitable infinity (in this case, a floating-point infinity). Although unlikely to occur in practice in Maple, it is easy to demonstrate underflow and overflow:

```
> Maple_floats(MIN_FLOAT)^2;  # underflow
```

$$0.$$

```
> Maple_floats(MAX_FLOAT)^2;  # overflow
```

$$\mathrm{Float}(\infty)$$

Overflow and underflow are much more serious problems in "conventional" programming languages such as C and FORTRAN, which have much smaller maximum exponent magnitudes.

Precision and accuracy

The other type of floating-point error is the error in representing the true value. The complexity of the representation is normally controlled by limiting the maximum number of digits in the mantissa, which corresponds to the maximum number of significant figures in the number. In Maple, the default number of digits in the mantissa of a float, and hence the number of significant figures, is determined by the value of the global reserved variable `Digits` (*with a capital D*). Initially this is set to 10, but it can be reassigned, e.g.,

```
> 'Digits' = Digits, 2/3 = 2./3.;
```

$$\mathit{Digits} = 10,\ \frac{2}{3} = .6666666667$$

```
> Digits := 5:  'Digits' = Digits, 2/3 = 2./3.;
```

$$\mathit{Digits} = 5,\ \frac{2}{3} = .66667$$

```
> Digits := 10:
```

It is sometimes necessary to change the value of Digits for some part of a computation, without changing its value for other parts of the computation. This is called a *local* change. In general, you cannot assume that the value was 10 before you changed it, because it may already have been changed from its default value. One simple and reliable way to change Digits locally is like this:

```
> OldDigits := Digits:  Digits := 5:
     2./3.;  # Do the computation
  Digits := OldDigits:
```

.66667

```
> Digits;
```

10

This technique for local reassignment applies to *any* variable, not just Digits. Another way to control the accuracy in the numerical evaluation of an expression is to use the function evalf with an index specifying the precision, e.g.,

```
> evalf[5](2/3);
```

.66667

```
> Digits;
```

10

However, non-trivial pieces of code are normally packaged as *procedures* or *modules* (to be discussed in detail later). Digits is one of a small number of pre-defined special variables called *environment* variables, which means that any reassignment within a procedure or module applies only while the procedure or module is executing and is automatically undone when the procedure or module terminates. (It is analogous to declaring the variable to be local, except that this does not work for pre-defined global variables such as Digits because only the global variable has any significance. ***Hence, variables such as* Digits *should not be declared local in procedures or modules!***) Using a module is the most elegant way to provide a local context for a computation, such as a numerical calculation that requires Digits to be changed, like this:

```
> module()
     local `2`, `3`;  `2` := 2.0;  `3` := 3.0;
     Digits := 15;
     print(`2`/`3`);  # Do the computation
  end module:
```

$$.666666666666667$$

```
> Digits;
```

$$10$$

Note that the value returned by a module is the module itself, so any output required from within a module must be performed explicitly, such as by calling the `print` function as above.

The digits of the mantissa are the significant figures of the approximation, so the maximum error *caused* by the floating-point representation is 0.5 in the last significant digit. The actual magnitude of the error depends also on the exponent and hence on the overall magnitude of the number, so the error in a floating-point representation is naturally a *relative error*; i.e., its significance is relative to the magnitude of the number. Once the number of significant figures used to represent floating-point approximations has been fixed then all floating-point approximations have the same fixed relative error.

If a number x has the floating-point representation:

$$x = mantissa \times 10^{exponent}$$

then the maximum *absolute* magnitude of the representation error is $e = .5 \times 10^{exponent}$ and the maximum *relative* magnitude of the representation error is

$$r = \frac{e}{|x|} = \frac{.5}{|mantissa|},$$

where $|mantissa| \approx 10^{Digits}$; i.e., $r \approx .5 \times 10^{-Digits}$. (By contrast, the number of *decimal places* of accuracy depends on the exponent. In a floating-point number this depends on the magnitude of the number and is not fixed, whereas in a fixed-point representation it is fixed, so the error in a fixed-point representation is naturally an *absolute error*.)

Infinite numbers in Maple

Maple understands the identifier `infinity` to represent the infinity normally denoted by ∞, and it has the correct arithmetic properties; e.g., for any finite k, $\infty + k = \infty$:

```
> k := - 123:
```

```
> infinity + k = infinity;
```

$$\infty = \infty$$

$\infty\, k = \infty\, \mathrm{sgn}(k)$:

```
> infinity*k = infinity*signum(k);
```

$$-\infty = -\infty$$

and $k/\infty = 0$:

```
> k/infinity = 0;
```

$$0 = 0$$

```
> k := 'k':
```

Extended IEEE arithmetic

At one time, details of floating-point computations varied among hardware manufacturers and software suppliers, especially at the limits of computation (i.e., for very large and very small quantities). In 1985 the Institute of Electrical and Electronics Engineers (IEEE) published an American National Standard (ANSI) for Binary Floating-Point Arithmetic, known as IEEE-754, and in 1987 they published an American National Standard for Radix-Independent Floating-Point Arithmetic, known as IEEE-854. (See the Maple online help topic `IEEE` for references.) Subsequently, hardware and software suppliers have tried to comply with these standards, leading to much better defined numerical computations. Numeric computation in Maple is based on these standards and their natural generalization to arbitrary precision and exact arithmetic.

One aspect of IEEE standard arithmetic is that zero has a sign, which can be important in the context of limits at discontinuities. Moreover, complex numbers with zero imaginary part are distinct from real numbers, and together with signed zeros this allows correct evaluation of functions that have branch cuts in the complex domain. IEEE arithmetic also defines infinities and undefineds (what IEEE-754 calls "NaN" — "Not a Number"), which Maple implements in both the exact and floating-point number domains. Here are simple examples of floating-point signed infinity and signed zero arising from over- and underflow respectively:

```
> -Maple_floats(MAX_FLOAT)^2;  # overflow
```

$$\mathrm{Float}(-\infty)$$

```
> -Maple_floats(MIN_FLOAT)^2;  # underflow
```

$$-0.$$

For most purposes it is not necessary to worry about the details of Maple's arithmetic, but it is comforting to know that it complies with modern standards. It is part of the move toward improving Maple's facilities for numerical computation by using the well-established Numerical Algorithms Group (NAG) library, the first major step of which is provision of the new facilities for numerical linear algebra that are the subject of the next chapter.

5.1.3 Number systems in other languages

"Conventional" programming languages that are not intended for "algebraic computation" do not normally support such flexible number systems as Maple. In particular, languages such as C, C++, and Java provide only a relatively small subset of the integers and fixed precision floating-point approximation of other real numbers. The restriction is principally that the whole representation of a number must fit into a fixed maximum amount of memory, which is usually at most 64 bits.

FORTRAN is a little more flexible because it is intended primarily for scientific and engineering computation; it provides floating-point approximation of complex numbers and recent versions of FORTRAN provide more flexible precision. Of course, these limitations can be circumvented by writing additional libraries of code to support more flexible number systems, and most modern systems for mathematical computation, such as Maple, are effectively written in C or C++.

Java is one of the most modern among widely used languages and was introduced by Sun in late 1995. Sun provides the main web site for Java at http://java.sun.com/ and a good succinct reference text is *Java in a Nutshell: A Desktop Quick Reference* [6]. Java provides four integer types using 8, 16, 32, and 64 bits and two IEEE 754 floating-point types using 32 and 64 bits. Compare this with Maple's arbitrary size integers and arbitrary precision floating-point types.

5.2 Floating-point computation in Maple

Some numerical computation is trivial; e.g., $1 + \frac{1}{2} + \frac{1}{3} + \frac{1}{4}$ implies a numerical computation that is nevertheless (in principle) exact. However, numerical computation normally involves *irrational* numbers, which implies *approximation.* The aim is to compute a result that is *sufficiently* accurate as *quickly* as possible. The theory behind this is called *numerical analysis.* Maple provides support for some relatively simple and standard numerical computations: e.g., evaluation, solving equations, integration, solving differential equations, and linear algebra. This section outlines the first four of these; linear algebra is deferred until the next chapter.

5.2.1 Floating-point evaluation: `evalf`, `evalhf`

Automatic floating-point evaluation

Normally, an arithmetic expression or a function application will be automatically evaluated using floating point to the current precision (`Digits` significant digits) if it contains at least one floating-point constant. Floating point is sometimes said to be contagious! This applies to algebraic operations and functions and *nearly all* transcendental functions, which includes trigonometric functions, e.g.,

```
> (2/3)^2 + sqrt(log(1.5));
```

$$1.081205866$$

But a "mixed" expression may not be fully evaluated in floating point by default, so floating point is not reliably contagious, e.g.,

```
> (2/3.0)^2 + sqrt(log(15/10));
```

$$.4444444445 + \sqrt{\ln(\frac{3}{2})}$$

```
> sqrt(2) + cos(tan(1.3));
```

$$\sqrt{2} - .8958260588$$

```
> sqrt(2.0) + cos(tan(13/10));
```

$$1.414213562 + \cos(\tan(\frac{13}{10}))$$

Explicit arbitrary-precision floating-point evaluation: `evalf`

Full floating-point evaluation (that is as complete as possible) of any *expression* can always be obtained by applying the function `evalf`, e.g.,

```
> evalf( (2/3)^2 + sqrt(log(15/10)) );
```

$$1.081205866$$

```
> evalf( sqrt(2) + cos(tan(13/10)) );
```

$$.5183875032$$

By default, the precision used is the current value of `Digits`, but if `evalf` is *indexed* with a positive integer then this specifies the precision to use *for this evaluation only*, e.g.,

```
> evalf(Pi);
```

$$3.141592654$$

```
> evalf[50](Pi);
```

$$3.1415926535897932384626433832795028841971693993751$$

Hardware floating-point evaluation: `evalhf`

Normal computer hardware can perform fast floating-point arithmetic on numbers using fixed precision (e.g., 32 or 64 bits), but arbitrary-precision floating-point arithmetic is currently performed by software (the Maple kernel) using the hardware to perform only *integer* arithmetic. Hence, the arbitrary-precision floating-point arithmetic normally used by Maple is significantly slower than hardware floating-point arithmetic. This is one reason why Maple is not normally used for serious large-scale numerical computation. (The other is that Maple programs are interpreted rather than compiled, which makes them portable but relatively slow.)

Because floating-point computation is so much faster using hardware than software, Maple provides the function `evalhf` as a hardware analogue of `evalf`. It takes a single argument that is the expression to be evaluated and uses (double-precision) hardware floating-point arithmetic if it is available. It is not possible to choose the precision. It is about 15 significant figures on many 32-bit architectures; the approximate value on any particular system can be found by evaluating the following expression, which under Microsoft Windows gives:

```
> trunc(evalhf(Digits));
```

$$14$$

The `evalhf` function is not easy to use and there are restrictions on the form of expression it can evaluate: it must evaluate to a *single* number and it cannot involve "symbolic" objects such as free variables or structures such as lists. Using `evalhf` naively will probably not save computation time because of the conversion overhead between Maple and hardware floating-point representations. A simple example of the use of `evalhf` is given in Chapter 12.

One major application that requires a great deal of relatively low-precision floating-point computation is graphics, and `evalhf` is used internally in the Maple plotting functions; e.g., see the source code for the plotting procedure `` `plot/adaptive` `` or do some plotting with a suitable value of `infolevel` (see also Chapter 11) set, such as:

```
> infolevel[plot] := 2:  plot(sin):
  plot/adaptive:   evalhf trying again
```

```
plot/adaptive:   evalhf succeeded

plot/adaptive:   produced    151.    output segments

plot/adaptive:   using    151.    function evaluations
```

```
> infolevel[plot] := 0:
```

5.2.2 Solution of algebraic equations: `solve`, `fsolve`

Equation solvers usually treat an expression *exprn* the same as the equation *exprn* = 0, so I will use the term "equation" to include the case of an expression that is assumed to be equated to zero. The Maple equation solvers normally accept either a single equation or a set (but not a list) of equations, and an optional variable or set of variables to solve for. If not specified, they try to solve for all the variables in the equations. The main solver for algebraic equations (and inequalities) is `solve`. It *always* attempts to find an *exact algebraic* solution and returns those that it can find, which may not be all solutions that exist; if it cannot find any solution at all then it returns nothing (equivalent to the empty sequence, `NULL`).

If a floating-point approximation to a purely numerical solution is required, perhaps because no exact solution can be found or because it is too complicated, then it is best to use `fsolve` instead (see below).

The general exact algebraic solver: `solve`

A single solution to a single equation is a single algebraic expression:

```
> solve(a*x + b = 0, x);
```

$$-\frac{b}{a}$$

Multiple solutions are returned as a *sequence* of expressions:

```
> solve((a*x + b)*(c*x + d), x);
```

$$-\frac{d}{c}, -\frac{b}{a}$$

If any argument to solve involves a set then each solution is returned as a *set of equations* with the variable(s) on the left and multiple solutions are returned as a sequence of such sets of equations:

```
> solve({2*x + 3});
```

$$\{x = \frac{-3}{2}\}$$

```
> solve((2*x + 3)*(4*x - 5) = 0, {x});
```

$$\{x = \frac{-3}{2}\}, \{x = \frac{5}{4}\}$$

```
> S := solve({x + y = 0, x - y = 2});
```

$$S := \{y = -1, x = 1\}$$

Using solutions: `subs`**,** `assign`

The best way to use the result returned by a solver is usually to assign it to a variable (as above), extract the desired solution if there is more than one, and then use `subs` to substitute the solutions into expressions as appropriate. This is also the best way to assign the solutions to other variables, e.g.,

```
> Solution_for_x := subs(S, x);
```

$$Solution\_for\_x := 1$$

```
> Solution_for_y := subs(S, y);
```

$$Solution\_for\_y := -1$$

Note that it is not safe (other than interactively) to extract elements of sets by their position in the set, because this is not fixed!

It is also possible to turn equations into assignments by applying the function `assign`, but often this is not the best approach because then the variables for which the equations were solved are no longer free.

Exact solutions: `RootOf`**,** `allvalues`

The function `solve` always returns *exact* solutions to exact problems. It will attempt to solve almost any equality (or inequality) *over the rational numbers.* Zeros that cannot be expressed exactly (and reasonably simply) using radicals (fractional powers) are represented using the Maple function `RootOf`. The first argument of a `RootOf` function returned by `solve` is always an expression in the variable `_Z` and the second is a tag to indicate the root to which this particular call of `RootOf` refers, e.g.,

```
> S := solve(x^4 + x - 1);
```

$$S := \mathrm{RootOf}(\_Z^4 + \_Z - 1, \mathit{index} = 1),$$
$$\mathrm{RootOf}(\_Z^4 + \_Z - 1, \mathit{index} = 2),$$
$$\mathrm{RootOf}(\_Z^4 + \_Z - 1, \mathit{index} = 3),$$
$$\mathrm{RootOf}(\_Z^4 + \_Z - 1, \mathit{index} = 4)$$

The function `evalf` can be used to compute a numerical approximation to such a solution:

```
> evalf(S);
```

$$.7244919590,\ .2481260628 + 1.033982061\, I,\ -1.220744085,$$
$$.2481260628 - 1.033982061\, I$$

Using one or more floats in the first argument to `solve` has an effect similar to applying `evalf` (except that the order of the solutions is different):

```
> solve(x^4 + x - 1.0);
```

$$-1.220744085,\ .2481260628 - 1.033982061\, I,$$
$$.2481260628 + 1.033982061\, I,\ .7244919590$$

The function `allvalues` will attempt to expand a tagged `RootOf` function in terms of radicals and an untagged `RootOf` function representing roots of a polynomial into the explicit sequence of exact roots that it represents. These expansions may be both complex and complicated — try it on the above example to see! To save space, I will illustrate a much simpler application:

```
> allvalues(RootOf(x^2+x+1));
```

$$-\frac{1}{2} + \frac{1}{2} I \sqrt{3},\ -\frac{1}{2} - \frac{1}{2} I \sqrt{3}$$

If `allvalues` cannot expand a `RootOf` exactly (in terms of radicals) then it will leave it essentially unchanged, e.g.,

```
> allvalues(RootOf(x^5 + x^2 - 1));
```

$$\mathrm{RootOf}(\%1, \mathit{index} = 1),\ \mathrm{RootOf}(\%1, \mathit{index} = 2),$$
$$\mathrm{RootOf}(\%1, \mathit{index} = 3),\ \mathrm{RootOf}(\%1, \mathit{index} = 4),$$
$$\mathrm{RootOf}(\%1, \mathit{index} = 5)$$
$$\%1 := \_Z^5 + \_Z^2 - 1$$

Approximate floating-point solution of equations: `fsolve`

There are algorithms to compute reliably all roots of single polynomial equations in one variable (e.g., using Sturm sequences; see Chapter 12) but computing reliably all roots of transcendental equations is much more difficult (and generally not possible). The function `fsolve` attempts to compute floating-point approximations to the roots of equations. (The equations must not contain any free parameters). *If you know that you need floating-point approximations to the roots of equations then it is best to use* `fsolve` *rather than* `solve`!

For *polynomial* equations, `fsolve` computes *all* real roots by default (see Chapter 12 for some idea of how it does this); for *transcendental* equations it *attempts* to compute any *one* real root. Like `solve`, it will try to deduce the variable(s) to solve for:

```
> fsolve(x^4 + x - 1);
```

$$-1.220744085, .7244919590$$

It accepts optional arguments *after* (or as part of) the solution variable specifications:

- The keyword `complex` causes `fsolve` to return complex roots instead of only real roots, e.g.,

```
> fsolve(x^4 + x - 1, x, complex);
```

$$-1.220744085, .2481260628 - 1.033982061\,I,$$
$$.2481260628 + 1.033982061\,I, .7244919590$$

- A range restricts the roots to lie in the closed interval specified, e.g.,

```
> fsolve(x^4 + x - 1, x = 0..1);
```

$$.7244919590$$

Ranges can be specified either alone or on the right of an equation with a variable on the left, much as for `plot`. To find several roots of a transcendental equation it is necessary to make several separate calls of `fsolve`, each with a different careful choice of interval. A suitable plot may help to choose intervals.

Example: find all non-negative roots of $10\sin(x) = x$

It is often best to plot two simple graphs rather than one more complicated graph, and a suitable interval for the plot will need to be found by trial and error in general, e.g.,

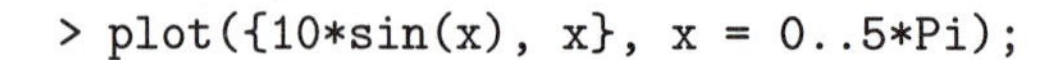

```
> plot({10*sin(x), x}, x = 0..5*Pi);
```

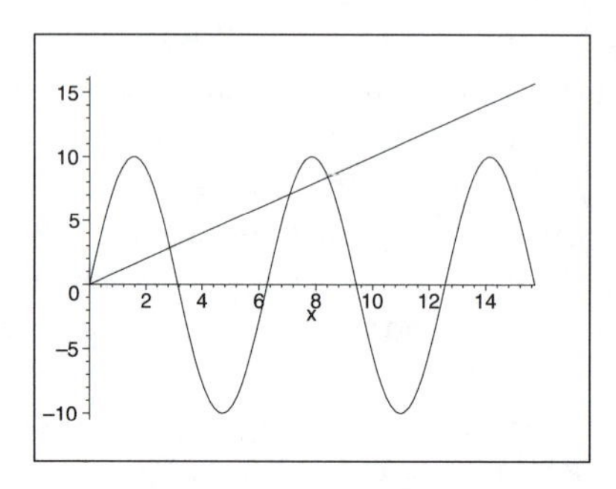

This shows that the required roots are all less than $3\,\pi$ (since we know that the zeros of $\sin(x)$ are at integer multiples of π). Let us see what root, if any, `fsolve` finds by default:

```
> eqn := (10*sin(x) = x):  fsolve(eqn);
```

$$0.$$

This gives only one root; in fact, it gives the trivial one at the origin. But the plot indicates *isolating intervals*, based on the known zeros and maxima of $\sin(x)$, that each contain (i.e., isolate) precisely one of the four required roots, which can therefore all be computed as an expression sequence like this:

```
> fsolve(eqn, x = -Pi/2..Pi/2),
  fsolve(eqn, x = Pi/2..Pi),
  fsolve(eqn, x = 2*Pi..5*Pi/2),
  fsolve(eqn, x = 5*Pi/2..3*Pi);
```

$$0., 2.852341894, 7.068174358, 8.423203932$$

Integer solutions: `isolve`, `msolve`

There are two other specialized solvers for finding integer solutions of polynomial equations over the integers (so-called *Diophantine* equations): `isolve` finds all solutions in $\mathbb{Z}$ and `msolve` finds all solutions in $\mathbb{Z}_m$.

5.2.3 Numerical integration (quadrature): `evalf/Int`

If the function `evalf` is applied to an unevaluated *definite* integral then *numerical integration* is performed, provided there are no symbolic parameters involved. An unevaluated integral arises *either* when the active function `int` is used but fails to evaluate the integral *or* when the inert function `Int` is used. If a floating-point approximation to the integral is *required* then `Int` should normally be used, because it will generally be faster and no less accurate to perform numerical integration directly than to perform symbolic integration followed by numerical evaluation, and the symbolic integration will fail in many cases anyway, just wasting time. In the case that the symbolic integration succeeds, the two numerical results may differ slightly; their difference indicates the true error in both of them and is simply a consequence of the different algorithms used in the two cases!

As an example, consider:

$$\int_0^\pi \sin(\sin(\sin(x)))\,dx.$$

A plot of the integrand confirms that this integral is well defined and visual inspection of the area under the curve suggests that the integral should have a value of about 1.5:

```
> sssx := sin(sin(sin(x)));
```

$$sssx := \sin(\sin(\sin(x)))$$

```
> plot(sssx, x=0..Pi);
```

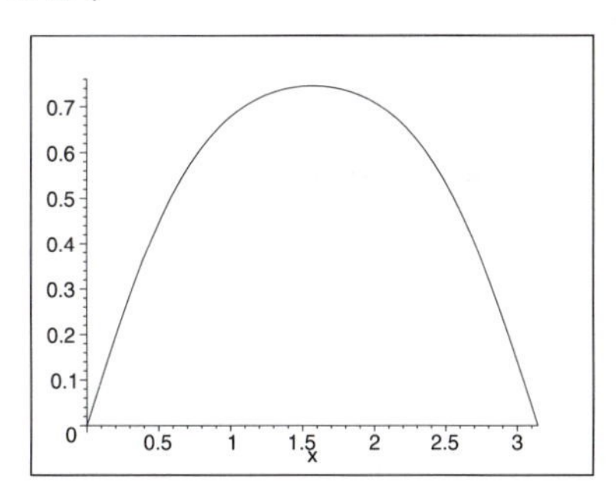

Maple cannot evaluate this integral exactly:

```
> int(sssx, x=0..Pi);
```

$$\int_0^\pi \sin(\sin(\sin(x)))\,dx$$

But it can evaluate the definite integral numerically, and the result looks about right:

```
> evalf(%);
```

$$1.642593493$$

The numerical integration can be performed more directly and more efficiently by applying `evalf` to an inert integral constructed using `Int`:

```
> evalf(Int(sssx, x=0..Pi));
```

$$1.642593493$$

5.2.4 Solution of differential equations: `dsolve`

An ordinary differential equation (often abbreviated to ODE) is an equation for an *unknown function* of one variable that involves its derivatives. For example,

$$\frac{dy(x)}{dx} = k\,y(x),$$

where k is any constant, is an ordinary differential equation for the function $y(x)$. (Note that Maple regards all derivatives as partial derivatives, and so always uses the curly partial derivative symbol to display them. This is entirely valid, although not usual mathematical typography.) This particular ODE has the solution

$$y(x) = A\,e^{kx}$$

for any constant A, as can easily be verified using Maple as follows:

```
> ODE := (diff(y(x),x) = k*y(x));
```

$$ODE := \frac{\partial}{\partial x}\,\mathrm{y}(x) = k\,\mathrm{y}(x)$$

```
> soln := (y(x) = A*exp(k*x));
```

$$soln := \mathrm{y}(x) = A\,\mathbf{e}^{(k\,x)}$$

```
> evalb(eval(ODE, soln));
```

$$true$$

The Maple function `dsolve` can solve a large class of ODEs exactly, e.g.,

```
> dsolve(ODE);
```

$$\mathrm{y}(x) = \_C1\,\mathbf{e}^{(k\,x)}$$

The name of the unknown function to solve for can optionally be provided as the second argument to `dsolve` (as for `solve`), and `dsolve` uses identifiers of the form *_C1* to indicate arbitrary constants in the solution. The arbitrary constants in the general solution of an ODE can be fixed by specifying additional conditions, which in Maple can be given as additional equations together with the ODE, e.g.,

```
> dsolve({ODE, y(0) = A});
```

$$\mathrm{y}(x) = A\,\mathbf{e}^{(k\,x)}$$

In general, it is not possible to solve an arbitrary ODE exactly but it is still possible to approximate the solution numerically, as with integrals. A problem can be solved numerically only if it contains no symbolic parameters; so, in the example we have been considering, it is necessary to give the parameters k and A specific values. Then `dsolve` will construct a numerical solution if it is given the optional argument `numeric`. By default, the solution returned is a procedure, which returns not (as one might expect) the values of $y(x)$ but a list of equations giving the values of x and $y(x)$. A more convenient form of solution is as a list of equations that have procedures on their right sides, which is obtained by giving the optional `output` argument shown below. (Note that, somewhat inconsistently, it is necessary to specify the variable to solve for when solving numerically.)

```
> soln :=
    dsolve({subs(k=1,ODE), y(0) = 1}, y(x),
    numeric, output = listprocedure);
```

$$soln := [x = (\mathbf{proc}(x) \ldots \mathbf{end\ proc}\ ), \mathrm{y}(x) = (\mathbf{proc}(x) \ldots\ \mathbf{end\ proc})]$$

The first procedure is the identity function, and the second corresponds to the unknown function $\mathrm{y}(x)$, which can be extracted as from a solution returned by `solve`:

```
> soln := subs(soln, y(x));
```

$$soln := \mathbf{proc}(x) \ldots \mathbf{end\ proc}$$

The identifier `soln` now represents a Maple function that can be used (only) in a numerical context just like any other Maple function (except that it may be slower). (In this particular example, *soln* is a numerical approximation to the exponential function.) For example, it can be plotted like this:

```
> plot(soln, -1..1);
```

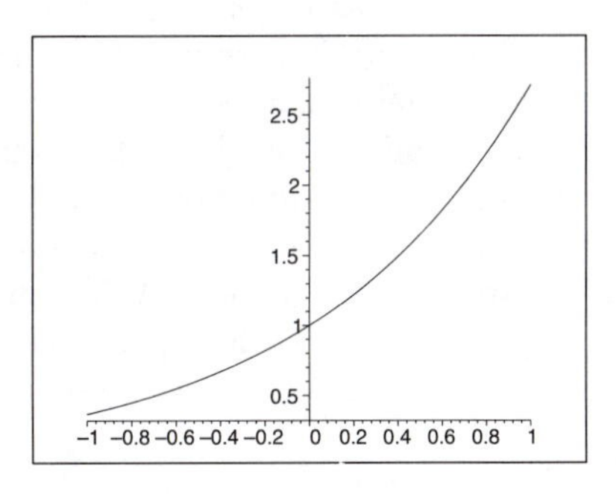

and tabulated like this:

```
> array([[x,y(x)],
    seq([0.1*i,soln(0.1*i)], i = 0..5)]);
```

$$\begin{bmatrix} x & \mathrm{y}(x) \\ 0. & 1. \\ .1 & 1.10517091795195954 \\ .2 & 1.22140275799013942 \\ .3 & 1.34985880735125074 \\ .4 & 1.49182469735216650 \\ .5 & 1.64872127033562310 \end{bmatrix}$$

Alternatively, tabulating such a function might be a good application of a Maple spreadsheet (using the spreadsheet *column fill* functions to do most of the work; see Chapter 13).

The `dsolve` function accepts other optional arguments and can solve much more complicated problems, including higher order ODEs and systems of coupled ODEs. There is also a solver called `rsolve` for solving *recurrence relations* or *difference equations*, which are closely related to differential equations, although the concept of numerical approximation does not really apply to discrete problems such as this. See the online help for further details of both `dsolve` and `rsolve`.

5.3 Multiple roots

When attempting to solve algebraic equations numerically, it is important to understand the nature of multiple roots, because they are numerically unstable and hence difficult to compute reliably. This is really a topic for a text on numerical analysis, but it is sufficiently important to merit a brief introduction here. There are two problems: the sign of a function does not change through a multiple root with even multiplicity, which makes it difficult to find, and because a multiple root is degenerate, small perturbations such as arise naturally in approximate solution methods can cause its nature to change. I will discuss multiple roots of *polynomial* equations further in the first section of Chapter 12 and show how to avoid the problems they cause.

If m roots of an equation or zeros of a function (of x) are made to coalesce (or merge) then the result is called a single root or zero of *multiplicity* m. When two zeros coalesce the resulting function is locally quadratic; i.e., with respect to an origin placed at the double root the function has the form x^2 + *higher degree terms*, as illustrated in Figure 5.1. When three zeros coalesce the resulting function is locally cubic; i.e., with respect to an origin placed at the triple root the function has the form x^3 + *higher degree terms*, as illustrated in Figure 5.2.

More generally, the Taylor series expansion of a function about a zero of multiplicity m has the form:

$$a\,x^m + b\,x^{m+1} + \cdots$$

and is said to be *of order m*. Equivalently, the first derivative of the function that does not vanish at the multiple root is the mth. A root of multiplicity

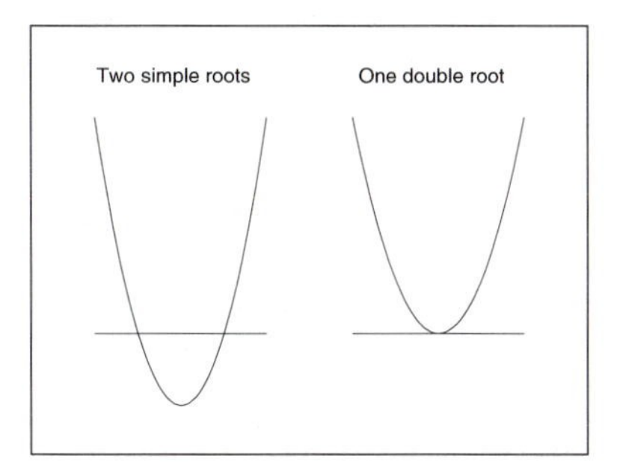

Figure 5.1: Graphs near and at a double root.

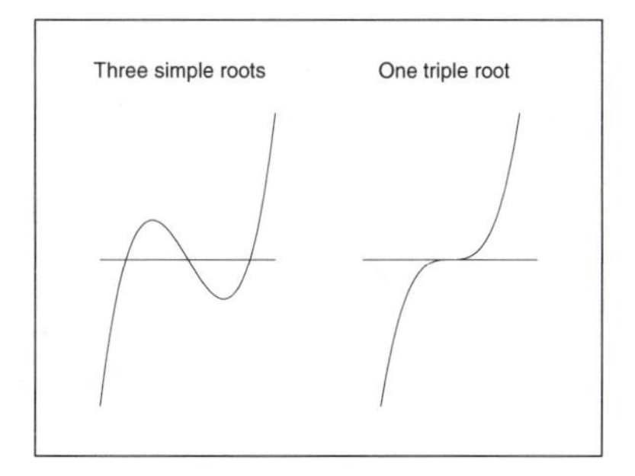

Figure 5.2: Graphs near and at a triple root.

one is called *simple*. The multiplicity is also the *maximum* number of roots (or equivalently the number of *simple* roots) that can be produced from a multiple root by a small perturbation of the equation or function, but beware that not all perturbations will split a multiple root into simple roots and it may require some thought to find a suitable perturbation to demonstrate this.

Hence, there are at least three ways to investigate a multiple zero of a function that are easy using Maple: plot the graph of the function and various small perturbations of it near the zero, evaluate successive derivatives at the zero or equivalently compute the Taylor series expansion about the zero; see the online help for the functions `taylor` or `series` for details.

5.4 Aliases and algebraic numbers

5.4.1 Alias

The `alias` mechanism allows an *alias* or alternative name to be given to an expression, such that the alias is used externally while the expression is used internally. This is similar to the effect of an assignment for input together with the inverse of the assignment for output. An alias may be defined for any Maple object *except* a numerical constant. The `alias` function takes an arbitrary sequence of equations of the form *`name = expression`* as arguments and returns a sequence of all the aliases currently defined; none is defined by default. Aliases can be used trivially to abbreviate long names. Here is an example that makes the identifier e represent the base of natural logarithms:

```
> alias(e = exp(1));
```

$$e$$

```
> evalf(e);
```

$$2.718281828$$

However, such use of `alias` is potentially misleading because, for example:

```
> e^x = exp(x);  evalb(%);
```

$$e^x = \mathbf{e}^x$$
$$\mathit{false}$$

```
> simplify(%%);  evalb(%);
```

$$\mathbf{e}^x = \mathbf{e}^x$$
$$\mathit{true}$$

There is no easy way to find out what an alias represents. An alias can be cleared by aliasing it to itself, e.g.,

```
> alias(e = e);
```

An important use of `alias` is to give names to *algebraic numbers* as explained in the next section; for any other purpose `alias` should be used with great care or not at all. A name can be given to a *numerical constant* (or any other Maple expression) for *input only* using the `macro` function; see the online help.

5.4.2 Algebraic numbers

An algebraic number is represented in Maple as a root of a polynomial — the primitive polynomial of lowest degree of which it is a root, which is called its *minimal* polynomial — using the `RootOf` function. For example, an algebraic number α defined to be a root of the polynomial x^2+x+1 might be represented in Maple as:

```
> alias(alpha = RootOf(x^2 + x + 1));
```

$$\alpha$$

Then expressions involving the constant α simplify correctly:

```
> alpha^2 + alpha + 1;  simplify(%);
```

$$\alpha^2 + \alpha + 1$$
$$0$$

```
> simplify(alpha^2);  simplify(alpha^3);
```

$$-1 - \alpha$$
$$1$$

More sophisticated evaluation of expressions involving algebraic numbers is performed by applying the algebraic evaluation function `evala`. This should normally be applied to expressions containing *inert* functions (whose names usually begin with a capital letter) and *inert* operators (whose names usually begin with an & symbol), to prevent them first being evaluated normally, e.g.,

```
> evala(Factor(x^2 + x + 1,alpha));
```

$$(x - \alpha)(x + 1 + \alpha)$$

This polynomial factorizes over the rationals extended by α (as it must be definition), whereas it is irreducible over the rationals:

```
> factor(x^2 + x + 1);
```

$$x^2 + x + 1$$

```
> alias(alpha = alpha);  # removes the alias
```

5.4.3 Implicit dependence

One final potential application of `alias` is to make a variable depend *implicitly* (i.e., in an unspecified way) on another variable. Suppose we want the variable y to depend on the variable x, so that its derivative remains symbolic. This is necessary if you want the independent variable in a differential equation to look like a variable rather than a function. By default, y does not depend on x, so

```
> diff(y,x);
```

$$0$$

But after defining this alias:

```
> alias(y = y(x));
```

$$y$$

it does:

```
> diff(y,x);
```

$$\frac{\partial}{\partial x} y$$

However, this technique can cause confusion because it is easy to forget that y is essentially an alias for itself, so it is not generally a good idea! For example, at first sight this result looks a little surprising:

```
> subs(x=3, y);
```

$$\mathrm{y}(3)$$

There is no *good* way to declare implicit dependence in Maple.

5.5 Computing with modular integers

Algebraic equations can be solved in $\mathbb{Z}_m$ by using the function `msolve`, as mentioned above.

An expression can be evaluated "modulo m" for some positive integer m by using the infix operator `mod` with the syntax:

exprn `mod` *m*

To prevent *exprn* being evaluated normally *before* it is evaluated modulo m, *inert* functions (whose names normally begin with a capital letter) and the *inert* power operator `&^` should be used (as when computing with algebraic numbers), e.g.,

```
> factor(x^2+1);
```

$$x^2 + 1$$

```
> Factor(x^2+1) mod 2;
```

$$(x + 1)^2$$

Suppose we want to find the last (unit) digit in the following large number:

```
> 2^(2^1000) mod 10;

    Error, integer too large in context
```

Although the direct approach fails, we can do it easily like this, by using the *inert* operator `&^`:

```
> 2&^(2^1000) mod 10;
```

$$6$$

If m is prime then $\mathbb{Z}_m$ is a (finite) field, in which division by any non-zero number is defined, e.g.,

```
> (1/2) mod 3;
```

$$2$$

Some matrix operations are supported directly by the `mod` operator, but not all. (See the online help for `mod` for details.) Determinant computation is supported directly, e.g.,

```
> A := matrix([[1,2],[1,1]]);
```

$$A := \begin{bmatrix} 1 & 2 \\ 1 & 1 \end{bmatrix}$$

```
> Det(A) mod 3;
```

$$2$$

But, generally, modular matrix computation must be performed over the rationals and the result then reduced by explicitly mapping the `mod` operation (as a function) over the resulting matrix, e.g.,

```
> map(`mod`, evalm(A^2), 3);
```

$$\begin{bmatrix} 0 & 1 \\ 2 & 0 \end{bmatrix}$$

By default, the "positive" representation $[0, m-1]$ is used for $\mathbb{Z}_m$, but if the assignment,

$$`\mathrm{mod}` := \mathrm{mods}$$

is made then the "symmetric" representation $[-\mathrm{iquo}(m-1, 2), \mathrm{iquo}(m, 2)]$ is used.

5.6 Exercises

Zeros of polynomials

Assign the expression $x^5 - 3x^3 - 1$ to the Maple variable `p` and plot a graph of p that shows *clearly* all its real zeros. Use Maple to prove that p has no rational zeros, i.e., that p has no linear factors (with coefficients in $\mathbb{Q}$). Use the Maple `solve` function to find the zeros of p *exactly* and assign the result to the variable `S`. Evaluate S using the floating-point evaluation function `evalf` to give an approximation to the zeros.

Use the Maple function `fsolve` to compute all the *real* zeros of p as floating-point approximations. Use `fsolve` to compute all five complex zeros of p as floating-point approximations.

Check that p evaluated at each of these complex zeros really is zero or very small, by using a *single* Maple expression to process the sequence of complex zeros in one step. (See the online help for `seq` for guidance on how best to do this.)

Zeros of transcendental functions

Plot the two graphs defined by $y = x$ and $y = \tan(x)$ superimposed so as to show *clearly* the five solutions nearest to 0 of the equation $x = \tan(x)$. Because the tangent function is discontinuous, you need to specify the plot option `discont = true`.

Let $f = \tan(x) - x$. Plot a graph of the expression f showing clearly its behaviour *near the zero at* $x = 0$ *but showing no other zeros.* From the graph, what would you expect to be the *multiplicity* of this zero (i.e., if f were changed slightly, what is the *maximum* number of zeros that could emerge from the zero at $x = 0$)? Demonstrate this by plotting a graph of f modified *slightly* so as to change its gradient at the origin; i.e., plot an expression of the general form $f + \alpha x$ for some appropriate *small* value of α. ("Small" means relative to the scale of the problem, which here is of order 1, so small means between about 1 and 0.1. Consider both positive and negative values of α.)

Suppose you conclude that the multiplicity is m. Prove this in two ways, by showing the following:

- The Taylor series expansion of f around $x = 0$ has order m (i.e., the term of lowest degree has degree m). (Use the online help as necessary.)
- The expression f and its first $m - 1$ derivatives all vanish at $x = 0$ but the mth derivative does not.

Find as many *explicit* zeros of f as possible by using the Maple function `solve` followed by `allvalues` and `evalf`. (Refer to the online help as necessary. Note that only exact representations are returned by `solve` and `allvalues`.)

By considering the plot in the first part of this exercise and the points of discontinuity of the tangent function, determine ranges (called isolating intervals) that each contain *precisely one* zero of f. Hence, construct a *Maple expression sequence* consisting of the four smallest non-negative zeros of f as floating-point approximations, by calling `fsolve` restricted to appropriate isolating intervals.

Numerical integration (quadrature)

Evaluate the integral

$$\int_0^2 e^{-x^2}\, dx$$

exactly, and *then* evaluate the resulting expression as a floating-point approximation. (Remember that in Maple the identifiers `e` and `E` have no special significance, so the exponential function can be input *only* using the functional syntax `exp(x)`.) Evaluate the original integral *directly* as a floating-point approximation *without* first attempting to evaluate it exactly. Explain briefly why the two results are (slightly) different. (Hint: Consider carefully how Maple has evaluated them.)

Repeat the above two steps for the integral

$$\int_0^2 e^{x-x^3}\, dx.$$

Explain briefly why the two results are exactly the same in this case.

The area and circumference of an ellipse

Let a curve C be specified parametrically as $(x(t),\, y(t),\, t = \alpha..\beta)$. The length of C and the area traced out by a straight line from the origin to a point that traverses C are given respectively by:

$$\mathit{Length} = \int_\alpha^\beta \sqrt{\left(\frac{dx(t)}{dt}\right)^2 + \left(\frac{dy(t)}{dt}\right)^2}\, dt$$

and

$$\mathit{Area} = \frac{1}{2}\int_\alpha^\beta x(t)\,\frac{dy(t)}{dt} - y(t)\,\frac{dx(t)}{dt}\, dt.$$

Hence find expressions for the exact total area and length of the general ellipse specified parametrically as $(a \cos(t),\, b \sin(t),\, t = 0..2\,\pi)$.

Verify that your results are correct for the special case when the ellipse is a circle of radius r. (Look at the online help for the `simplify` function to find out how to use the fact that the radius is by definition positive in order to *fully simplify* the expression for the circumference.)

Find expressions for the exact total area and circumference of the ellipse in the special case $a = 2$, $b = 1$. Finally, evaluate these expressions as floats accurate to 15 significant figures.

Numerical solution of ODEs

Find the exact solution of the ordinary differential equation

$$\frac{dy(x)}{dx} = e^{-x^3}$$

subject to the condition that $y(0) = 0$, and plot it on the interval $[-1, 1]$. Now solve the same problem purely numerically.

You should have noticed that the plot of the exact solution is missing for $x < 0$ (because the solution is *apparently* complex), and that plotting the numerical solution is much faster than plotting the exact solution. Confirm this by timing the two plots. Use the online help for the `time` function to find out how Maple can time evaluations. This illustrates the observation that it is often better to solve a problem using a suitable numerical method from the beginning than to solve it exactly and then numerically evaluate the exact result. Experiment with some other ODEs of your choice.

5.7 Appendix: code to plot the figures

```
> restart;
```

```
> plots[setoptions](colour = black, axes = NONE);
```

Graphs near and at a double root (Figure 5.1)

```
> plots[display](array([
    plot({x->x^2-1, 0}, -2..2,
       title = "Two simple roots"),
    plot({x->x^2, 0, [[0,-1]]}, -sqrt(3)..sqrt(3),
       # [[0,-1]] is just for alignment!
       title = "One double root")]));
```

Graphs near and at a triple root (Figure 5.2)

```
> plots[display](array([
    plot({x->x^3-2*x, 0}, -2..2,
       title = "Three simple roots"),
    plot({x->x^3, [[-2,0],[2,0]]}, -4^(1/3)..4^(1/3),
       title = "One triple root")]));
```

Chapter 6

Numerical Linear Algebra

The intention of this chapter is twofold: partly to introduce the general facilities for linear algebra that are new in Maple 6, but mainly to focus on the new facilities for numerical linear algebra that use the NAG library internally. These facilities allow sophisticated numerical computations to be performed automatically, with very little user effort, and much of the programming that would be needed in other languages is simply unnecessary. They complement the facilities for numerical solution of equations, numerical integration, etc., that were introduced in the previous chapter.

The first two sections introduce linear algebra in Maple and numerical linear algebra in general. The next two sections present basic facilities for constructing, manipulating, and viewing matrices, which are intended primarily for interactive use. Section 6.5 introduces the new `LinearAlgebra` package and the next three sections present some of its main facilities for solving systems of linear equations, constructing special matrices, and solving eigenvalue problems. The last two sections discuss three classes of problems in linear algebra that are particularly difficult to solve numerically, namely ill-conditioned, under- and over-determined problems.

6.1 Linear algebra in Maple

Maple provides two representations of matrices, as either *tables* or *rtables*, and two packages to support linear algebra, which are called `linalg` and `LinearAlgebra`, based, respectively, on matrices represented as tables and rtables. The rtable data structure, so named because it represents a *rectangular table*, and the `LinearAlgebra` package, are new in Maple 6. The `linalg` package provides slightly better support for abstract linear algebra in general, whereas the new `LinearAlgebra` package is more closely integrated with the rest of Maple and provides much better support for numerical linear algebra by incorporating code developed by the Numerical Algorithms Group (NAG).

Hence, I will focus on rtable-based matrices and the `LinearAlgebra` package in this chapter. (However, in the final chapter I will focus on the `linalg` package as a basis for supporting block matrices.) The standard Maple function `convert` allows easy conversion between the table and rtable representations of matrices and vectors.

A lot of the linear algebra support in Maple applies equally well to symbolic and numeric values, although I will focus on problems involving numeric, and in particular approximate floating-point, values. Maple automatically chooses the appropriate low-level support to use.

One could probably write an entire book on numerical linear algebra in Maple and this chapter will not be more than a first introduction. I will outline only simple use of the main interactive functions in the `LinearAlgebra` package that do not require much background in numerical linear algebra to understand. However, most of these functions accept many optional arguments, which I will not discuss, and there are a lot of low-level facilities that are intended primarily for use in other procedures by experts, which I also will not discuss.

From now on in this chapter I will assume rtable-based Matrices and Vectors. Note the convention that the *rtable* data-type names *Matrix* and *Vector* are capitalized to distinguish them from the related *table* data-type names *matrix* and *vector*.

6.2 Numerical linear algebra

Numerical linear algebra involves two main problems, the solution of linear algebraic equations and eigenvector/eigenvalue problems, which are described in most textbooks on numerical methods. A good introduction to numerical algorithms and their implementation is *Numerical Recipes: The Art of Scientific Computing* by Press *et al.* [23] and a good presentation of the theoretical background is provided by *Introduction to Numerical Analysis* by Stoer and Bulirsch [25]. Problems in linear algebra require sophisticated numerical methods only if they involve fairly large systems of equations or matrices, otherwise they can be solved by various *ad hoc* methods. Nevertheless, I will use only fairly small problems as examples, in order to save space and avoid getting lost in inessential complexity. Tasks such as inverting a matrix and computing determinants are essentially special cases of the solution of linear algebraic equations.

Numerical computation becomes difficult when the problem is *ill conditioned*, which means that the solution is very sensitive to small changes or perturbations in the problem specification. Unfortunately, the unavoidable errors implicit in numerical approximation cause such perturbations, and obtaining accurate solutions to ill-conditioned problems is one of the major tasks of numerical analysis. The *condition* of a matrix problem can be quantified by the condition number of the matrix: the larger the condition number the worse the condition of the problem. Matrices that are singular or nearly so are particularly difficult to deal with, in which case the technique of singular value decomposition (SVD) may help.

The task of numerical computation (in general) is to produce a solution that is sufficiently accurate as quickly as possible; thus, the first concern is accuracy and the second is speed. But to some extent there is a trade-off between these two requirements: obtaining higher accuracy may require using more steps or a more sophisticated algorithm, either of which will probably take more time. Using higher precision arithmetic will probably require more memory, which may cause the computation to take more time because, for example, the process may have to wait longer for resources or it may have to use slower memory.

Often, the mathematical model from which a problem in numerical linear algebra arises leads to some special structure in the problem; for example, the coefficient matrix may be sparse, triangular, tridiagonal, banded, or block-structured, and numerical methods have been developed to take advantage of such structure. The appropriate method is normally chosen automatically by the `LinearAlgebra` package; nevertheless, it is useful to have some idea of the issues involved.

6.3 Interactive linear algebra

Maple 6 provides a succinct angle-bracket syntax for interactively constructing row and column vectors and matrices, and the standard algebraic operators are all automatically overloaded to perform matrix algebra. Angle brackets containing values separated by commas construct a *column* structure, in the simplest case a single column vector:

```
> < a , b , c >;
```

$$\begin{bmatrix} a \\ b \\ c \end{bmatrix}$$

whereas angle brackets containing values separated by vertical bars construct a *row* structure, in the simplest case a single row vector:

```
> < a | b | c >;
```

$$[a, b, c]$$

The input spacing is optional, and I am inserting more spaces than usual here to emphasize the syntax and its symmetry.

The values used in these constructors can themselves be either row or column structures (matrices in general) provided their sizes are compatible (conformable): a column of rows or a row of columns constructs a matrix by stacking the columns or rows:

```
> < < a | b > , < c | d > >;
```

$$\begin{bmatrix} a & b \\ c & d \end{bmatrix}$$

```
> < < a , c > | < b , d > >;
```

$$\begin{bmatrix} a & b \\ c & d \end{bmatrix}$$

whereas a column of columns or a row of rows just concatenates the columns or row:

```
> < < a , b > , < c , d > >;
```

$$\begin{bmatrix} a \\ b \\ c \\ d \end{bmatrix}$$

```
> < < a | b > | < c | d > >;
```

$$\begin{bmatrix} a & b & c & d \end{bmatrix}$$

Moreover, the whole process is completely dynamic, the constructors evaluate their operands, and matrices and vectors can be stacked together in arbitrary ways (provided they are conformable):

```
> M := < < a | b > , < c | d > >;
```

$$M := \begin{bmatrix} a & b \\ c & d \end{bmatrix}$$

```
> R := < e | f >;
```

$$R := [e, f]$$

```
> < M , R >;
```

$$\begin{bmatrix} a & b \\ c & d \\ e & f \end{bmatrix}$$

```
> < M | R >;
```

```
Error, (in Matrix/MakeInit) initializer parameter is
non-conformant: (1,2)
```

Addition and subtraction of conformable matrices and vectors are supported by the usual operators + and -, e.g.,

```
> M+M, M-M;
```

$$\begin{bmatrix} 2a & 2b \\ 2c & 2d \end{bmatrix}, 0$$

```
> R+R, R-R;
```

$$[2e, 2f], 0$$

Note the appearance of the scalar 0 here where one would expect a zero matrix or vector. Maple interprets a numerical scalar used in matrix addition as a diagonal matrix of whatever size and shape is appropriate, e.g.,

```
> M+3, M-3;
```

$$\begin{bmatrix} a+3 & b \\ c & d+3 \end{bmatrix}, \begin{bmatrix} a-3 & b \\ c & d-3 \end{bmatrix}$$

```
> R+3;
```

```
Error, (in rtable/Sum) invalid arguments
```

Multiplication of *conformable* matrices and vectors is supported by the standard dot operator representing non-commutative multiplication:

```
> R . M;
```

$$[e\,a + f\,c, e\,b + f\,d]$$

```
> M . R;
```

```
Error, (in LinearAlgebra:-Multiply)
LinearAlgebra:-MatrixVectorMultiply expects its 2nd argument, v,
to be of type Vector[column], but received Vector[row](2, [...],
datatype = anything, storage = rectangular, order =
Fortran_order, shape = [])
```

The dot operator also supports (commutative) multiplication of matrices and vectors by scalars, for which the * operator can also be used:

```
> 3 . M = 3*M;
```

$$\begin{bmatrix} 3a & 3b \\ 3c & 3d \end{bmatrix} = \begin{bmatrix} 3a & 3b \\ 3c & 3d \end{bmatrix}$$

```
> R . 3 = R*3;
```

$$[3\,e,\,3\,f] = [3\,e,\,3\,f]$$

But remember that the dot operator must not be followed immediately (with no space) by a natural number, which would ambiguously imply floating-point decimal notation:

```
> 3. M;
```

```
Error, missing operator or `;`
```

Square matrices can be raised to arbitrary powers using the standard ^ operator:

```
> M^2;
```

$$\begin{bmatrix} a^2 + b\,c & a\,b + b\,d \\ c\,a + d\,c & b\,c + d^2 \end{bmatrix}$$

This includes symbolic powers:

```
> M^p;
```

$$\begin{bmatrix} a & b \\ c & d \end{bmatrix}^p$$

and negative powers, provided the matrix is non-singular:

```
> M^(-1);
```

$$\begin{bmatrix} \dfrac{d}{a\,d - b\,c} & -\dfrac{b}{a\,d - b\,c} \\ -\dfrac{c}{a\,d - b\,c} & \dfrac{a}{a\,d - b\,c} \end{bmatrix}$$

Note that arrays that use the rtable representation do *not* obey the last name evaluation that applies to table-based structures:

```
> M;
```

$$\begin{bmatrix} a & b \\ c & d \end{bmatrix}$$

6.4 Viewing large matrices and vectors

Clicking the right mouse button (*Option*-clicking on Macintosh platforms) on any Maple output produces a context menu that offers a number of operations that depend on the precise nature of the output. If the output exists only as "static" text in the worksheet but not as an "active" Maple data structure (because the statements that produced it have not been executed in the current session) then the context menu offers only the option to copy the output, but for an "active" data structure it normally offers many more operations. Most of these operations can also be performed by executing commands. However, one important interactive facility is the Structured Data Browser. While this can be used with any rtable (only), it is most important for large ones. Small rtables are displayed *in situ*, as we have already seen, but here is another example, which also illustrates direct use of the rtable function:

```
> rtable([[1,2],[3,4]]);
```

$$\begin{bmatrix} 1 & 2 \\ 3 & 4 \end{bmatrix}$$

However, large rtables (and all rtables having more than two dimensions) are displayed only in an abbreviated form, e.g.,

```
> rtable(1..100, 1..100, (i,j)->1/(i+j-1));
```

$$\left[\begin{array}{l} \text{1..100 x 1..100 2-D Array} \\ \text{Data Type: anything} \\ \text{Storage: rectangular} \\ \text{Order: Fortran\_order} \end{array}\right]$$

The context menu obtained by *right*-clicking on either a displayed or an abbreviated rtable (provided it is active) offers the option to *Browse*. Selecting this option brings up a separate window showing a graphical representation of the rtable. You can zoom in on a sub-range of the elements by either double clicking or dragging the mouse. When the sub-range is small enough the actual values of the elements in the sub-range are displayed. The browser can also be used to change elements and other properties of an rtable interactively. The best way to learn how to use this facility is by experiment.

The definition of large for purposes of choosing to abbreviate rtables is controlled by the interface option `rtablesize`, which on my system has the following default value:

```
> interface(rtablesize);
```

$$10$$

although it can be changed via the `interface` function.

This abbreviation facility does not apply to table-based structures. The following statement generates the table-based equivalent of the above rtable:

```
matrix(100, 100, (i,j)->1/(i+j-1))
```

If you execute it you will see why the abbreviation facility is useful! Incidentally, this example constructs a Hilbert matrix, a topic to which I will return in Section 6.9.

6.5 Using the LinearAlgebra package

Almost all the functions used in this chapter are in the `LinearAlgebra` package, with the exception of the `Matrix` and `Vector` constructor functions, which are in the main Maple library. The `LinearAlgebra` package can be used most conveniently after executing:

```
> with(LinearAlgebra):
```

which makes all its functions available via their short names, as I will assume in the rest of this chapter. The names of the functions in this package all begin with a capital letter, to distinguish them from similar functions in the older `linalg` package. Beware that this violates the loose Maple convention that functions whose names begin with capital letters are inert!

Maple determines automatically what kind of arithmetic to use for linear algebra. By default, it uses the rule that if only numerical values are involved, at least one of which is a float, then hardware floating-point arithmetic is used; otherwise, normal Maple symbolic computation is used. Hence, in this chapter I will use mainly float constants. Hardware floating-point arithmetic is accurate to about 15 significant digits on 32-bit architectures such as those I am using to run examples. If higher precision is required then software floating-point arithmetic can be used by setting the environment variable `UseHardwareFloats` to `false`; by default, it is `true`.

For testing (and illustrating) numerical linear algebra it is convenient to be able to generate random matrices and vectors, and the `RandomMatrix` and `RandomVector` functions in the `LinearAlgebra` package provide these facilities, e.g.,

```
> RandomMatrix(2);
```

$$\begin{bmatrix} 62 & -71 \\ -79 & 28 \end{bmatrix}$$

The general constructor functions in the `LinearAlgebra` package allow a great deal of control over the structure of the matrices and vectors that they generate, such as the data type of the entries. Here is an example of how to generate a matrix with float entries from an existing matrix:

```
> Matrix(%, datatype=float);
```

$$\begin{bmatrix} 62. & -71. \\ -79. & 28. \end{bmatrix}$$

Many functions, such as `RandomMatrix` and `RandomVector`, allow control information to be passed on via an optional argument of the form

`outputoptions = [`*constructor_options*`]`

Here is an example of how to generate a random matrix directly with float entries:

```
> RandomMatrix(2, outputoptions = [datatype=float]);
```

$$\begin{bmatrix} -56. & -50. \\ -8. & 30. \end{bmatrix}$$

Most functions in the `LinearAlgebra` package accept optional arguments that provide a great deal of additional control over their operation but which I will not discuss here; see the online help for details.

6.6 Solving systems of linear equations

The function-call `LinearSolve(A,B)` returns the solution of the system of linear equations presented in matrix form as $A \cdot x = B$, where A is a Matrix and B is a Matrix or column Vector. For example:

```
> A := RandomMatrix(3, outputoptions = [datatype=float]);
```

$$A := \begin{bmatrix} -36. & -7. & -62. \\ -41. & 16. & -90. \\ 20. & -34. & -21. \end{bmatrix}$$

```
> B := RandomVector(3, outputoptions = [datatype=float]);
```

$$B := \begin{bmatrix} -65. \\ 5. \\ 66. \end{bmatrix}$$

```
> x := LinearSolve(A, B);
```

$$x := \begin{bmatrix} 4.54400699912510841 \\ 1.84234470691163565 \\ -1.79807524059492540 \end{bmatrix}$$

The following matrix expression, which is called the *residual vector*, should be very small:

```
> A.x - B;
```

$$\begin{bmatrix} .142108547152020037\,10^{-13} \\ .284217094304040074\,10^{-13} \\ 0. \end{bmatrix}$$

The error can be expressed as a single number, called the *residual*, by taking the *norm* of this vector. There is a whole family of norms defined on vectors and matrices; let us use the Euclidean or 2-norm, which represents the geometrical length of a vector:

```
> 'error' = Norm(%, 2);
```

$$error = .317764371615650958\,10^{-13}$$

This error is consistent with the expectation of a precision of about 15 significant digits. (To use the word *error* as an identifier it must be escaped with backquotes because it is a Maple keyword.)

A system of linear equations can be put into matrix form by using the function `GenerateMatrix` and a system of equations can be recovered from a matrix form by using the function `GenerateEquations`.

A square matrix A can be inverted provided it is non-singular, meaning that its *determinant* is non-zero, e.g.,

```
> Determinant(A);
```

$$74295.$$

Hence, the square matrix A defined above is indeed non-singular. (In fact, a *random* square matrix will be non-singular almost always.)

A non-singular square matrix A can be inverted by solving the matrix equation $A \cdot x = U$, where U is the identity or unit matrix of the same size as A. (I will denote the unit matrix by U rather than by I, which is perhaps more common, because by default Maple uses I to represent the imaginary unit $\sqrt{-1}$.) We can, of course, construct a 2×2 floating-point unit matrix like this:

```
> U := <<1.0|0.0|0.0>, <0.0|1.0|0.0>, <0.0|0.0|1.0>>;
```

$$U := \begin{bmatrix} 1.0 & 0. & 0. \\ 0. & 1.0 & 0. \\ 0. & 0. & 1.0 \end{bmatrix}$$

However, it is easier (and generally better for reasons that I will explain later) to do it like this:

```
> U := IdentityMatrix(3, outputoptions = [datatype=float]);
```

$$U := \begin{bmatrix} 1.0 & 0. & 0. \\ 0. & 1.0 & 0. \\ 0. & 0. & 1.0 \end{bmatrix}$$

Then, this is one way to compute the inverse of A:

```
> AI := LinearSolve(A, U);
```

$AI :=$
$[-.0457096709065212964\,,\ .0263947775758799249\,,$
$.0218318863988155276]$
$[-.0358166767615586482\,,\ .0268658725351638636\,,$
$-.00939497947371963418]$
$[.0144558853220270526\,,\ -.0183592435560939460\,,$
$-.0116158557103438980]$

Now let us check it:

```
> A . AI = AI . A;
```

$[.99999999999999988\,,\ .222044604925031308\,10^{-15}\,,$
$.111022302462515654\,10^{-15}]$
$[.222044604925031308\,10^{-15}\,,\ 1.\,,\ 0.]$
$[.555111512312578272\,10^{-16}\,,\ 0.\,,\ 1.] =$

$$\begin{bmatrix} 1.00000000000000022\,,\ 0.\,,\ .116573417585641436\,10^{-14} \\ .222044604925031308\,10^{-15}\,,\ 1.\,,\ .721644966006351751\,10^{-15} \\ 0.\,,\ .555111512312578272\,10^{-16}\,,\ .99999999999999978 \end{bmatrix}$$

Both sides of the above equation should be equal to the unit matrix, which they are *to the expected accuracy.* Nevertheless, the fact that the two sides are clearly different reminds us that approximate matrices may well not commute even when their exact counterparts do.

Of course, a more direct way to invert a matrix is like this:

```
> A^(-1);
```

$[-.0457096709065213034\,,\ .0263947775758799422\,,$
$.0218318863988155344]$
$[-.0358166767615586482\,,\ .0268658725351638706\,,$
$-.00939497947371963766]$
$[.0144558853220270526\,,\ -.0183592435560939460\,,$
$-.0116158557103438980]$

which agrees with the inverse computed above, but again only *to within the expected accuracy.*

6.7 Special matrices

Matrices represented as rtables can have various *shape* identifiers; see the online help for the full set. Specifying the shape explicitly via the `shape` option has a number of advantages. It makes it easier to initialize such matrices because only the independent elements need be initialized. It means that Maple need not store all the elements since it can deduce some of them, and Maple can potentially manipulate them more efficiently because it knows precisely what their shape is and does not need to try to deduce it from the values of their elements, which is unreliable when they are floating-point approximations.

The shape identifiers *identity* and *scalar* are special cases of *diagonal* and require storage of only the diagonal elements, the other elements all being zero. For the shapes *triangular*, *symmetric*, *antisymmetric*, *hermitian*, and *antihermitian*, only about half the elements need to be stored. The other elements are either zero or related to those stored. The diagonal elements of an antisymmetric or antihermitian matrix must be zero and those of a complex hermitian matrix must be real, and Maple enforces that, e.g.,

```
> A := Matrix(2, shape=antisymmetric);
```

$$A := \begin{bmatrix} 0 & 0 \\ 0 & 0 \end{bmatrix}$$

```
> A[1,2] := 2:  A;
```

$$\begin{bmatrix} 0 & 2 \\ -2 & 0 \end{bmatrix}$$

```
> A[1,1] := 1;

   Error, attempt to assign non-zero to antisymmetric diagonal
```

A symmetric matrix is equal to its transpose and an antisymmetric matrix is equal to minus its transpose, which we can easily verify for appropriate random matrices, even when they have floating-point elements:

```
> S := RandomMatrix(2, outputoptions =
      [shape=symmetric, datatype=float]):
  S = Transpose(S);  Equal(op(%));
```

$$\begin{bmatrix} 13. & 26. \\ 26. & 68. \end{bmatrix} = \begin{bmatrix} 13. & 26. \\ 26. & 68. \end{bmatrix}$$

true

It is necessary to use the function `Equal` in the `LinearAlgebra` package, rather than `evalb`, to compare rtables reliably:

```
> AS := RandomMatrix(2, outputoptions =
      [shape=antisymmetric, datatype=float]):
  AS = -Transpose(AS);  Equal(op(%));
```

$$\begin{bmatrix} 0. & 55. \\ -55. & 0. \end{bmatrix} = \begin{bmatrix} 0. & 55. \\ -55. & 0. \end{bmatrix}$$

$$true$$

Hermitian and antihermitian matrices are complex-valued analogues of symmetric and antisymmetric matrices. The hermitian conjugate or *hermitian transpose* of a matrix is the complex conjugate of its transpose. By default, random matrices have integer elements in the range $[-99, 99]$. To generate non-trivially complex matrices, we need to be able to generate analogous complex elements, namely random *Gaussian* integers, with real and imaginary parts in the same range. To do this, let us define a complex random number generator called `crand`:

```
> rand(-99..99):  crand := % + I*%:
```

A hermitian matrix is equal to its hermitian transpose and an antihermitian matrix is equal to minus its hermitian transpose, which we can now easily verify:

```
> H := RandomMatrix(2, generator = crand, outputoptions =
      [shape=hermitian, datatype=complex[8]]):
  H = HermitianTranspose(H);  Equal(op(%));
```

$$\begin{bmatrix} 0.+0.I & -37.-35.I \\ -37.+35.I & 0.+0.I \end{bmatrix} = \begin{bmatrix} 0.+0.I & -37.-35.I \\ -37.+35.I & 0.+0.I \end{bmatrix}$$

$$true$$

```
> AH := RandomMatrix(2, generator = crand, outputoptions
      = [shape=antihermitian, datatype=complex[8]]):
  AH = -HermitianTranspose(AH);  Equal(op(%));
```

$$\begin{bmatrix} 0.+0.I & 49.+63.I \\ -49.+63.I & 0.+0.I \end{bmatrix} = \begin{bmatrix} 0.+0.I & 49.+63.I \\ -49.+63.I & 0.+0.I \end{bmatrix}$$

$$true$$

A related property of matrices, which matters computationally although it does not relate to any particular theory in abstract linear algebra, is *sparseness*. A matrix is *sparse* if most of its elements are zero, in which case it is more efficient to store only its non-zero elements together with their indices rather than to store all its elements, as is necessary for a general dense matrix. The ultimate sparse matrix is a zero matrix:

```
> Z := ZeroMatrix(2);
```

$$Z := \begin{bmatrix} 0 & 0 \\ 0 & 0 \end{bmatrix}$$

But the `ZeroMatrix` function generates a constant matrix that cannot be changed:

```
> Z[1,1] := 1;
```

```
Error, attempt to assign non-zero to zero Matrix entry
```

A sparse matrix is different in that it defaults to a zero matrix but it can be changed, and the usual change to make would be to set a small proportion of its elements to non-zero values, e.g.,

```
> M := Matrix(2, storage=sparse);
```

$$M := \begin{bmatrix} 0 & 0 \\ 0 & 0 \end{bmatrix}$$

```
> M[1,1] := 1:  M;
```

$$\begin{bmatrix} 1 & 0 \\ 0 & 0 \end{bmatrix}$$

6.8 Eigenvalue problems

The *eigenvalue problem* for a square matrix A is to find (column) vectors x and scalars λ such that $A \cdot x = \lambda \cdot x$. The significance of a solution is that the effect of A on any vector in the direction of an *eigenvector* x is simply to change the length of the vector by a factor of the corresponding *eigenvalue* λ. Thus, eigenvectors represent distinguished directions for a matrix. There can be up to n distinct eigenvectors x and corresponding eigenvalues λ if A is an $n \times n$ matrix.

The function-call `Eigenvectors(A)` returns the complete solution to the simple eigenvalue problem for the matrix A as an expression sequence consisting of a Vector Λ of the eigenvalues of A followed by a Matrix X whose columns are the eigenvectors of A. The ith column of X is an eigenvector associated with the ith eigenvalue in the Vector Λ. (The related but generally less useful function-call `Eigenvalues(A)` returns only the Vector Λ of eigenvalues.) In general, eigenvectors and eigenvalues are complex, even for a real matrix:

```
> A := RandomMatrix(2, outputoptions = [datatype=float]);
```

$$A := \begin{bmatrix} 38. & -82. \\ 97. & -66. \end{bmatrix}$$

```
> Lambda, X := Eigenvectors(A);
```

$$\Lambda, X := \begin{bmatrix} -14. + 72.4568837309471974\, I \\ -14. - 72.4568837309471974\, I \end{bmatrix},$$
$$[.394630784700000014 + .549879170800000038\, I,$$
$$.394630784700000014 - .549879170800000038\, I]$$
$$[.736138194499999954 + 0.\, I,$$
$$.736138194499999954 + 0.\, I]$$

Now let us check this result:

```
> x := Column(X, 1);   lambda := Lambda[1];
```

$$x := \begin{bmatrix} .394630784700000014 + .549879170800000038\, I \\ .736138194499999954 + 0.\, I \end{bmatrix}$$
$$\lambda := -14. + 72.4568837309471974\, I$$

```
> A . x = lambda . x;
```

$$\begin{bmatrix} -45.3673621303999966 + 20.8954084904000014\, I \\ -10.3059347210999946 + 53.3382795676000044\, I \end{bmatrix} =$$
$$\begin{bmatrix} -45.3673621300044161 + 20.8954084920865634\, I \\ -10.3059347230000001 + 53.3382795680986206\, I \end{bmatrix}$$

The two sides are not identical, so let us represent the error as a single number:

```
> `error` = Norm(lhs(%)-rhs(%), 2);
```

$$error = .261908653999467274\, 10^{-8}$$

For the other eigenvector/eigenvalue pair we get:

```
> x := Column(X, 2);   lambda := Lambda[2];
```

$$x := \begin{bmatrix} .394630784700000014 - .549879170800000038\, I \\ .736138194499999954 + 0.\, I \end{bmatrix}$$
$$\lambda := -14. - 72.4568837309471974\, I$$

```
> A . x = lambda . x;
  `error` = Norm(lhs(%)-rhs(%), 2);
```

$$\begin{bmatrix} -45.3673621303999966 - 20.8954084904000014\, I \\ -10.3059347210999946 - 53.3382795676000044\, I \end{bmatrix} =$$
$$\begin{bmatrix} -45.3673621300044161 - 20.8954084920865634\, I \\ -10.3059347230000001 - 53.3382795680986206\, I \end{bmatrix}$$

$$error = .261908653999467274\,10^{-8}$$

The error in this approximate numerical solution is of the order of 10^{-8} for both eigenvector/eigenvalue pairs, which is much larger than for the solution of a pair of linear equations but probably still small enough for most practical purposes.

Another way of checking the solution in the case where the maximal number of linearly independent eigenvectors exists is to leave them in the matrix X but to make the eigenvalues the elements of a diagonal matrix, e.g.,

```
> Lambda := DiagonalMatrix(Lambda);
```

$$\Lambda := \begin{bmatrix} -14. + 72.4568837309471974\,I & 0 \\ 0 & -14. - 72.4568837309471974\,I \end{bmatrix}$$

Then the following single matrix equation is equivalent to stacking the individual eigenvalue equations in the form $A \cdot (x_i) = (x_i) \cdot (\lambda_i)$ side by side:

```
> A . X = X . Lambda:
```

and again we get a similar estimate of the error:

```
> `error` = Norm(lhs(%)-rhs(%), 2);
```

$$error = .3841968646\,10^{-8}$$

The above matrix equation $A \cdot X = X \cdot \Lambda$ shows how to transform a matrix into a diagonal form, provided a full set of linearly independent eigenvectors can be found, in which case the matrix X must be invertible:

```
> X^(-1) . A . X = Lambda;
```

$$[-13.9999999999999930 + 72.4568837309471974\,I,$$
$$.258103938222120632\,10^{-8} + .162454227847774746\,10^{-8}\,I]$$
$$[.258103227679384872\,10^{-8} - .162454583119142626\,10^{-8}\,I,$$
$$-14. - 72.4568837309471832\,I] =$$
$$\begin{bmatrix} -14. + 72.4568837309471974\,I & 0 \\ 0 & -14. - 72.4568837309471974\,I \end{bmatrix}$$

Within the numerical error that we have come to expect, this result confirms that we can use the matrix of eigenvectors to transform a matrix into a diagonal matrix whose elements are the eigenvalues.

In general, eigenvectors and eigenvalues are complex, even for a real matrix. However, if A is a real symmetric or complex hermitian matrix then its eigenvalues are real, e.g.,

```
> A := RandomMatrix(2, outputoptions =
     [shape=symmetric, datatype=float]);
```

$$A := \begin{bmatrix} -75. & 5. \\ 5. & 25. \end{bmatrix}$$

```
> Lambda, X := Eigenvectors(A);
```

$$\Lambda,\, X := \begin{bmatrix} -75.2493781056044498 \\ 25.2493781056044533 \end{bmatrix},$$
$$\begin{bmatrix} .998758526924798940 & .0498137018801597594 \\ -.0498137018801597594 & .998758526924798940 \end{bmatrix}$$

Moreover, its eigenvectors are orthogonal and can be normalized (if necessary) to be real and of unit length. This makes the matrix of eigenvectors X an orthogonal matrix, which means that its transpose is equal to its inverse. In fact, Maple does the normalization automatically and moreover knows that the matrix is in principle orthogonal, even though in practice this is true only approximately. This is one of the advantages of formally specifying the matrix A to be real-symmetric. For example:

```
> IsOrthogonal(X);
```

true

```
> X . Transpose(X);
```

$$\begin{bmatrix} .99999999999999978 & 0. \\ 0. & .99999999999999978 \end{bmatrix}$$

The analogue for a complex hermitian matrix is that the matrix of eigenvectors is unitary, which means that its hermitian transpose is equal to its inverse, e.g.,

```
> A := RandomMatrix(2, generator = crand, outputoptions =
    [shape=hermitian, datatype=complex[8]]);
```

$$A := \begin{bmatrix} 0. + 0.\, I & -93. + 92.\, I \\ -93. - 92.\, I & 0. + 0.\, I \end{bmatrix}$$

```
> Lambda, X := Eigenvectors(A);
```

$$\Lambda,\, X := \begin{bmatrix} -130.816665604960264 \\ 130.816665604960264 \end{bmatrix},$$

$[.502695358777401791 - .497290032338935206\, I,$
$-.502695358777401791 + .497290032338935206\, I]$
$[.707106781186547572 + 0.\, I,$
$.707106781186547572 + 0.\, I]$

```
> IsUnitary(X);
```

$$true$$

```
> X . HermitianTranspose(X);
```

$$\begin{bmatrix} 1. + 0.\,I & 0. + 0.\,I \\ 0. + 0.\,I & 1.00000000000000022 + 0.\,I \end{bmatrix}$$

6.9 Ill-conditioned problems

A *Hilbert matrix* is a matrix whose (i, j)-element is $1/(i + j - 1)$ and Hilbert matrices are notoriously ill conditioned; they become worse conditioned as their size increases because their elements become more similar for larger values of i and j. The function HilbertMatrix returns a Hilbert matrix of a specified size, e.g.,

```
> HM := HilbertMatrix(4);
```

$$HM := \begin{bmatrix} 1 & \frac{1}{2} & \frac{1}{3} & \frac{1}{4} \\ \frac{1}{2} & \frac{1}{3} & \frac{1}{4} & \frac{1}{5} \\ \frac{1}{3} & \frac{1}{4} & \frac{1}{5} & \frac{1}{6} \\ \frac{1}{4} & \frac{1}{5} & \frac{1}{6} & \frac{1}{7} \end{bmatrix}$$

The *condition number* of a matrix is a measure of its condition, and the larger its condition number the worse its condition, e.g.,

```
> ConditionNumber(HM);
```

$$28375$$

The following table shows how the condition numbers of Hilbert matrices increase with their size. (I use a table-based array only because an rtable-based array of this size is just too big to display by default.)

```
> array([['n','Cond'],
      seq([n,evalf(ConditionNumber(HilbertMatrix(n)))],
        n = 1..10)]);
```

$$\begin{bmatrix} n & Cond \\ 1 & 1. \\ 2 & 27. \\ 3 & 748. \\ 4 & 28375. \\ 5 & 943656. \\ 6 & .29070279\,10^{8} \\ 7 & .9851948865\,10^{9} \\ 8 & .3387279110\,10^{11} \\ 9 & .1099654541\,10^{13} \\ 10 & .3535743925\,10^{14} \end{bmatrix}$$

If the reciprocal of the condition number of a matrix is comparable to the relative accuracy of the arithmetic to be used in an approximate numerical solution then the problem is seriously ill conditioned and large numerical errors can be expected. For example, we see from the above table that the reciprocal of the 10×10 Hilbert matrix is comparable with the expected relative accuracy of hardware floating-point computations (on a 32-bit machine). Hence, if we try to invert this Hilbert matrix numerically we should expect the large errors that we get:

```
> HM := HilbertMatrix(10, outputoptions = [datatype=float]):

> HMI := MatrixInverse(HM):

> HM . HMI - IdentityMatrix(10):

> 'error' = Norm(%,2);
```

$$error = .01185581530$$

What are the solutions? One possible solution (at least in Maple) is to try to solve the problem using exact rational arithmetic, which works well for this artificial problem:

```
> HM := HilbertMatrix(10):

> HMI := MatrixInverse(HM):

> HM . HMI - IdentityMatrix(10):

> 'error' = Norm(%,2);
```

$$error = 0$$

Another possible solution (again, at least in Maple) is to try to solve the problem using higher precision floating-point arithmetic. Computer hardware usually supports a floating-point representation using at most two machine words; to use higher precision than that requires software floating point.

The Maple LinearAlgebra package uses software instead of hardware floating point if the environment variable UseHardwareFloats is set to false (by default, it is true) and then floating-point arithmetic respects the value of Digits. Such changes can be kept local by making them inside a module (or procedure), which I will do in order to avoid changing my default computational environment.

The default value of Digits, namely 10, is less accurate than hardware floating point and gives an appalling result for this trial problem, but a value of 20 is more accurate and gives a somewhat better result:

```
> module()
    local HM, HMI;
    UseHardwareFloats := false;
    Digits := 20;  print('Digits' = Digits);
    HM := HilbertMatrix(10,
       outputoptions = [datatype=float]);
    HMI := MatrixInverse(HM);
    HM . HMI - IdentityMatrix(10);
    print(`error` = Norm(%,2))
  end module:
```

$$Digits = 20$$

$$error = .14800178766130261091\,10^{-5}$$

Setting Digits to 30 uses approximately twice as many significant decimal digits as hardware floating-point arithmetic and gives a respectable result:

```
> module()
    local HM, HMI;
    UseHardwareFloats := false;
    Digits := 30;  print('Digits' = Digits);
    HM := HilbertMatrix(10,
       outputoptions = [datatype=float]);
    HMI := MatrixInverse(HM);
    HM . HMI - IdentityMatrix(10);
    print(`error` = Norm(%,2))
  end module:
```

$$Digits = 30$$

$$error = .833726016901030448241066077062\,10^{-16}$$

Singular value decomposition can be a useful tool to analyse (and perhaps solve) problems involving ill-conditioned and singular matrices. The singular values of a matrix A are equal to the square roots of the (real) eigenvalues of the product of A with its (hermitian) transpose. Since this product is either real-symmetric or hermitian, and positive semi-definite, the eigenvalues are all real and non-negative, and so their square roots are also purely real. The function-call `SingularValues(A)` returns the singular values of Matrix A as a column Vector. Although this function will attempt to compute the exact singular values of an exact matrix, the result can be so complicated as to be of little use, whereas the floating-point approximations to the singular values of a floating-point matrix are much more intelligible, e.g.,

```
> HM := HilbertMatrix(10, outputoptions = [datatype=float]):
  SingularValues(HM);
```

$$\begin{bmatrix} 1.75191967028317342 \\ .342929548480457813 \\ .0357418162776099932 \\ .00253089075296064658 \\ .000128749619730106912 \\ .472966824446632998\,10^{-5} \\ .122910711243663234\,10^{-6} \\ .214521311681321278\,10^{-8} \\ .218435361531967234\,10^{-10} \\ .556447199604206512\,10^{-13} \end{bmatrix}$$

The fact that the singular values become very small is a symptom of the ill condition of this Hilbert matrix. In fact, any $m \times n$ matrix A with $n \leq m$ can be represented in the form $A = U \cdot SS \cdot (V^T)$ where U is an $m \times n$ column-orthogonal matrix, V^T is the transpose of an $n \times n$ orthogonal matrix, and SS is a diagonal $n \times n$ matrix whose diagonal elements are the singular values of A. Hence, SVD generalizes the diagonalization of a square matrix and the singular values capture the essence of a matrix as do the eigenvalues of a square matrix. The condition number of a matrix can be defined as the ratio of the largest to the smallest singular value, which for this Hilbert matrix gives:

```
> %[1] / %[-1];
```

$$.3148402348\,10^{14}$$

which is essentially the same value that we obtained by direct computation earlier.

The problem caused by small singular values is perhaps easiest to explain for a square $n \times n$ matrix A, for which all the matrices in the SVD are square $n \times n$ matrices and the matrices U and V are both orthogonal. Then the inverse of A can be expressed in terms of its SVD as $A^{-1} = V \cdot SS^{-1} \cdot U^T$. If $SS = \text{diag}(s_j)$ then $SS^{-1} = \text{diag}(1/s_j)$. Hence, if any of the singular values of the matrix are so small that they get "lost in the numerical approximation error" then the numerical solutions of problems involving the matrix become swamped by numerical error and are very inaccurate. The ultimate ill condition occurs when a singular value is zero and the inverse of the matrix is undefined; such a matrix is singular. However, in problems that are only artificially singular but which do in principle have a solution, that solution can often be found by replacing the reciprocal of any zero singular values by zero, thereby essentially avoiding infinities by replacing them with zero. We will use this method in the next section. *Numerical Recipes* gives a nice introduction to this approach, which I have largely followed in this presentation.

6.10 Under- and over-determined problems

A system of linear equations $A \cdot x = b$ represented by an $m \times n$ matrix A is said to be *under*-determined if $n < m$ and *over*-determined if $m < n$. An under-determined system cannot be solved uniquely and its solution will typically involve $m - n$ parameters; an over-determined system has a solution only if $n - m$ of the equations are linear combinations of the rest, so a *random* over-determined system typically has no solution.

6.10.1 Under-determined problems

In a purely numerical computational system, under-determined systems cannot be handled directly because there is no way to represent the parameters, but in Maple this is no problem and by default Maple just invents parameters as necessary (although some control over the names used is available), e.g.,

```
> A := RandomMatrix(2,3, outputoptions = [datatype=float]);
```

$$A := \begin{bmatrix} 1. & -82. & -69. \\ -42. & 59. & 23. \end{bmatrix}$$

```
> b := RandomVector(2, outputoptions = [datatype=float]);
```

$$b := \begin{bmatrix} 24. \\ 65. \end{bmatrix}$$

```
> x := LinearSolve(A, b);
```

$$x := \begin{bmatrix} -1.992909896 - .6454948295\,\_t13_1 \\ -.3169867059 - .8493353024\,\_t13_1 \\ \_t13_1 \end{bmatrix}$$

```
> A . x - b;
```

$$\begin{bmatrix} -.2\,10^{-7} - .3\,10^{-7}\,\_t13_1 \\ -.2\,10^{-7} \end{bmatrix}$$

The magnitude of this residual vector is clearly small relative to the magnitude of x for all values of the parameter, although the error is much larger than when A is a non-singular square matrix.

6.10.2 Over-determined problems

Over-determined systems typically have no solution and Maple just reports this as an error, e.g.,

```
> A := RandomMatrix(3,2, outputoptions = [datatype=float]);
```

$$A := \begin{bmatrix} 22. & -2. \\ -81. & -98. \\ -99. & -41. \end{bmatrix}$$

```
> b := RandomVector(3, outputoptions = [datatype=float]);
```

$$b := \begin{bmatrix} 53. \\ 61. \\ -70. \end{bmatrix}$$

```
> x := LinearSolve(A, b);
```

```
Error, (in LinearAlgebra:-LA_Main:-LinearSolve) inconsistent
system
```

A solution to such a 3×2 matrix equation exists only if one of the equations is a linear combination of the others. As an example, suppose we start with a random 3×2 problem and then include a linear combination of the two equations as a third equation. It is convenient to use the augmented matrix representation, in which the constant vector b is attached to the right of the coefficient matrix A, e.g.,

```
> Ab := RandomMatrix(2,3);
```

$$Ab := \begin{bmatrix} -90 & -11 & 61 \\ -66 & 35 & 96 \end{bmatrix}$$

```
> x := LinearSolve(Ab);
```

$$x := \begin{bmatrix} \dfrac{-3191}{3876} \\ \dfrac{769}{646} \end{bmatrix}$$

Now let us make this problem randomly but consistently over-determined:

```
> rand(-99..99):
  Ab := <Ab, %()*Row(Ab,1) + %()*Row(Ab,2)>;
```

$$Ab := \begin{bmatrix} -90 & -11 & 61 \\ -66 & 35 & 96 \\ -11286 & 1463 & 11033 \end{bmatrix}$$

```
> x := LinearSolve(Ab);
```

$$x := \begin{bmatrix} \frac{-3191}{3876} \\ \frac{769}{646} \end{bmatrix}$$

Maple has no difficulty solving this problem when specified exactly, but what about if we make the same problem numerical:

```
> Ab := Matrix(Ab, datatype=float);
```

$$Ab := \begin{bmatrix} -90. & -11. & 61. \\ -66. & 35. & 96. \\ -11286. & 1463. & 11033. \end{bmatrix}$$

```
> x := LinearSolve(Ab);
```

```
Error, (in LinearAlgebra:-LA_Main:-LinearSolve) inconsistent
system
```

It appears that numerical errors that are too small to be apparent in the input above have already made this problem numerically inconsistent.

6.10.3 Solution by singular value decomposition

One way to proceed is to use SVD. First, let us put the problem back into the conventional form $A \cdot x = b$.

```
> A := SubMatrix(Ab, 1..-1, 1..-2);
```

$$A := \begin{bmatrix} -90. & -11. \\ -66. & 35. \\ -11286. & 1463. \end{bmatrix}$$

```
> b := Column(Ab, 3);
```

$$b := \begin{bmatrix} 61. \\ 96. \\ 11033. \end{bmatrix}$$

We can obtain the full SVD (only available for a floating-point matrix) by specifically requesting the output of all three matrices involved in the decomposition of A:

```
> U,S,Vt := SingularValues(A, 'output=[U,S,Vt]');
```

$U, S, Vt :=$
$[-.00771806403822983180, .650788665733404104,$
$-.759219694186364702]$
$[-.00614636961645026286, -.759258839726383860,$
$-.650759737874035071]$
$[-.999951325629422350, -.000356160469855825694,$
$.00985999602839437994], \begin{bmatrix} 11380.9829962690637 \\ 34.5403913481611654 \\ 0. \end{bmatrix},$

$$\begin{bmatrix} .991702645624878865 & -.128552956638949638 \\ -.128552956638949638 & -.991702645624878865 \end{bmatrix}$$

The zero singular value (in Vector S) is a consequence of the singular nature of this problem and it causes the matrix V to be missing the corresponding column (and hence its transpose V^T to be missing the corresponding row). We need formally to put this row back before we can use the decomposition in order to ensure conformability, although the actual values in the replaced row of V^T do not matter since they will all be multiplied by the zero singular value (which is why they were omitted), so let us use 0 for each missing element:

```
> Vt := <Vt, <0|0>>;
```

$$Vt := \begin{bmatrix} .991702645624878865 & -.128552956638949638 \\ -.128552956638949638 & -.991702645624878865 \\ 0 & 0 \end{bmatrix}$$

Now we can check that we can recover a reasonable approximation to A:

```
> U.DiagonalMatrix(S).Vt:
  A = %;  'error' = Norm(A-%%, 2);
```

$$\begin{bmatrix} -90. & -11. \\ -66. & 35. \\ -11286. & 1463. \end{bmatrix} =$$

$$\begin{bmatrix} -89.9999999999988348 & -11.0000000000004316 \\ -66. & 34.9999999999999858 \\ -11285.9999999999964 & 1462.99999999999910 \end{bmatrix}$$

$$error = .3948365673\,10^{-11}$$

To use the SVD to solve the over-determined system of equations we compute a (generalized) inverse AI of (the rectangular matrix) A by inverting the SVD representation, replacing the reciprocal of the zero singular value by zero (i.e., replacing infinity by zero). We can replace the matrix S by its "inverse" using the `Map` function from the `LinearAlgebra` package, which maps in place rather than making a copy, as follows:

```
> Map(x->if x=0 then 0 else 1/x end if, S);
```

$$\begin{bmatrix} .000087865872390000018 \\ .028951611750000016 \\ 0. \end{bmatrix}$$

```
> AI := Transpose(Vt) . DiagonalMatrix(S) . Transpose(U);
```

$AI :=$
$[-.002422787734179024\,90, .002825285583613414\,84,$
$-.000085807013297148\,9454]$
$[-.018684959989595378\,8, .021799446059528073\,6,$
$.000021520730027115\,8972]$

Now the numerical solution to the consistent over-determined problem and its numerical error relative to the exact solution are:

```
> AI.b;  'error' = Norm(x-%, 2);
```

$$\begin{bmatrix} -.823271413465477053 \\ 1.19040247673854658 \end{bmatrix}$$

$$error = .365591340466306320\,10^{-9}$$

6.10.4 Least squares solution

Another way to solve problems such as this numerically is not to seek the exact solution, which makes $A \cdot x - b$ zero, but to seek the solution that makes $A \cdot x - b$ as small as possible. The function-call `LeastSquares(A,b)` returns a value for x that minimizes `Norm(A.x-b, 2)`. For our current problem this approach gives a very accurate result, which is somewhat better than that obtained by SVD:

```
> LeastSquares(A, b);  'error' = Norm(x-%, 2);
```

$$\begin{bmatrix} -.823271413828692400 \\ 1.19040247678016218 \end{bmatrix}$$

$$error = .237268453326604004\,10^{-13}$$

6.11 Exercises

These exercises should all be performed using small rtable-based vectors and matrices with random floating-point elements, and the `LinearAlgebra` package should be used where appropriate. (You could also try using small matrices with integer or rational elements, or in some cases even symbolic elements, and you could try using table-based vectors and matrices with the older `linalg` package.)

1. Verify for both a diagonal and a triangular matrix that its determinant is the product of its diagonal elements.

2. Let V be an $n \times r$ matrix with $r \leq n$ and maximal rank (i.e., rank r, which means that its r columns are linearly independent). Then $M = V^T \cdot V$ is a symmetric positive definite matrix. Verify this as follows. Construct a suitable random matrix V and check that it has maximal rank. Construct M, check explicitly that it is symmetric and apply the function `IsDefinite` to check that it is symmetric positive definite. Verify that the determinant of M is positive. Positive definite means that $0 < x^T \cdot M \cdot x$ for all non-zero r-dimensional column vectors x. Verify this for a few suitable random vectors x.

3. Construct a matrix representation of a system of linear equations (which is easiest in the form of a single augmented matrix) and solve it using `LinearSolve`. Convert the system to an explicit set of equations and solve it using `fsolve`. Compare the solutions.

4. Construct a random quadratic form $q(x, y) = a\,x^2 + b\,x\,y + c\,y^2$ using the matrix representation $X \cdot Q \cdot X^T$ where $X = (x, y)$ is a row vector and Q is a *symmetric positive-definite* 2×2 matrix. (You can either use the theory in an earlier exercise to construct a random symmetric positive-definite matrix or just generate random symmetric matrices until one is positive definite.) Find the eigenvectors and eigenvalues of Q and interpret them geometrically as follows. Plot the ellipse defined by $q(x, y) = 1$ and plot each eigenvector as a line through the origin extending in both directions a distance that is the reciprocal of the square root of the corresponding eigenvalue. These two lines should be the axes of the ellipse.

5. Construct a matrix equation of the form $A \cdot x = b$, where A is an $n \times n$ Hilbert matrix and b is a suitable random vector. Solve the equation using the `LinearSolve` function and compute the residual of the solution. Solve the problem again using the `LeastSquares` function and compare the residuals obtained using the two methods. Repeat this comparison for several values of n.

Chapter 7

Logic and Control Structures

Programming is about computers making decisions. The features of programming languages that support this are called control structures and the theoretical framework is that of logic. This chapter develops these ideas: first the simple aspects of logic and Boolean algebra that underlie all computer programming and then their application in Maple control structures. Control structures can be "nested", which means that one is contained within another, and sophisticated programs can be expected to contain quite deeply nested control structures that are reminiscent of the leaves of an onion or a set of "Russian dolls". I will illustrate nested control structures briefly at the end of this chapter, and then we will see them in action in subsequent chapters.

Most logical expressions arise from the need to compare two quantities and the first section introduces the Maple comparison operators. Boolean algebra provides a basis for building more complicated logical expressions; the next two sections introduce the background and the Maple Boolean operators. The principal application of Boolean algebra in computing is to provide conditional execution, which is introduced in Section 7.4. A simple application of conditional execution is piecewise-defined functions, which Maple supports quite well, as explained in the next section. One of the main applications of conditional execution is to control code repetition or *loops*, which is the topic of the final section.

7.1 Relational operators: <, <=, >, >=, =, <>

In computing it is often necessary to *compare* two quantities, which is done using the *relational operators* that are normally denoted in mathematics as:

$$<, \ \leq, \ >, \ \geq, \ \neq$$

In Maple they are input as, respectively, the binary operators `<`, `<=`, `>`, `>=`, `=`, and `<>`, which do nothing more than construct inert *data structures* that can then be interpreted by other Maple functions and statements. However, Maple always automatically simplifies relations into a canonical form that uses only `<`, `<=`, `=`, and `<>`, although by default it performs no further evaluation, e.g.,

```
> 3 > 2;  a >= b;  # automatic "simplification"
```

$$2 < 3$$
$$b \leq a$$

Relational data structures can be interpreted by the `solve` function; the most commonly used relation in this context is `=`, but `solve` will attempt to solve (or simplify) other relations, e.g.,

```
> solve(x^2 <= 1);
```

$$\mathrm{RealRange}(-1, 1)$$

Algebra on relations is supported, e.g.,

```
> (x + a >= b) - (a = a);
```

$$-a + b \leq x$$

```
> -(x + a >= b);
```

$$-x - a \leq -b$$

All normal operations on data structures, such as the `op` and `map` functions, may be applied to relational data structures. In addition, the functions `lhs` (left-hand side) and `rhs` (right-hand side) that are specific to binary data structures may be used instead of `op`, e.g.,

```
> lhs(a = b);
```

$$a$$

Beware that element or operand extraction by position is unreliable on data structures that are automatically simplified, such as sets and inequalities, e.g.,

```
> lhs(a >= b);  # you might expect a, but you get ...
```

$$b$$

Relational data structures are treated as logical propositions and *automatically* evaluated to Boolean (logical) values when required by the context, such as when they appear in control statements, e.g.,

```
> if 3 > 2 then `3 > 2` else `3 <= 2` end if;
```

$$3 > 2$$

```
> 4 > 3 and 3 > 2;
```

$$true$$

However, beware that Boolean expressions are simplified before they are evaluated, and the simplification may remove the Boolean context; e.g., in the following example Maple discards the redundant "and true" and hence the Boolean context, so the expression is not evaluated:

```
> 3 > 2 and true;
```

$$2 < 3$$

Boolean evaluation can always be forced by applying the function `evalb`. This is not usually necessary within Maple programs (except sometimes in functions that must return a Boolean value), but it can be useful for interactive experimentation:

```
> evalb(3 > 2);
```

$$true$$

But note that Boolean evaluation never causes algebraic simplification, and as usual in Maple this must be explicitly requested by applying the appropriate algebraic simplification functions. Hence,

```
> evalb(x*(x+1) = x^2+x);
```

$$false$$

```
> evalb(expand(x*(x+1) = x^2+x));
```

$$true$$

(The above example works because `expand` automatically maps itself over the two sides of the relation. By contrast, `simplify` does not!)

Note, however, that the Boolean-valued function `is`, which is really part of the `assume` system (see the online help for details), provides a more sophisticated interpretation of mathematical propositions:

```
> is(x*(x+1) = x^2+x);
```

$$true$$

The Boolean *constants* are denoted `true` and `false`. (There is also a pseudo-Boolean constant `FAIL`, meaning that Maple cannot determine the correct Boolean value, which is usually treated the same as `false`. Maple actually uses three-valued logic, but the existence of the third value `FAIL` can be ignored most of the time.)

The order-relation operators `<`, `<=`, `>`, `>=` can be evaluated to Boolean constants only when *both* operands evaluate to *explicit real numerical values.* If this requires floating-point approximation then you must *explicitly* apply a floating-point evaluation function, e.g.,

```
> evalb(Pi > 3);
```

$$-\pi < -3$$

```
> evalb(evalf(Pi) > 3);
```

$$true$$

But you must be aware that floating-point comparison is *unreliable* and depends on the precision used, e.g.,

```
> evalb(evalf(Pi) = 3.141592654);   # REALLY ?????
```

$$true$$

```
> proc() Digits := 20;
      evalb(evalf(Pi) = 3.141592654)
  end proc();
```

$$false$$

Floating-point approximations should only be compared for equality to within some tolerance that represents the allowed numerical error; e.g., a truncated decimal representation implies an error of less than 0.5 in the last digit, and using this as the tolerance gives a meaningful numerical comparison:

```
> proc() Digits := 20;
      evalb( abs(evalf(Pi) - 3.141592654) < 5e-10 )
  end proc();
```

$$true$$

7.2 Boolean algebra

Boolean algebra is named after the British mathematician and logician George Boole (1815–1864), who invented an "algebra of logic" that is related to ordinary algebra. Logic (from the Greek word *logos*, meaning "word" or "reason") is a branch of philosophy dealing with the principles of valid reasoning and argument.

In its simplest form, as required for computer programming, Boolean algebra is an algebra on a set $\mathbb{B}$ of two elements, usually denoted {`true`, `false`}(or sometimes $\{1, 0\}$) equipped with two binary operations called `and` and `or` and one unary operation called `not`, i.e., the following mappings:

$$\text{and} : \mathbb{B}^2 \to \mathbb{B}, \quad \text{or} : \mathbb{B}^2 \to \mathbb{B}, \quad \text{not} : \mathbb{B} \to \mathbb{B}.$$

Informally, the meanings of these Boolean operators reflect their normal usage in English: "not" inverts or negates a truth value; the proposition "a and b" is true if and only if both a and b are true; the proposition "a or b" is true if either a or b is true (or both are true). A function or mapping can be defined as a set of pairs of points in the domain and corresponding values in the range. If the domain and range are small discrete sets then this set of pairs can be conveniently represented as a table. In Boolean algebra such a table is called a *truth table*. The truth table for "not" is the simplest:

a	**not** a
true	false
false	true

For the binary operators, a slightly more complicated truth table structure (like that of a multiplication table) is required, which has this general form:

operator name •	values of operand b
values of operand a	value of $a \bullet b$

The truth tables for "and" and "or" are as follows:

and	true	false
true	true	false
false	false	false

or	true	false
true	true	true
false	true	false

7.3 Boolean operators in Maple: `and`, `or`, `not`

The Maple Boolean operators `and`, `or`, and `not` all cause automatic Boolean evaluation of their operands (unless automatic simplification removes the operator; see above).

7.3.1 Truth tables in Maple

Truth tables can be constructed in Maple like this (although there are more elegant ways to do it):

```
> array([[a,      `|`, not a],
        [`====`, `+`, `====`],
        [true,   `|`, not true],
        [false,  `|`, not false]]);
```

$$\begin{bmatrix} a & | & \textbf{not } a \\ ==== & + & ==== \\ \mathit{true} & | & \mathit{false} \\ \mathit{false} & | & \mathit{true} \end{bmatrix}$$

```
> array([[`and`, `|`, true,           false],
        [`====`, `+`, `====`,         `====`],
        [true,   `|`, true and true,  true and false],
        [false,  `|`, false and true, false and false]]);
```

$$\begin{bmatrix} \mathit{and} & | & \mathit{true} & \mathit{false} \\ ==== & + & ==== & ==== \\ \mathit{true} & | & \mathit{true} & \mathit{false} \\ \mathit{false} & | & \mathit{false} & \mathit{false} \end{bmatrix}$$

```
> array([[`or`,  `|`, true,          false],
        [`====`, `+`, `====`,        `====`],
        [true,   `|`, true or true,  true or false],
        [false,  `|`, false or true, false or false]]);
```

$$\begin{bmatrix} \mathit{or} & | & \mathit{true} & \mathit{false} \\ ==== & + & ==== & ==== \\ \mathit{true} & | & \mathit{true} & \mathit{true} \\ \mathit{false} & | & \mathit{true} & \mathit{false} \end{bmatrix}$$

Note that an array must have the structure of a Maple matrix in order to be printed as a two-dimensional table as above — a general array is not printed in this way. The array initialization used here ensures that the arrays are automatically matrices. (Remember also that if an array is assigned to a variable then that variable must be *explicitly* `eval`ed or `print`ed in order to see the array, since array identifiers evaluate to themselves by default.)

It is necessary to make the special symbols used to "draw lines" through the tables into identifiers by enclosing them in backquotes and it is also necessary similarly to escape the Maple keywords `and`, `or`, and `not` in order to use them as identifiers. This is not because they contain special symbols but because they have operator syntax. (I have used "=" signs here to draw horizontal lines only because in the default display font on my system the "–" sign looks rather insubstantial and does not line up well with the "+" sign.) We reconsider this technique for displaying tabular data in Maple in more detail as an application of double loops at the end of this chapter.

7.3.2 Boolean expressions

Boolean expressions are constructed using Boolean operators, relational expressions and variables that have Boolean values. The syntax of these programming constructs is similar in all programming languages. It must be followed precisely, which often requires careful interpretation of the colloquial or even the mathematical formulation of a logical proposition or condition. For example, the proposition $2 < x < 5$, which might be read as "x is greater than 2 and less than 5", must be expanded to "x is greater than 2 and ***x is*** less than 5" and programmed as:

```
x > 2 and x < 5
```

It *cannot* be programmed as either "`2 < x < 5`" or "`x > 2 and < 5`", both of which are syntax errors. These remarks apply to *all* relational operators.

The Boolean operators `and` and `or` in Maple obey "conditional", "McCarthy",[1] or "short-circuit" evaluation rules, as they do in other modern languages such as C, C++, and Java. What this means is that they evaluate their left operands first and only evaluate their right operands if necessary to determine the result; this is the sense in which they are conditional or in which they "short-circuit". Hence, conditional Boolean operators in programming languages are not strictly symmetrical in the way that their mathematical counterparts are. Here is an example of how to use this feature and why it is important:

```
> x := 0;
```

$$x := 0$$

```
> x > 0 and log(x) > 1;
```

false

```
> log(x) > 1 and x > 0;
```

```
Error, (in ln) numeric exception: division by zero
```

[1] "McCarthy" here refers to John McCarthy, who began the development of LISP in the Artificial Intelligence Group at MIT in 1959. LISP was one of the first languages for symbolic computation and many features of Maple can be traced back to LISP.

```
> x := 'x':
```

It is generally bad style to write explicit Boolean constants within Boolean expressions, and a particularly common example is explicitly to compare a Boolean expression with a Boolean constant. This is inelegant and should be avoided when programming Boolean expressions. (Maple, and other systems, will automatically simplify some such redundant constructs, but it is still bad practice to use them in the first place.) In particular:

- "`x = true`" is the same as "`x`";
  "`x = false`" is more elegantly expressed as "`not x`";
- "`x and true`" is the same as "`x`";
  "`x and false`" is always "`false`";
- "`x or true`" is always "`true`";
  "`x or false`" is the same as "`x`".

7.4 Conditional execution: `if`

Maple can be made to execute different sequences of statements depending on *conditions*, which are Boolean expressions (that are automatically evaluated to Boolean values). This is rarely of direct interactive use, but it is very useful within loops and procedures and can be used to define functions piecewise.

7.4.1 Simple conditional execution: `if`

In the simplest case, a statement sequence is executed if some condition is true and is not executed otherwise. The syntax is

`if` *condition* `then` *statement_sequence* `end if`

Here is a trivial (non-mathematical) example:

```
> WeekEnd := Day ->
    if Day = Sunday then
       print(`Take it easy!`);
       print(`But back to work tomorrow.`)
    end if:
```

```
> WeekEnd(Saturday):   # prints nothing
```

```
> WeekEnd(Sunday):
```

Take it easy!
But back to work tomorrow.

Here is the same example, implemented as a function that returns a value rather than outputting it as a side-effect via `print`:

```
> WeekEnd := Day ->
     if Day = Sunday then
        'Take it easy! But back to work tomorrow.'
     end if:
```

```
> WeekEnd(Saturday);  # returns nothing
```

```
> WeekEnd(Sunday);
```

Take it easy! *But back to work tomorrow.*

The words `if`, `then`, and `end` are Maple *keywords*. This means that they have a defined purpose in the Maple language and they cannot be used as identifiers (unless they are escaped with backquotes). Every statement that begins with `if` must end with `end if`. (This way of grouping statement sequences follows the language Modula, and is a development of the "`begin ... end`" grouping used in its predecessors such as Algol 60 and Pascal. It is an alternative to the "`{ ... }`" grouping used in languages based on C, C++, Java, etc. Maple also accepts the keyword `fi` in place of `end if` for backward compatibility; this use of reversed keywords to terminate control structures came from Algol 68.) The statements between `then` and `end if` are all executed in the order they are written if the condition between `if` and `then` evaluates to true; otherwise, none of them is executed.

The statements in a statement sequence should be separated by semicolons (colons are also allowed — it makes no difference which is used, but semicolons are better because they are used in most other modern languages); it is optional in Maple whether the last statement is terminated. (Maple is fairly flexible about the presence of redundant statement terminators.) A statement sequence can, of course, consist of a single statement. If a top-level `if` statement itself is terminated with a semicolon then the result of the last statement executed under its control, if any, is displayed.

7.4.2 Alternative execution: `else`

To execute one of *two* alternative statement sequences the `if` syntax is extended by the introduction of the `else` keyword. The syntax is

```
if condition then true_statement_sequence
   else false_statement_sequence end if
```

The statements between `then` and `else` are all executed in the order they are written if the condition between `if` and `then` evaluates to true; otherwise, the statements between `else` and `end if` are all executed in the order they are written. Here is another trivial non-mathematical example:

```
> WeekEnd := Day ->
    if Day = Saturday or Day = Sunday then
       print('Enjoy the weekend!')
    else
       print('Work hard!')
    end if:
```

```
> WeekEnd(Saturday):
```

Enjoy the weekend!

```
> WeekEnd(Monday):
```

Work hard!

7.4.3 Multiple alternative execution: `elif`

To execute one of *many* alternative statement sequences the `if` syntax is extended again by the introduction of the `elif` keyword. The syntax is

`if` *condition_1* `then` *true_statement_sequence_1*
`elif` *condition_2* `then` *true_statement_sequence_2* ...
`else` *false_statement_sequence* `end if`

The statement sequence corresponding to the first condition that evaluates to true is executed, but if no condition evaluates to true then the statement sequence between `else` and `end if` is executed, if there is one. This is the general form of the `if` statement; there can be any number (including zero) of `elif` clauses, and the final `else` clause is optional. If there is no appropriate default action then leave it out completely. For example:

```
> WeekEnd := Day ->
    if Day = Saturday then
       print('The start of the weekend!')
    elif Day = Sunday then
       print('Take it easy!');
       print('But back to work tomorrow.')
    elif Day = Monday then
       print('Back to work!')
    else
       print('Keep working!')
    end if:
```

```
> WeekEnd(Saturday):
```

The start of the weekend!

```
> WeekEnd(Sunday):
```

Take it easy!
But back to work tomorrow.

```
> WeekEnd(Monday);
```

Back to work!

```
> WeekEnd(Tuesday):
```

Keep working!

Compiled languages often provide an efficient statement (e.g., `switch` in C, C++, and Java; `case` in Algol, Pascal, and Modula) for performing multiple alternative execution based on the value of an expression (which is less general than the conditions allowed in `if` statements). In Maple, no such special statement exists, perhaps because the efficiency gain would be minimal in an interpreted language, and the general form of the `if` statement should be used instead.

Note that it is allowed to nest `if` statements, but if the purpose of the nesting is to implement a multi-alternative decision then using the `elif` keyword is better than explicit nesting. For example, instead of:

```
if condition_1 then
   true_statement_sequence_1
else
   if condition_2 then
      true_statement_sequence_2
   else
      false_statement_sequence
   end if
end if
```

it is better to use:

```
if condition_1 then
   true_statement_sequence_1
elif condition_2 then
   true_statement_sequence_2
else
   false_statement_sequence
end if
```

In the first form it is difficult to pair up the `if` and `end if` delimiters correctly and if the code is consistently indented it wanders further over to the right the more alternatives there are. The second form is much clearer and avoids both of these problems.

7.5 Piecewise-defined functions

It is very rare that any kind of conditional statement is useful as a "top-level" statement, and in practice they are nearly always used as part of the definition of a function, procedure, or loop. One of the simplest applications is in piecewise-defined functions.

7.5.1 Piecewise-defined functions using `if`

A *piecewise-defined mathematical function* is one that is defined using different formulae in different regions of its domain, which is assumed to be some continuous set, in particular the set of all real numbers or some real interval. A piecewise-defined function may or may not be discontinuous.

An important discontinuous example is the unit step (or Heaviside) function, which can be defined by:

$$H(x) = \begin{cases} 1 & x > 0, \\ 0 & x < 0. \end{cases}$$

This function is undefined at $x = 0$, its point of discontinuity. The definition could be modified to give it some (arbitrary) value there, most easily by making one of the inequalities non-strict, e.g.,

$$H(x) = \begin{cases} 1 & x > 0, \\ 0 & x \leq 0. \end{cases}$$

This makes no difference in the contexts in which it is normally used, namely within integrals.

The unit step function is provided in Maple with the name `Heaviside`. Its graph looks like this:

```
> plot(Heaviside);
```

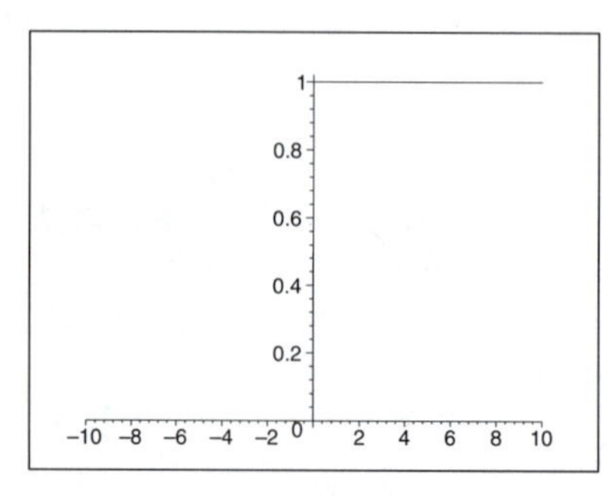

The second definition of $H(x)$ above could be implemented using conditional execution like this:

```
> H := x -> if x > 0 then 1 else 0 end if;
```

$$
\begin{array}{l}
H := \textbf{proc}(x) \\
\textbf{option}\ \textit{operator},\ \textit{arrow}; \\
\quad \textbf{if}\, 0 < x\, \textbf{then}\, 1\ \textbf{else}\, 0\, \textbf{end if} \\
\textbf{end proc}
\end{array}
$$

The way that Maple prints such mappings with conditional bodies as special procedures (which they are) is unfortunate, but the definition works, at least for simple purposes:

```
> plot(H);
```

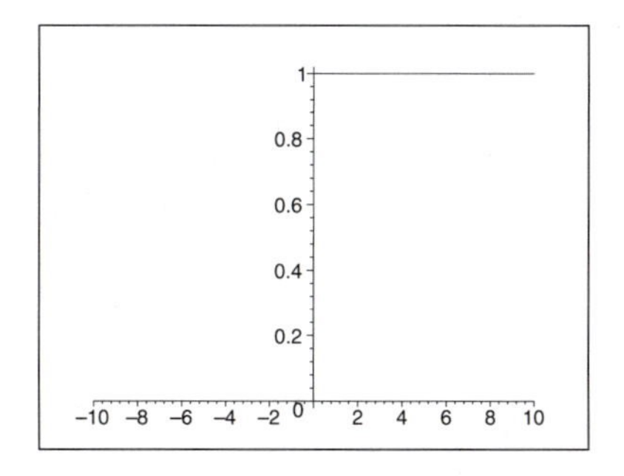

Another important related discontinuous example is the "top-hat" function:

$$T(x) = \begin{cases} 1 & |x| < 1, \\ 0 & |x| \geq 1. \end{cases}$$

It can be implemented using conditional execution like this:

```
> T := x -> if abs(x) < 1 then 1 else 0 end if;
```

$$
\begin{array}{l}
T := \textbf{proc}(x) \\
\textbf{option}\ \textit{operator},\ \textit{arrow}; \\
\quad \textbf{if}\, \mathrm{abs}(x) < 1\, \textbf{then}\, 1\, \textbf{else}\, 0\, \textbf{end if} \\
\textbf{end proc}
\end{array}
$$

```
> plot(T, -2..2);
```

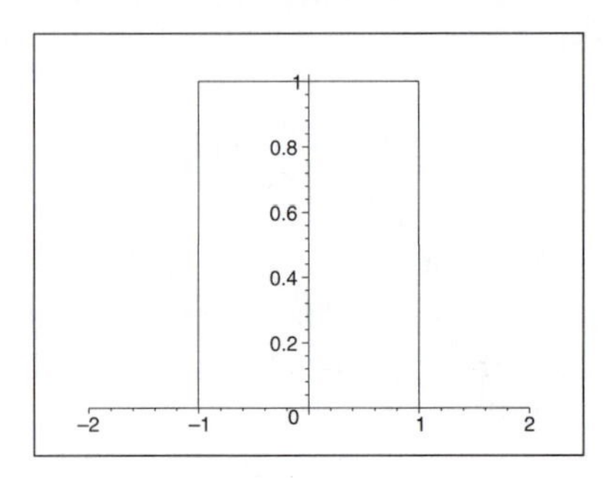

Continuous functions can also be piecewise-defined, e.g.,

$$f(x) = \begin{cases} x^2 & x > 0, \\ 0 & x \le 0. \end{cases}$$

```
> f := x -> if x > 0 then x^2 else 0 end if;
```

$$\begin{array}{l} f := \textbf{proc}(x) \\ \textbf{option}\ operator,\ arrow; \\ \quad \textbf{if}\, 0 < x\, \textbf{then}\, x^2\, \textbf{else}\, 0\, \textbf{end if} \\ \textbf{end proc} \end{array}$$

Maple functions defined this way can be plotted and evaluated without difficulty, provided their arguments are always *explicitly numerical*:

```
> plot(f, axes = FRAME);
```

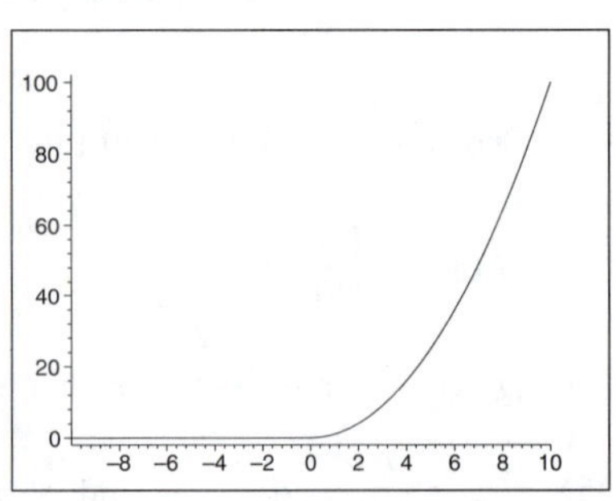


```
> seq(f(x), x = -2..2);
```

$$0, 0, 0, 1, 4$$

However, they cannot be used with symbolic arguments, even if they are implicitly numerical, e.g.,

```
> plot(f(x), x, axes = FRAME);
  Error, (in f) cannot evaluate boolean: -x < 0
```

unless evaluation of the function is deferred (by unevaluation) until after the argument has been given a numerical value, e.g.,

```
> plot('f(x)', x, axes = FRAME);
```

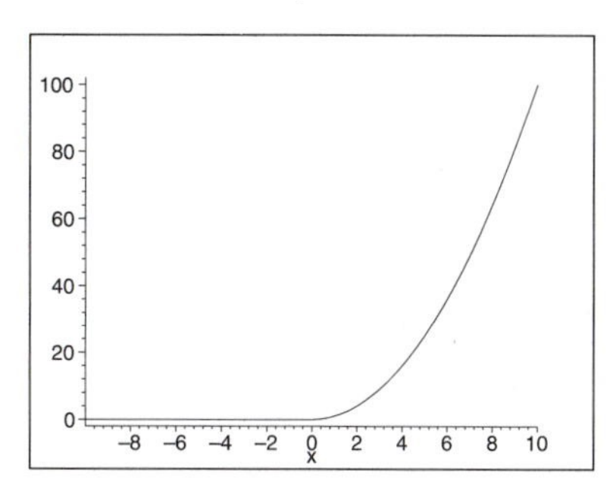

Consequently, this simple method of piecewise definition cannot be used with symbolic operations such as differentiation and integration:

```
> diff(f(x), x);

    Error, (in f) cannot evaluate boolean: -x < 0

> int(f(x), x);

    Error, (in f) cannot evaluate boolean: -x < 0
```

One solution is to modify the definition to allow for non-numerical arguments, but generally the best solution is to use the `piecewise` function which is intended for this purpose.

7.5.2 Piecewise-defined functions using `piecewise`

The `piecewise` function accepts a variable number of arguments of the following form:

`piecewise(`*cond_1*`, `*f_1*`, `*cond_2*`, `*f_2*`, ..., `*cond_n*`, `*f_n*`,`
f_otherwise`)`

which appear in the same order and have the same meaning as they do in the following analogous `if` statement:

`if` *cond_1* `then` *f_1* `elif` *cond_2* `then` *f_2* `... elif` *cond_n* `then` *f_n*
`else` *f_otherwise* `end if`

For every successive pair of arguments, the first, *cond_i*, must be a Boolean expression. The last expression, *f_otherwise*, can be omitted and its default value is 0. The `piecewise` function works exactly like an `if` statement but with the added advantage that Maple prints it in the conventional form, allows it to remain unevaluated or symbolic, and knows how to differentiate, integrate, and simplify it, e.g.,

```
> f := x -> piecewise(x > 0, x^2);
```

$$f := x \to \text{piecewise}(0 < x,\, x^2)$$

```
> f(x);
```

$$\begin{cases} x^2 & 0 < x \\ 0 & \textit{otherwise} \end{cases}$$

```
> plot(f(x), x, axes = FRAME);
```

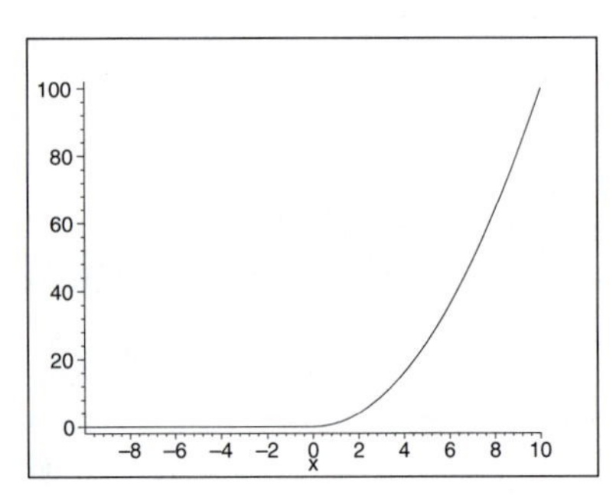


```
> seq(f(x), x = -2..2);
```

$$0, 0, 0, 1, 4$$

```
> diff(f(x), x);
```

$$\begin{cases} 0 & x \leq 0 \\ 2\,x & 0 < x \end{cases}$$

```
> int(f(x), x);
```

$$\begin{cases} 0 & x \leq 0 \\ \dfrac{1}{3}\,x^3 & 0 < x \end{cases}$$

7.5.3 Continuity class

We have seen that piecewise-defined functions can be either continuous or discontinuous, but in fact there is a finer classification that can be important in mathematical applications. This can often be seen by plotting a piecewise-defined function together with its first few *derivatives*; e.g., the following code plots the function $f(x)$ defined above together with its first three derivatives and adds labels at the right:

```
> plot({f(x), seq(diff(f(x),x$n), n = 1..3)},
     x = -1..3, discont = true,
     axes = FRAME, color = BLACK):
  plots[textplot]( # labels for the graphs
     {[3,9," f"],[3,6," f1"],[3,2," f2"],[3,0," f3"]},
     align = RIGHT):
  plots[display]({%%,%}); # plot the graphs with labels
```

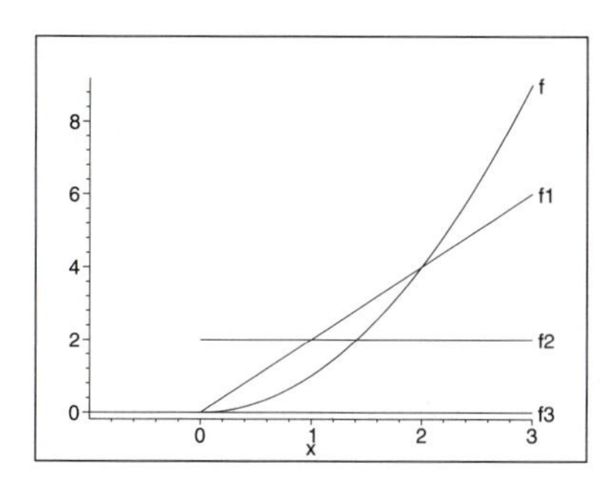

We see that both the function (labelled f) and its first derivative (labelled f1) are continuous, but the second derivative (labelled f2) is discontinuous (and higher derivatives are all zero).

The equivalent plot of derivatives using mappings (see Chapter 9) instead of functions can be produced as follows (omitting the labelling, which could be done exactly as before):

```
> plot({f, D(f), (D@@2)(f), (D@@3)(f)}, -1..3,
    discont = true, axes = FRAME, color = BLACK):
```

The continuity class of a function is the number of derivatives that are continuous everywhere (i.e., over the whole domain). It determines how smoothly the pieces of a piecewise-defined function join together. If a function f has continuity class n this is denoted:

$$f \in C^n.$$

If all derivatives of f are continuous everywhere then f is called "smooth", which is denoted:

$$f \in C^\infty$$

and if its Taylor series converges everywhere then f is said to be "analytic", which is denoted:

$$f \in C^\omega.$$

For the example considered above $n = 1$, i.e.,

$$f \in C^1.$$

Differentiating a function decreases its continuity class by 1 down to a minimum of -1 and integrating it *may* increase it by 1. If

$$f \in C^{-1}$$

then f is discontinuous and all its derivatives are also discontinuous.

7.5.4 Conditional evaluation: the `if` function

It is occasionally useful to be able to write an expression (as opposed to a statement), the value of which is conditional. The `piecewise` function provides precisely this functionality and is appropriate in explicitly mathematical contexts, but Maple also provides an `if` *function* with the syntax:

```
`if`(condition, true_value, false_value)
```

which may be preferable in some programming contexts, as I will illustrate now and again later in this book. It is analogous to the statement:

```
if condition then true_value else false_value end if
```

except that it is an expression and so can be used within another expression, e.g.,

```
> a + `if`(3>2, 3, 2);
```

$$a + 3$$

```
> seq(`if`(isprime(i),1,0), i=1..10);
```

$$0, 1, 1, 0, 1, 0, 1, 0, 0, 0$$

The function name `if` must be enclosed in backquotes because it is a Maple keyword. The `if` function always requires *precisely three arguments*; the first is evaluated and if it evaluates to true then the second argument is evaluated and returned; otherwise, the third argument is evaluated and returned. Hence, the argument that is not returned is also not evaluated. Languages such as C, C++, Perl, and Java all provide a ternary operator `?:` that provides equivalent conditional evaluation. The `if` function is similar to, but not the same as, the `piecewise` function; generally `piecewise` is preferable in a symbolic contexts.

7.6 Loops: `do`, `while`, `for`

A *loop* is a section of code that is executed (or iterated) several times. Loops are classified according to what stops them — a loop that has no way to stop itself is called an *infinite* loop and is to be avoided! The basic loop or repetition statement in Maple has the form:

```
do statement_sequence end do
```

The word do and phrase end do are the Maple delimiters that bracket the sequence of statements that are repeatedly executed within the loop, which is often called the *loop body*. (Maple also accepts the keyword od in place of end do for backward compatibility.) As for conditional execution, the statements in the statement sequence should be separated by semicolons (again, colons are also allowed — it makes no difference which is used) and it is optional whether the last statement in the statement sequence is terminated. The statement sequence can, of course, consist of a single statement (or even no statement at all). If the do-loop itself is terminated with a semicolon then the values of (top-level) intermediate expressions are displayed; usually this is not desirable, in which case the do-loop itself should be terminated with a colon.

7.6.1 The general controlled loop: while

This basic do-loop is an infinite loop, which therefore *always* requires some additional control code. The keyword do can be preceded by various control clauses (in any order, but with no more than one of each type). The most general type of controlled loop has the form:

```
while condition do statement_sequence end do
```

The condition must be a Boolean expression, which is *automatically* evaluated to a Boolean value. The statement sequence that forms the *body* of the loop is executed repeatedly while the condition remains true. If the condition is initially false then the loop body is not executed at all.

There are two requirements for a while-loop to work correctly.

- The loop body must be preceded by code that sets the initial value of the condition (usually to be true) and sets the initial state for the computation to be performed (if necessary). This is called *loop initialization.*
- The loop body statements must (eventually) change the value of the condition.

Very many programming errors are caused by incorrect or missing loop initialization, so ***always check that every loop is correctly initialized!*** Also, always ensure that the initialization statements are in the same execution group as the loop itself so that the initialization is automatically executed whenever the loop is executed.

Here is an example of a very simple while-loop used to find the smallest non-negative integer power of 2 that is not less than a given positive number N:

```
> N := 123: # for example
```

Let p represent the integer power of 2. Its minimum value is $2^0 = 1$, so initialize p to 1:

```
> p := 1:  # initialize the loop
```

Now if p is less than N we increase the power by 1 (by multiplying p by 2), and while p remains less than N we keep doing this:

```
> while p < N do p := 2*p end do:
```

Finally, we display the value of p that results:

```
> p;
```

$$128$$

As an independent check on this result, a completely different way of looking at the same problem is to observe that $N = 2^{\log_2(N)}$. If we replace $\log_2(N)$ by the smallest integer not less than $\log_2(N)$ (see below) then we get the required smallest integer power of 2:

```
> 2^ceil(evalf(log[2](N)));
```

$$128$$

This computation is less satisfactory because it uses floating-point approximation to solve a problem that is expressed exactly and purely in terms of integers.

Aside: computing the integer part of a number

The smallest integer not less than a number x is called its *ceiling* and the largest integer not greater than a number x is called its *floor*. They are denoted, respectively,

$$\lceil x \rceil \quad \text{and} \quad \lfloor x \rfloor$$

and in Maple they can be computed by using the functions `ceil` and `floor`. These function can be thought of as converting a number to the nearest integer above or below, respectively. There are three related functions: `trunc` truncates a number to the nearest integer towards 0; `round` rounds a number to the nearest integer; `frac` returns the fractional part of a number.

7.6.2 Nested control structures

Nested means one inside another. Control structures can be nested but they cannot overlap. For example, this nested control structures is allowed:

```
for i do
   if ... then
      for j do ... end do
   end if
end do
```

but this overlapped control structure is not allowed:

```
for i do
   if ... then ... end do
end if
```

We will return to nested loops later. Almost any non-trivial computation will probably require some kind of nested control structure. As a simple but non-trivial example of a `while`-loop, let us compute the prime factorization of an integer n, assumed to be greater than 1, as a sequence of prime factors in increasing order but allowing repeats. The algorithm we will use is to divide n as many times as possible by successively larger primes p, starting with the smallest prime, namely 2. We test divisibility of n by p using the criterion `irem(n,p)=0`. If we succeed in dividing n by a prime p then we replace n by its quotient by p and include p in the factor sequence f; otherwise, we update p to be the next smallest prime. We repeat this process until it is no longer possible to perform the division, which is certainly the case if $p > n$. Here is an application of the algorithm to factor the integer 12:

```
> n := 12;
  p := 2:  f := NULL:
  while p <= n do
     if irem(n,p) = 0 then
        n := iquo(n,p);
        f := f,p
     else
        p := nextprime(p)
     end if
  end do:
  'f' = f;
```

$$n := 12$$
$$f = (2,\ 2,\ 3)$$

This result is consistent with that returned by the standard Maple integer factorizer `ifactor`:

```
> ifactor(12);
```

$$(2)^2\ (3)$$

7.6.3 Iterating over an integer range: `for ... to`

To repeat exactly the same statement sequence a fixed number of times n a do-loop can be controlled by a `to` clause like this:

```
to n do statement_sequence end do
```

For example, a multiple indefinite integral (the inverse of a multiple derivative) can be computed by repeatedly re-integrating in a loop as in the following example, which computes the third-order multiple integral of an arbitrary function $g(x)$:

```
> y := g(x):
  to 3 do y := int(y,x) end do:
  y;
```

$$\int\int\int \mathrm{g}(x)\,dx\,dx\,dx$$

A named loop or iteration counter i that is *automatically* incremented for each successive execution of the loop body and whose value can be used within the body is defined by using the `for` keyword like this:

`for` *i* `do` *statement_sequence* `end do`

By default, the variable i is automatically initialized to 1 and is incremented by 1 each time around the loop. But this is an infinite loop, so it should normally only be used together with a control clause, i.e., either `to`:

`for` *i* `to` *n* `do` *statement_sequence* `end do`

or `while`:

`for` *i* `while` *condition* `do` *statement_sequence* `end do`

or occasionally both:

`for` *i* `to` *n* `while` *condition* `do` *statement_sequence* `end do`

For example, the following code will print the first three derivatives of a function *f*, where the value of the automatic loop counter i is used to determine the order of the derivative. The inert derivative `Diff` together with the activation function `value` is used here to produce meaningful output:

```
> f := sin(sin(x)):
  for i to 3 do
     Diff(f, x$i); print(% = value(%))
  end do:
```

$$\frac{\partial}{\partial x}\sin(\sin(x)) = \cos(\sin(x))\cos(x)$$

$$\frac{\partial^2}{\partial x^2}\sin(\sin(x)) = -\sin(\sin(x))\cos(x)^2 - \cos(\sin(x))\sin(x)$$

$$\frac{\partial^3}{\partial x^3}\sin(\sin(x)) = -\cos(\sin(x))\cos(x)^3 + 3\sin(\sin(x))\cos(x)\sin(x) - \cos(\sin(x))\cos(x)$$

In non-trivial control structures (conditional and loop statements) it is good practice to indent the code, rather in the way that paragraphs of English text are often indented, to help make the logical structure of the code clear from its textual layout. I will illustrate this where it is appropriate. A good amount to indent by is 3 spaces; much less makes the indentation less clear and much more wastes too much space.

To have the counter in a `for`-loop start from m instead of 1 add a `from` clause, and to have the counter automatically incremented in steps of s instead of 1 add a `by` clause, like this:

```
for i from m by s to n do statement_sequence end do
```

The counter can be made to count downwards by making the step s negative, e.g.,

```
> print(`Countdown commences!`):
  for i from 3 by -1 to 1 do print(i) end do:
  print(`Blast off!`):
```

Countdown commences!

3

2

1

Blast off!

Never assign to a `for`-loop variable (e.g., `i` above) inside the loop body. The values of a `for` variable are set automatically by Maple, and reassigning them is not prevented but will cause confusion and probably disaster! The value of a `for` variable after the end of a loop will be the first value that exceeds the bound set by the `to` clause or the value when the condition set by the `while` clause fails. This value can be used in subsequent computation. Indeed, the loop body can be *empty* if all the necessary computation can be performed by the `for` and `while` clauses, and the only purpose of the loop it to increment the `for` variable to some particular value for later use. For example, if there were no `prevprime` function in Maple then we could still, for example, find the largest prime not greater than 1000 like this:

```
> for p from 1000 by -1 while not isprime(p) do
  end do:
  p;
```

$$997$$

```
> prevprime(1000);  # check!
```

$$997$$

7.6.4 Iterating over a structure: `for ... in`

Another very common type of loop involves repeating the same code for each element of some data structure, such as a sequence, list, set, sum, or product. To repeat exactly the same statement sequence for each element of a structure a `do`-loop can be controlled by an `in` clause like this:

`in` *structure* `do` *statement_sequence* `end do`

Since the data structure iterated over must be finite this loop will always terminate, but it can additionally be controlled by a `while` clause. To give access to each element of the structure via a named variable a `for` clause can be added like this to give the usual form of this loop structure:

`for` *i* `in` *structure* `do` *statement_sequence* `end do`

Here is an example of a very simple `for`...`in`-loop used to find the maximum element of a sequence S of numbers:

```
> S := 1, 3, -5, 7:  # for example
```

Let m represent the maximum. Its minimum possible value is $-\infty$, so initialize m to this value:

```
> m := -infinity:  # initialize the loop
```

For each element *el* in the sequence, compare the element with m and if it is bigger then reset m to that element:

```
> for el in S do
     if el > m then m := el end if
  end do:
```

Finally, display the value of m that results:

```
> m;
```

$$7$$

This is obviously correct, but another way to check it is to use the built-in Maple function `max` that returns the maximum of its argument sequence:

```
> max(S);
```

$$7$$

We could use a `for`...`in`-loop with a `while` clause and an empty body to find the first negative value in the sequence S like this:

```
> for el in S while not el < 0 do end do:  el;
```

$$-5$$

However, this algorithm is not reliable if there is no negative value, e.g.,

```
> for el in 1,2,3 while not el < 0 do end do:  el;
```

$$3$$

One solution to that problem is to add a *sentinel* to the end of the data, which should be some value that cannot be confused with real data. We can then either test for the sentinel or accept it as the default value, as in the following code. It uses `NULL` as the sentinel, which is a reasonable default when failing to find a negative value, but it must be unevaluated in order to prevent it being discarded from the end of the input sequence. This code also illustrates the use of `or` as a *conditional* Boolean operator; without the `el = NULL` test the code fails. In the case of a more complicated loop, such as this has become, it can be helpful to indicate the intentional lack of a loop body by a comment:

```
> for el in S, 'NULL' while not(el = NULL or el < 0) do
      # nothing
  end do:
  el;
```

$$-5$$

```
> for el in 1,2,3, 'NULL' while not(el = NULL or el < 0) do
      # nothing
  end do:
  el;
```

7.6.5 Other ways to control loops: `break`, `next`

Loops can also be controlled by conditional statements within their bodies. Two statements specific to loops are these:

break — terminate the loop (as in C, etc.).

next — skip the rest of the body and continue the loop (like `continue` in C, etc.).

A `while`-loop has its test "at the top"; i.e., a test is made before the loop body is ever executed. Sometimes, it is more convenient to execute the loop body before making the first test, particularly when the loop body needs to be executed in order to establish the test condition. Some languages provide a "repeat ... until" or "repeat ... while" construct, but Maple does not. However, these two constructs can be emulated in Maple as follows:

- repeat *until* condition is true:

```
do
   statement_sequence;
   if condition then break end if
end do
```

- repeat *while* condition is true:

```
do
   statement_sequence;
   if not condition then break end if
end do
```

We will see various examples of the use of these control structures in later chapter.

7.6.6 The for ... to-loop implemented as a while-loop

The general for-loop

```
for i from m by s to n while condition do
      statement_sequence end do
```

is just a special case of the while-loop

```
while condition do statement_sequence end do
```

because a simple for-loop can be implemented like this (provided $s > 0$; if $s < 0$ the inequality must be reversed):

```
i := m:
while i <= n and condition do
   statement_sequence;
   i := i + s
end do
```

If the loop body (*statement_sequence*) contains a next statement then the increment statement (i := i + s) needs to be copied into the loop body immediately before the next statement.

7.6.7 Double loops

A very common example of a nested control structure is a double loop, which is an inner loop contained entirely within the body of an outer loop. It often arises when processing data in a two-dimensional table, such as a two-dimensional array or matrix. Usually one of the loops runs over the rows and the other over the columns. If the elements of the table all have similar significance (or the significance is symmetric about the leading diagonal, which runs from top-left to bottom-right) then it probably does not matter which loop does which, i.e., whether the table is processed "by rows" or "by columns". Otherwise, it may be much simpler and/or more efficient to process the table in an order that reflects its structure. The convention is that the first index denotes the row and the second denotes the column, and it is common but certainly not obligatory to call them i and j.

For example, let us generalize our earlier construction of Boolean truth tables and construct a small multiplication table. In general mathematical terms, this could be for any group, and the table would be symmetric if the group is Abelian but not otherwise. However, even if the table is not strictly symmetric the *significance* of its elements is symmetric, so it makes no difference whether we process it by rows or by columns. To be specific, let us construct the conventional multiplication table for the integers from 1 to 3, but so that the upper bound can be changed easily let us make it the value of a variable, which I will call mx to indicate that it is the *maximum* multiplicand (since the identifier max is reserved for a built-in function).

```
> mx := 3:
```

Maple will automatically display a table in conventional two-dimensional layout only if the table has the structure of a matrix, which means that it is indexed by two integer ranges starting from 1. This is the default when a two-dimensional array is initialized without specifying the index ranges (which we relied upon earlier for our truth tables). Hence, for automatic tabular display it is necessary always to construct a table that has the structure of a matrix. I will call it MT since it will represent a *multiplication table*:

```
> MT := array(1..mx, 1..mx);
```

$$MT := \mathrm{array}(1..3,\, 1..3,\, [])$$

Maple assignments always display as assignments rather than as the value assigned, which in some cases (such as tables and plots) prevents the normal display of the value. Hence, it is often best to suppress display of the assignment itself and instead immediately display the value of the variable that has been assigned, remembering that a table or array must be explicitly evaluated or printed in order to see the value of the variable, like this:

```
> MT := array(1..mx, 1..mx):  print(MT):
```

$$\begin{bmatrix} MT_{1,1} & MT_{1,2} & MT_{1,3} \\ MT_{2,1} & MT_{2,2} & MT_{2,3} \\ MT_{3,1} & MT_{3,2} & MT_{3,3} \end{bmatrix}$$

This is not yet a multiplication table; it just shows how Maple displays unassigned matrix elements of a matrix accessed by name. A suitable double loop to assign the elements, using the variables i and j to denote the row and column indices and letting Maple do the multiplication to compute the elements, looks like this. This loop processes the array "by rows" by fixing the index for each row in turn and varying the column index within each row, so it assigns all the elements in the first row, then the second row, then the third row. It is worth remembering (in general) that the control variable in an inner loop varies faster than that in an outer loop, and an inner loop scans a sub-structure specified by an outer loop.

```
> for i to mx do  # for each row
     for j to mx do  # for each element in the row
        MT[i,j] := i*j
     end do
  end do:
  print(MT);
```

$$\begin{bmatrix} 1 & 2 & 3 \\ 2 & 4 & 6 \\ 3 & 6 & 9 \end{bmatrix}$$

This is of limited use as a multiplication table, because it does not show what is multiplied by what. To add row and column labels, and also horizontal and vertical "rules" to distinguish labels from values, we need to make the array bigger. To do this while preserving the matrix structure unfortunately means sacrificing the direct relationship between multiplicand and index. Here is a more informative multiplication table:

```
> mx := 3:  shift := 2:
  MT := array(1..mx+shift, 1..mx+shift,
     [(1,1)=`*`, (1,2)=`|`, (2,1)=`---`, (2,2)=`+`]):
  for j to mx do  # label the columns
     MT[1, j+shift] := j;
     MT[2, j+shift] := `---`
  end do:
  for i to mx do  # label the rows
     MT[i+shift, 1] := i;
     MT[i+shift, 2] := `|`
  end do:
  for i to mx do  # for each row
     for j to mx do  # for each element in the row
        MT[i+shift, j+shift] := i*j
     end do
  end do:
  print(MT);
```

$$\begin{bmatrix} * & | & 1 & 2 & 3 \\ --- & + & --- & --- & --- \\ 1 & | & 1 & 2 & 3 \\ 2 & | & 2 & 4 & 6 \\ 3 & | & 3 & 6 & 9 \end{bmatrix}$$

This is about the most convincing tabular display possible in Maple. It is a moot point whether it is worth introducing the variable *shift* to represent the displacement of the data from the top-left corner, but I think it makes the code a little clearer. I have used array initialization to set the top-left elements of the table that do not fit into the general repeating structure and two single loops to label the columns and rows. The multiplication symbol in the top-left corner is just to indicate the meaning of the table. I have used identifiers such as `` `|` `` instead of the corresponding strings such as `"|"` in order to avoid the string quotes being displayed; the following print statement illustrates the difference:

```
> print(`|`, "|"):
```

$$|, \text{“|”}$$

As another example of a double loop, a 3×3 *unit* (or identity) matrix could be defined explicitly using a double `for`-loop like this:

```
> U := array(1..3, 1..3):
  for i to 3 do
     for j to 3 do U[i,j] := 0 end do;
     U[i,i] := 1
  end do:
  eval(U);
```

$$\begin{bmatrix} 1 & 0 & 0 \\ 0 & 1 & 0 \\ 0 & 0 & 1 \end{bmatrix}$$

An alternative, and perhaps more obvious, algorithm that uses a conditional statement nested inside a double `for`-loop is this:

```
> U := array(1..3, 1..3):
  for i to 3 do
     for j to 3 do
        if i = j then U[i,j] := 1 else U[i,j] := 0 end if
     end do
  end do:
  eval(U);
```

$$\begin{bmatrix} 1 & 0 & 0 \\ 0 & 1 & 0 \\ 0 & 0 & 1 \end{bmatrix}$$

However, the second algorithm is less elegant, requires more code, and makes nine tests in order to save three assignments, and so is probably slower!

7.7 Exercises

Truth tables in Boolean algebra

The Boolean operator "implies" can be defined either by a truth table or in terms of standard Boolean operators using the fact that (a **implies** b) is equivalent to (b **or not** a). Use this equivalence to implement "implies" as a Boolean-valued function of two arguments (i.e., a mapping), such that "**implies**(a, b)" means "a **implies** b".

Use your function to construct and print the truth table for "implies" in the form of a Maple array, and confirm that it is as follows:

$$\left[\begin{array}{c|cc} \textbf{implies} & \text{true} & \text{false} \\ \hline \text{true} & \text{true} & \text{false} \\ \text{false} & \text{true} & \text{true} \end{array}\right].$$

Beware that this truth table is *not* symmetrical: the values of the left operand (a) are in the left column and the values of the right operand (b) are in the top row.

By using the online help (or looking ahead to Chapter 9), repeat the above exercise using an *infix* operator called "`&implies`" instead of a prefix operator (function).

By using the online help (or looking ahead to Chapter 13), construct the truth table as a Maple *spreadsheet* instead of an array, but still using **implies** as either a function or operator to generate the table entries.

Piecewise-defined functions

Let the function f be defined mathematically as follows:

$$f(x) = \begin{cases} \frac{1}{2}x^2 + x + 1 & x < 0, \\ e^x & \text{otherwise.} \end{cases}$$

Define f as a Maple mapping in terms of an `if` statement and plot it on the domain $[-4, 4]$ with a vertical range from 0 to 5 to check that the definition works correctly.

Redefine f as a Maple mapping in terms of a `piecewise` function and compute a sequence consisting of the first three derivatives of f.

Plot the graphs of f and its first three derivatives together (superimposed on one plot) on the domain $[-4, 4]$. Use a vertical range of $[-1, 5]$ and *frame axes*, and *avoid plotting a vertical line at any discontinuities.* What is the continuity class of f; in other words, how many of its derivatives are everywhere continuous (i.e., have joined up graphs)? (You may find it helpful to plot the derivatives separately to answer this question.) Annotate the plot using the `legend` option; see Chapter 3 and/or the online help.

Conditional execution and loops

Do not use the functions `nextprime`, `prevprime`, or `ithprime` in this exercise (except to check your solutions), but you may use the function `isprime`. *All related lines of code should be within the same execution group.*

1. Use *one* `for`-loop and an appropriate conditional statement to split the positive integers not greater than 20 into a sequence P of prime numbers and a sequence Q of non-prime numbers. Display the results in the form of equations like this:

$$P = (2,\ 3,\ 5,\ 7,\ 11,\ 13,\ 17,\ 19)$$

$$Q = (1,\ 4,\ 6,\ 8,\ 9,\ 10,\ 12,\ 14,\ 15,\ 16,\ 18,\ 20)$$

2. Use a `for...while`-loop (with an empty body) to compute the smallest positive prime number greater than 1000, and display its value.

3. Use some kind of `do`-loop or nested `do`-loops and conditional statements to compute and display an expression sequence, assigned to the variable P, consisting of the 10 smallest positive prime numbers.

4. Write a `for...in` loop that constructs from a sequence P of positive integers a new sequence consisting of each element n in P divided by $n + 1$. Apply it to the sequence P defined in exercise 3 immediately above and display the result.

5. Construct and display the first 10 elements of the Fibonacci sequence F_n defined such that $F_1 = 1$, $F_2 = 1$, and $F_n = F_{n-1} + F_{n-2}$ for $n > 2$.

Double loops

Display the multiplication table for the positive elements of the finite ring Z_m that consists of the non-negative integers modulo the integer $m > 1$ for a few small values of m. (The multiplicands should be the integers in the range $[1, m - 1]$.) Observe that when m is prime then every non-zero element has a unique multiplicative inverse (i.e., the number 1 appears precisely once in each row and in each column), and that there are no "zero-divisors" (i.e., zero does not arise as the product of two non-zero numbers), but when m is not prime neither of these two properties holds.

Chapter 8

Procedures and Recursion

Serious programming in any language involves writing procedures. (In some languages these have other names, such as "function" or "subroutine".) A *procedure* is a piece of a program (a sub-program) that is written to be used later. It is (normally) *not* executed immediately after it is written. Its execution can usually be controlled by supplying *parameters* or *arguments* using a functional notation when the procedure is executed or *called.* Procedures provide a mechanism for implementing the subtasks into which a programming task should be divided.

This chapter focuses on the technical aspects of designing, writing, and executing procedures. The next chapter focuses on applications of procedures as operators and functions, including both those built into Maple and those that can be written by the user. Recursion refers to an object being defined in terms of itself. It is a topic of major importance, both to the practice of writing procedures and in their applications, and is discussed in both this and the next chapter.

The first two sections set out the basics of defining and using procedures and Section 8.3 introduces the concept of the scope of variables. In Maple, procedure arguments are handled somewhat idiosyncratically, as explained in Section 8.4. Maple procedures can be called with a variable number of arguments and can return copies of themselves via mechanisms described in Section 8.5, and Section 8.6 explains how to control what is returned by a procedure and how to handle errors. Section 8.7 presents an extended example consisting of a case study showing the development of a simple integer factorizer. The next three sections discuss some technicalities of displaying and using procedures and further ramifications of using local variables. Section 8.11 introduces the theory and practice of recursive procedures. The next three sections are devoted to the use of procedures for plotting and give case studies of two slightly different approaches to plotting recursive or fractal structures. The last section gives a brief illustrated outline of how to approach program design, development, and testing.

8.1 Defining and using procedures

One of the main uses of procedures, which is especially relevant in Maple, is to implement mathematical functions. Any action of a procedure apart from *returning* a value, such as printing output, is called a *side-effect.* Usually, side-effects should be avoided and a procedure should interact with the rest of a program only via its arguments and return value.

Procedure definitions in Maple generalize the arrow syntax for defining mappings. The following are identical except for notation:

- mapping definition

```
> (x,y) -> x^2 + y^2;
```

$$(x, y) \to x^2 + y^2$$

- procedure definition

```
> proc(x,y) x^2 + y^2 end proc;
```

$$\mathbf{proc}(x,\, y)\, x^2 + y^2\, \mathbf{end\ proc}$$

Normally, both expressions would be assigned to variables in order to use them later, i.e., to give names to the mapping or procedure.

To cause a procedure (or mapping) to be executed it must be *applied* to an argument sequence *in parentheses*, either directly, e.g.,

```
> proc(x) x^2 end proc (a);
```

$$a^2$$

or (usually) indirectly via its name, e.g.,

```
> sqr := proc(x) x^2 end proc;  sqr(a);
```

$$sqr := \mathbf{proc}(x)\, x^2\, \mathbf{end\ proc}$$

$$a^2$$

8.1.1 Returning versus printing values

Superficially, the following two procedures appear to have the same effect:

```
> ReturnIt := proc(a) a end proc:
```

```
> ReturnIt(x+y);
```

$$x + y$$

```
> PrintIt := proc(a) print(a) end proc:
```

```
> PrintIt(x+y);
```

$$x + y$$

The results look similar because Maple automatically displays the values of expressions that are evaluated at "top level" (if they are terminated by semicolons) in the same way that `print` displays its arguments. The difference is that a value output by printing cannot be used as input to another stage of a computation and a procedure that outputs only by printing does not act as a function in the mathematical sense, e.g.,

```
> ReturnIt(x+y) + z;
```

$$x + y + z$$

```
> PrintIt(x+y) + z;
```

$$x + y$$

```
Error, invalid terms in sum
```

A procedure that outputs its main result by returning it is much more flexible than one that outputs only by printing. Hence, procedures should normally output their main values by returning them, either implicitly as illustrated above or explicitly via the `return` statement to be described below, unless they are specifically required to produce printed output. (However, printing from within a procedure can be useful to display subsidiary information, such as details of the computation being performed, which may help when *debugging* or finding and correcting errors. Debugging is described further in Chapter 11.)

8.2 General procedure definition syntax

Procedures are more general than mappings (or functions). In particular, a procedure body can consist of any number of statements whereas a mapping (using arrow syntax) can be defined only by a single expression (or a single conditional statement). The general procedure definition has the form:

```
proc(arguments)
    local variables;
    global variables;
    options ...;
    description ...;
    body_statements
end proc
```

All parts of a procedure definition are optional, although a procedure definition with no body statements serves no useful purpose! The statements before the body are *declarations* that provide information to Maple. The keywords `local` and `global` are each followed by a sequence of variables that are intended to be either local or global, respectively, to this procedure, as explained below. The keyword `options` (or `option`) is followed by a sequence of options, some of which will be described later. When a procedure is called, the value it returns by default is the value of the last expression executed in the body (i.e., the value that the ditto operator `%` would return). The statements within a procedure definition should be *indented* in some consistent way to show the extent of the procedure definition. As for control structures, an indentation step of about 3 spaces seems to work fairly well.

The body of the procedure consists of a sequence of statements, which must be separated by either semicolons or colons (it makes no difference which). A statement terminator after the last body statement (i.e., immediately before the `end` keyword) is optional, and I will normally omit it. The `proc(...)` header should *not* be followed by a terminator, because it is not required and generally it is *not allowed.*

The terminator used *after* `end proc` determines whether the procedure definition itself is displayed. It is usually best to let Maple display a procedure when it is first defined in order to check for errors. The reformatting used by Maple can give useful clues to errors in control structures in non-trivial procedure bodies. Once the definition is correct, the version displayed by Maple can be deleted, or the terminator changed to a colon and the definition re-executed.

8.3 Variable scope

The *scope* of a variable is the part of a program in which the variable has the same meaning (but not necessarily the same value). A variable whose meaning is limited to some region is said to be *local* to that region and a variable whose meaning is the same inside and outside some region is said to be *global* to that region. By default, variables are usually global to an entire program (Maple session). To use a variable *temporarily* without affecting any use it may have elsewhere it must be declared to be local to a procedure (or module), and ***this should be done whenever possible.*** If a variable used in a procedure body *must* be global then it should be declared global. It is good practice to declare all variables used in a procedure body *except the procedure arguments* to be either local or global. This includes `for`-loop variables. Arguments should not be declared — they are treated specially in Maple and are not variables at all. (There is no need to declare the control variables used in `seq`, `add`, and `mul` functions, because these are automatically local to those functions.)

The following procedure returns the maximum value of a list (or set) of explicit numbers; the algorithm used in the procedure body was described in the previous chapter:

```
> MAX := proc(L)  # Return maximum of a list (or set)
      local m, el;
      m := -infinity;
      for el in L do
         if el > m then m := el end if
      end do;
      m
  end proc;
```

$$\begin{array}{l} MAX := \mathbf{proc}(L) \\ \mathbf{local}\, m,\ el; \\ \quad m := -\infty\,; \\ \quad \mathbf{for}\ el\ \mathbf{in}\ L\ \mathbf{do}\ \mathbf{if}\ m < el\ \mathbf{then}\ m := el\ \mathbf{end\ if}\ \mathbf{end\ do}\,; \\ \quad m \\ \mathbf{end\ proc} \end{array}$$

Note that when a procedure definition is displayed by Maple it is automatically reformatted and simplified, and any comments are discarded. This is because it is Maple's internal representation that is being displayed rather than the input text.

Procedure `MAX` works like this:

```
> MAX( [1, 3, -5, 7] );
```

$$7$$

One could argue about the maximum value of the elements of an empty list or set; this implementation returns $-\infty$.

```
> MAX( [ ] );
```

$$-\infty$$

Note that the values of the global variables `m` and `el` have not been affected by the procedure:

```
> m, el;
```

$$m,\ el$$

A local variable is completely independent and different from a global variable with the same name, and it is important to remember this.

A procedure can be defined and executed immediately, solely in order to provide a local environment (as introduced earlier in the context of setting a local value for `Digits`), e.g.,

```
> proc() local f;  global x;
    f := sin(sin(x));
    [seq(diff(f, x$i), i = 1..3)]
  end proc ();
```

$$[\cos(\sin(x))\cos(x), -\sin(\sin(x))\cos(x)^2 - \cos(\sin(x))\sin(x),$$
$$-\cos(\sin(x))\cos(x)^3 + 3\sin(\sin(x))\cos(x)\sin(x)$$
$$- \cos(\sin(x))\cos(x)]$$

However, often a better alternative is to use a module (see later, especially Chapter 15), which involves very similar syntax apart from the need to explicitly print any output required:

```
> module() local f;  global x;
    f := sin(sin(x));
    print([seq(diff(f, x$i), i = 1..3)])
  end module:
```

$$[\cos(\sin(x))\cos(x), -\sin(\sin(x))\cos(x)^2 - \cos(\sin(x))\sin(x),$$
$$-\cos(\sin(x))\cos(x)^3 + 3\sin(\sin(x))\cos(x)\sin(x)$$
$$- \cos(\sin(x))\cos(x)]$$

These mechanisms provide a form of "block structure" with local variables, as in C and other languages. In both cases, the variables `f`, `i` used internally are completely independent of their use anywhere else:

```
> f, i;
```

$$f, i$$

In the case of `i` this is because `seq` implicitly makes its control variable local (as do the related functions `add` and `mul`), so it is unnecessary (but would do no harm) to explicitly declare `i` to be local to the procedure or module.

You may be tempted to try to declare a reserved variable such as `Digits` to be local, but this will almost certainly not have the desired effect, because Maple will continue to use the global value to determine the precision for numerical evaluation. (This is a consequence of the fact that Maple variable scoping is *static*; i.e., it is applied when procedures are *defined* rather than when they are *executed.* This means that local variables and procedure arguments can be accessed only by code that is defined *inside* the procedure [or module] that declares the variables, which works because Maple uses *lexical* variable scoping. This contrasts with the programming model supported by some other languages, such as those based on LISP, which use dynamic scoping.) However, this problem is partly overcome by the use of *environment* variables, which behave *as if* they were implicitly declared local but should *not* be explicitly declared local. `Digits` is an environment variable, as are the ditto "pseudo-variables" (which can also be regarded as nullary operators) and any variable name beginning with "_Env". A small number of other environment variables are predefined in Maple; see the online help for further details.

Note that the value returned by a procedure should never normally involve free local variables because they are different from global variables with the same names, which can cause very strange effects. Hence, the variable x in the above procedure was not declared local (and in fact was explicitly declared global) to ensure that the value returned by the procedure involves only global objects.

8.4 Procedure arguments

These are automatically local but they are *not variables.* The *actual arguments* to which the procedure is applied are normally *evaluated* (by default) and then *every* occurrence of each *formal argument* used in the procedure definition is *replaced* by its value. Hence, all traces of formal arguments have disappeared before a procedure body is executed. This behaviour is different from most other languages, in which formal arguments are genuine local variables that are automatically initialized to the actual argument values.

Any assignment to a formal argument will attempt to assign to the *value* of the actual argument. This means that formal arguments cannot be freely used as local variables within procedure definitions in Maple, and it is often necessary to copy them to local variables at the start of a procedure body. The following procedure definitions illustrate this. (Note that this procedure is specifically intended to be executed for its side effect of printing and is not intended to be used as a function.)

```
> CountDown := proc(n)
     while n > 0 do
        print(n);
        n := n - 1
     end do;
     NULL # return nothing
  end proc:
```

```
> CountDown(3);
```

$$3$$

```
Error, (in CountDown) illegal use of a formal parameter
```

The solution is to use a local variable in place of the argument within the procedure body, like this:

```
> CountDown := proc(N)
     local n;  n := N;
     while n > 0 do
        print(n);
        n := n - 1
     end do;
     NULL # return nothing
  end proc:
```

```
> CountDown(3);
```

$$3$$
$$2$$
$$1$$

(The purpose of explicitly returning NULL is to avoid the value of n that fails the while condition (i.e., 0) being returned, which is what would happen by default and could be confusing.) Note that copying procedure arguments is not necessary if either you do not intend to assign to them or you intend to assign to them as a way of returning information from the procedure. This is often regarded as bad practice and it is certainly contrary to the principles of structured programming (and of functional programming) because it is a side-effect of the procedure call; nevertheless, it is sometimes convenient in practice and it is used quite extensively by the standard Maple library functions.

Finally, here is a version of the procedure that returns the countdown as a Maple *sequence* rather than printing it on separate lines. This version acts as a mathematical function and is more in keeping with good programming principles.

```
> CountDown := proc(N)
     local n, S;
     n := N;  S := NULL;
     while n > 0 do
        S := S, n;
        n := n - 1
     end do;
     S
  end proc:
```

```
> CountDown(3);
```

$$3, 2, 1$$

But note that this is not an efficient way to generate a trivial decreasing sequence of numbers! Here is a better way to solve *this particular problem*; the seq control variable i is implicitly local to the seq function so there is no need explicitly to declare it local to the procedure:

```
> CountDown := proc(N)
     seq(N-i, i=0..N-1)
  end proc:

> CountDown(3);
```

$$3, 2, 1$$

8.5 Special identifiers in procedures

Maple sometimes uses identifiers that superficially appear to be variables, but in fact do not behave like variables and so are better referred to as *special identifiers*. They act as *place-holders* for the information they represent; their replacement by this information does not rely on evaluation and cannot be prevented in any way (such as by unevaluation quotes). We have already seen that formal arguments in procedure definitions fall into this class. In addition, the special identifiers `nargs`, `args`, and `procname` are reserved as special identifiers within procedures.

8.5.1 Variable numbers of arguments: nargs, args

A Maple procedure may be called with a number of actual arguments different from the number of formal arguments with which it was defined. (Not all languages allow this, and in some it can be awkward to use.)

The special identifiers `nargs` and `args` represent, respectively, the number of *actual* arguments and the *actual* argument sequence to which the procedure was applied. (They are essentially the same as the parameters `argc` and `argv` available to C programs run under UNIX and most other modern operating systems.) They can be used to access all or some of the actual arguments, and are often used to handle optional arguments. They are the only way to handle an unspecified number of arguments (i.e., to access the whole actual argument sequence as a single object). The only simple way to determine whether an optional actual argument is present is via the number of actual arguments provided by `nargs`. (The alternative is to examine the *values* of the arguments, but this requires *keyword* rather than *positional* arguments as used by many Maple library functions, such as the plotting functions.)

Here are a couple of implementations of a division procedure that can accept either two arguments or just one argument and interprets the latter as a request to return its reciprocal. The first implementation uses two formal arguments and uses `nargs` to determine how many actual arguments were supplied:

```
> DIV := proc(p, q)  # Unary or binary division
     if nargs = 1 then 1/p else p/q end if
  end proc:
```

```
> DIV(x), DIV(x, y);
```

$$\frac{1}{x}, \frac{x}{y}$$

The second implementation uses both **nargs** and **args** and no formal arguments:

```
> DIV := proc()  # Unary or binary division
     if nargs = 1 then 1/args[1] else args[1]/args[2] end if
  end proc:
```

```
> DIV(x), DIV(x, y);
```

$$\frac{1}{x}, \frac{x}{y}$$

In this case, the first implementation is probably better. Generally, it is a matter of taste and convenience whether to use formal arguments or **args**, but usual practice is to use explicit formal arguments for all the *required* arguments (i.e., the arguments that will *always* be accessed regardless of how the procedure is executed). Often, optional arguments are accessed as elements of the **args** sequence.

If a procedure takes an arbitrary number of arguments (that usually all have the same status), and so effectively treats its arguments as a single sequence, then the arguments *must* be accessed as elements of the **args** sequence. A procedure that returns the maximum of all its arguments is a typical example. Here is another version of `procedure MAX` that maximizes its argument sequence:

```
> MAX := proc()  # Return maximum of arguments
     local m, arg;
     if nargs > 0 then
        m := args[1];
        for arg in args[2..nargs] do
           if arg > m then m := arg end if
        end do;
        m
     end if
  end proc:
```

Note that the identifier **arg** has no special significance in Maple, but it is a descriptive name for a local variable that will represent each argument of a procedure in turn under the control of a `for`-loop, as here. (This procedure could alternatively have been implemented by using explicit indexing and a loop of the form "`for i from 2 to nargs do`".)

```
> MAX(1, 3, -5, 7);
```

$$7$$

This version of procedure MAX (implicitly) returns NULL (nothing) if called with no arguments, because no expression is evaluated if the condition in the if statement is false, which probably makes more sense than returning $-\infty$ as we did before:

```
> MAX();
```

8.5.2 Returning unevaluated: procname

A procedure can return an unevaluated version of itself (or any other procedure), which is extremely useful in algebraic computing. Here is a trivial example:

```
> dummy := proc() 'dummy(args)' end proc:
```

```
> dummy(a,b,c);
```

$$\mathrm{dummy}(a,\, b,\, c)$$

Note that it is *essential* to use unevaluation quotes as above. If you do not, you will cause infinite recursion that will lead to a stack overflow error when the procedure is executed, e.g.,

```
> dummy := proc() dummy(args) end proc:
```

```
> dummy(a,b,c);
  Error, (in dummy) too many levels of recursion
```

It makes no difference whether the special identifier args is inside or outside of the unevaluation quotes, because args is not a variable and does not acquire its special value via evaluation. What is important is to prevent evaluation of the procedure name if the procedure is to return itself unevaluated!

A variable to which a procedure has been assigned evaluates to itself by default. Hence, it is possible to set up a chain of assignments of one variable to another, with only the last evaluating directly to a procedure. The special identifier procname represents the *last name* in any chain of assignments by which a procedure was called. It is possible to use any specific identifier when returning an unevaluated function, but the use of procname returns the name by which the procedure was actually called, which is consistent with the fact that in Maple the name given to a procedure is not part of the procedure definition, but is just the name of a variable to which the procedure definition has been assigned. Consider this procedure definition:

```
> DUMMY := proc() 'procname(args)' end proc:
```

The following assignment sets up a (short) chain of assignments:

```
> dummy := DUMMY;
```

$$dummy := DUMMY$$

The last name is `DUMMY`:

```
> dummy(a,b,c);
```

$$\mathrm{DUMMY}(a,\, b,\, c)$$

The following statement assigns to `dummy` the *value* of `DUMMY`, which is a procedure definition:

```
> dummy := eval(DUMMY);
```

$$dummy := \mathbf{proc}()\ \text{'procname(args)'}\ \mathbf{end\ proc}$$

The last name is now `dummy`, and the intermediate identifier `DUMMY` has been lost:

```
> dummy(a,b,c);
```

$$\mathrm{dummy}(a,\, b,\, c)$$

It would probably be wrong (and very confusing) if this procedure returned a completely different unevaluated procedure.

Here is a slightly less trivial example — a division procedure that remains symbolic if it is not possible to perform the division:

```
> DivideIfPossible := proc(p, q)
    if q = 0 then 'procname(p, q)' else p/q end if
  end proc:
```

```
> DivideIfPossible(x, y);
```

$$\frac{x}{y}$$

```
> DivideIfPossible(x, 0);
```

$$\mathrm{DivideIfPossible}(x,\, 0)$$

This procedure illustrates the fact that formal arguments can be used in place of `args` when a procedure returns unevaluated, and moreover that they do not acquire their values via evaluation, as shown by the fact that they are within unevaluation quotes.

8.6 Terminating procedure execution

By default, a procedure executes all the statements in its body and returns the value of the last expression executed. However, sometimes it is convenient to stop execution earlier and to return a different value. The `return` statement stops execution and returns the value of the expression sequence that follows the keyword `return` as the value of the procedure. It can be particularly useful within a loop. For example, the following procedure picks the first prime number in its argument sequence efficiently, in that it stops looking as soon as it has found a prime:

```
> FirstPrime := proc()
     local arg;
     for arg in args do
        if isprime(arg) then return arg end if
     end do
  end proc:
```

```
> seq(i^3+2, i = 2..10);  FirstPrime(%);
```

$$10, 29, 66, 127, 218, 345, 514, 731, 1002$$

$$29$$

If an error occurs so that no meaningful value can be returned then execution of a Maple program (but not Maple itself) can be terminated by executing the `error` statement. The string following the keyword `error` is printed after a special error message indicating the value of `procname`, like this:

```
> DivideIfPossible := proc(p, q)
     if q = 0 then
        error "attempt to divide by zero."
     else p/q end if
  end proc:
```

```
> DivideIfPossible(x, y);
```

$$\frac{x}{y}$$

```
> DivideIfPossible(x, 0);

   Error, (in DivideIfPossible) attempt to divide by zero.
```

More generally, the error message string can be followed by an expression sequence of values that are to be printed as part of the error message, under control of numbered parameter locations of the form %n or %-n embedded in the string, where n is a positive decimal integer that refers to the nth argument after the message string. This mechanism allows values to be extracted from a sequence in any order, should that be necessary. If the form %-n is used then the nth argument must be a non-negative integer i and is printed in ordinal form as ith (e.g., 1st, 2nd, 3rd, etc.). Values are printed linearly as they would be by `lprint` rather than pretty-printed as they would be (by default) by `print`. The special parameter `%0` prints the entire sequence of arguments after the message string separated by a comma and a space. Here is a slightly contrived example that tries to divide its first argument by all subsequent arguments. It also illustrates argument access by explicit indexing:

```
> DivideAll := proc(p)
     local R, i;  R := p;
     for i from 2 to nargs do
        if args[i] = 0 then
           error "%-2 arg is zero, previous arg was %1.",
              args[i-1], i
        else R := R/args[i] end if
     end do;
     R
  end proc:
```

```
> DivideAll(a, b, c, d);
```

$$\frac{a}{b\,c\,d}$$

```
> DivideAll(a, b, c, 0);
   Error, (in DivideAll) 4th arg is zero, previous arg was c.
```

It is important to use `error` appropriately when procedures call each other in a program, so that any failure is detected as soon as possible and is not allowed to propagate and cause other errors. Errors can be *trapped* and handled by using the `try` statement, which also provides a mechanism for making long (i.e., non-local) jumps and is discussed in Chapter 11.

8.7 Examples: integer factorization

In Chapter 7, as an example of the `while` loop, we considered a simple algorithm to compute the prime factorization of an integer n, assumed to be greater than 1, as a sequence of prime factors in increasing order but allowing repeats. Here is exactly the same code packaged as a procedure, which makes it much easier to test:

```
> ifac := proc(N)
     local n, p, f;
     n := N;  p := 2;  f := NULL;
     while p <= n do
        if irem(n,p) = 0 then
           n := iquo(n,p);
           f := f,p
        else
           p := nextprime(p)
        end if
     end do;
     f
  end proc:
```

```
> ifac(12);
```

$$2, 2, 3$$

```
> ifac(13);
```

$$13$$

There are various ways to improve this procedure. It is dangerous to make assumptions without checking them, so the next version checks that the value of its argument N is really a positive integer via the type declaration `N::posint`, to which we will return in Chapter 10. It also allows N to be 1, in which case it returns the factorization of 1 as simply 1. Hence, the next version should behave reliably for all values of the argument N.

The previous version is very inefficient because it attempts many more divisions than is really necessary. If N is divisible by a positive integer p then it must have the form $N = p\,q$ where q is another positive integer. The possible values of p, q must be symmetric about $p = q = \sqrt{N}$, so if N has a factor $p < \sqrt{N}$ then the other factor $q > \sqrt{N}$. Therefore, if we test for divisibility by successively increasing values of p then we need not go above $p = \sqrt{N}$; i.e., we need consider divisibility only by values of p that satisfy $p^2 \leq N$. (This form of the condition has the advantage that it involves only integers and does not require floating-point computations.)

To see the efficiency gain, consider trying to factorize a moderately large prime N, e.g., the smallest prime not less than $100^2 = 10000$, which is

```
> N := nextprime(10000);
```

$$N := 10007$$

Clearly, trying to divide it only by primes not greater than $\sqrt{N}$, of which (with help from the Maple number theory package) there are

```
> numtheory[pi](100);
```

$$25$$

is much faster than trying to divide it by all primes not greater than N, of which there are

```
> numtheory[pi](N);
```

$$1230$$

The efficiency gain is greater the larger is N.

However, using this form of the `while` condition has two consequences. The smallest value of N that will cause the loop to be executed is 4 (since the smallest prime is 2 and $2^2 = 4$). Therefore, we need to treat $N \leq 3$ as a special case, which is easy because 2 and 3 are prime and we have already decided to define the prime factorization of 1 to be 1, so we simply return N. If we do not divide N by all possible factors then the last factor remains in N as the value of n in our algorithm, which must now be explicitly appended to the sequence of prime factors. Here is the improved procedure followed by a few tests:

```
> ifac := proc(N::posint)
     local n, p, f;
     if N <= 3 then return N end if;
     n := N;  p := 2;  f := NULL;
     while p^2 <= n do
        if irem(n,p) = 0 then
           n := iquo(n,p);
           f := f,p
        else
           p := nextprime(p)
        end if
     end do;
     f,n
  end proc:
```

```
> ifac(1);
```

$$1$$

```
> ifac(12);
```

$$2, 2, 3$$

```
> ifac(13);
```

$$13$$

```
> ifac(12345);
```

$$3, 5, 823$$

```
> ifactor(12345);
```

$$(3)\ (5)\ (823)$$

A common alternative output format for factorization is to group repeated factors together in some way, as do the standard Maple factorizers. The following version of procedure `ifac` returns a sequence of *lists*, where each list contains a unique prime factor followed by its *multiplicity*, which is the maximum *positive* integer power to which it occurs as a factor, and again the factors are in increasing order. (It may not be immediately obvious, but this is the same concept of multiplicity as introduced in Chapter 5.) The input 1 is now treated as a separate special case and returned unchanged (which is a somewhat arbitrary but common design decision). The implementation adds another local variable `mult` to store the multiplicity of each prime factor. This is reset to 0 for each new prime and incremented by 1 each time that prime divides N. When a prime ceases to divide N, it is appended to the sequence of factors together with its multiplicity provided its multiplicity is non-zero. Care is required at the end of the loop to ensure that the correct value is returned. As a minor improvement to the efficiency, this version performs each division once only and allows the call to `irem` to return also the quotient via a new local variable `q` rather than calling `iquo` explicitly:

```
> ifac := proc(N::posint)
     local n, p, mult, f, q;
     if N = 1 then return 1 end if;
     if N <= 3 then return [N,1] end if;
     n := N;  p := 2;  mult := 0;  f := NULL;
     while p^2 <= n do
        if irem(n,p,'q') = 0 then
           n := q;  mult := mult + 1
        else
           if mult > 0 then
              f := f,[p,mult];  mult := 0
           end if;
           p := nextprime(p)
        end if
     end do;
     if n = p then return f,[p,mult+1] end if;
     if mult > 0 then f := f,[p,mult] end if;
     f,[n,1]
  end proc:
```

```
> ifac(1);
```

$$1$$

```
> ifac(12);
```

$$[2, 2], [3, 1]$$

```
> ifac(13);
```

$$[13, 1]$$

```
> ifac(12345);
```

$$[3, 1], [5, 1], [823, 1]$$

```
> ifactor(12345);
```

$$(3)\ (5)\ (823)$$

8.7.1 Aside: `ifactor`

We have now written an integer factorizer that behaves in a very similar way to the standard Maple function `ifactor`, although the algorithm used by `ifactor` is different (and presumably much more efficient). However, as we see from the above examples, `ifactor` uses a different output format. Because it is impossible to stop Maple from simplifying any arithmetic expression involving only numbers, e.g.,

```
> '2 * 3^2 * 4^3';
```

$$1152$$

Maple represents each integer factor as the argument of an inert function, the name of which is the empty identifier ``, so that only the arguments and their enclosing parentheses are displayed. Maple does not simplify expressions involving this function by default:

```
> ``(2)*``(3)^2*``(4)^3;
```

$$(2)\ (3)^2\ (4)^3$$

although `expand` replaces a function of the form ``(x) by its argument `x`, thereby allowing Maple to simplify the expression:

```
> expand(%);
```

$$1152$$

Users are free to use this trick for their own purposes if they wish.

8.8 Displaying Maple procedures

A procedure written in the Maple language is just an expression that can be displayed like any other expression, e.g.,

```
> proc(x) x^2 end proc;
```

$$\mathbf{proc}(x)\, x^2\, \mathbf{end\ proc}$$

If a procedure is given a name by assigning it to a variable, then that variable evaluates by default only to itself, e.g.,

```
> sqr := %:  sqr;
```

$$sqr$$

However, explicit application of either `eval` or `print` to the name of a normal procedure written by a Maple user will (by default) evaluate it to the procedure body, and either return or print it respectively, e.g.,

```
> eval(sqr);
```

$$\mathbf{proc}(x)\, x^2\, \mathbf{end\ proc}$$

```
> print(sqr):
```

$$\mathbf{proc}(x)\, x^2\, \mathbf{end\ proc}$$

This is not the case with the procedures in the Maple library, even though they are all written in the Maple programming language. By default, their bodies are not displayed, e.g.,

```
> print(simplify);
```

$$\mathbf{proc}(s)\ \ldots\ \mathbf{end\ proc}$$

The interface parameter `verboseproc` controls the display of procedures. Interface parameters are accessed via the function `interface`, which accepts either parameter names or *equations* (not assignments) with parameters on the left. The default value of `verboseproc` is 1, but if this is increased to 2 (or more) then the bodies of library procedures are displayed, e.g.,

```
> interface(verboseproc = 2);
```

Now the command

```
print(simplify);
```

produces a display that is too long to be worth including here. (Note that it is the value of `verboseproc` in effect when a worksheet is *displayed*, rather than when it was written, that determines whether procedure bodies are displayed.)

Complicated operations such as simplification normally use not one procedure but a hierarchy of sub-procedures. In Maple, such sub-procedures usually have names containing `/` characters that indicate the hierarchical structure. (This is exactly the same convention that is used in Maple to represent file hierarchies, and its use here is based on the way that the Maple library was originally implemented as a UNIX file hierarchy.) Each sub-procedure can itself be displayed (once a suitable value of `verboseproc` has been set), e.g.,

```
> print(`simplify/normal`);
```

proc(x)
option
`*Copyright (c) 1992 by the University of Waterloo. All rights reserved.*`;
  if hastype(x, '*float*') **then** evalf(normal(convert(x, '*rational*', '*exact*')))
  else normal(x)
  end if
end proc

Remember that *any* identifier containing special characters, such as `/`, must be enclosed in backquotes.

Most Maple system commands and functions (such as `simplify`) are implemented in the Maple programming language as procedures in the Maple libraries. However, a few of the most critical ones are implemented in the Maple kernel (written essentially in C), and it is not possible to display the bodies of these procedures.

8.9 Procedure options and remember tables

Most procedure options are used either by the Maple developers or automatically by Maple itself. For example, "`option builtin`" means that the procedure is implemented in the Maple kernel rather than in a Maple library and the body of such a procedure is just a reference (or pointer) to internal code, e.g.,

```
> print(diff);
```

proc() **option** *builtin*, *remember*; 145 **end proc**

```
> print(expand);
```

proc() **option** *builtin*, *remember*; 157 **end proc**

When a mapping is defined using the arrow syntax, it is represented internally as a procedure with options "`operator, arrow`", although the pretty-printer usually hides this, e.g.,

```
> x -> x^2;  lprint(%);
```

$$x \to x^2$$

```
proc (x) options operator, arrow; x^2 end proc
```

However, this information is displayed when a mapping whose body is an `if` statement is displayed, e.g.,

```
> x -> if x > 0 then 1 else 0 end if;
```

proc(x)
option *operator*, *arrow*;
  if $0 < x$ **then** 1 **else** 0 **end if**
end proc

The procedure option `remember` means that a procedure *remembers* the arguments with which it has been called, together with the corresponding return value, in its *remember table.* When a procedure is called, Maple first looks in its remember table. If it finds a matching actual argument sequence then it returns the return value stored in the remember table without actually executing the procedure; otherwise, it executes the procedure and, if `option remember` is set, it stores the return value in its remember table. Returning a remembered return value is very fast, but remembering does not allow a procedure to be controlled reliably by global variables, only by its arguments. This is why it is sometimes possible to change a global control parameter (such as `Digits` or `Order`), recompute some value, and not get what you expected. Remembering allows some recursive algorithms to be implemented efficiently but very simply, although it must be used with caution. Moreover, this particular facility is specific to Maple and is not directly available in most other languages.

8.9.1 Remember tables

A procedure's remember table is stored internally as its fourth operand and so can be accessed like this:

```
> sqr := proc(x)
     option remember;
     x^2
  end proc;
```

sqr := **proc**(x) **option** *remember*; x^2 **end proc**

```
> op(4,eval(sqr));
```

Initially, the remember table does not exist, but after executing the procedure it does:

```
> sqr(x);
```

$$x^2$$

```
> op(4,eval(sqr));
```

$$\mathrm{table}([x = x^2])$$

The setting

```
> interface(verboseproc = 3);
```

causes a procedure's remember table to be displayed as a comment after its body when the procedure definition is displayed, e.g.,

```
> print(sqr):
```

proc(x) **option** *remember*; x^2 **end proc** # (x) = x^2

Remember tables can be cleared (reset) by the function `forget`. (They can also be manipulated directly.)

Assignment to a *functional form*, as opposed to a name, works by putting values in the remember table of a procedure that is created automatically if it does not already exist, e.g.,

```
> sqr := 'sqr':  sqr(a) := a^2;
```

$$\mathrm{sqr}(a) := a^2$$

```
> print(sqr):
```

proc() **option** *remember*; 'procname(args)' **end proc** # (a) = a^2

This mechanism is very different from normal procedure definition. It can be used to define a mapping on a discrete domain, but not a continuous mapping as is usually required. Beware that assigning a procedure definition to an identifier will reset any remember table associated with that identifier:

```
> sqr := proc(x)
    option remember;
    x^2
  end proc:
```

```
> print(sqr):
```

$$\textbf{proc}(x)\,\textbf{option}\ \textit{remember};\ x^2\ \textbf{end proc}$$

Hence, any point mapping definitions must be made *after* the continuous definition; e.g., let us define the following somewhat unusual discontinuous function:

$$\mathrm{DELTA}(x) = \begin{cases} 1 & x = 0 \\ 0 & \text{otherwise} \end{cases}$$

without using any conditional code:

```
> DELTA := proc() 0 end proc:  DELTA(0) := 1:
```

```
> print(DELTA):
```

$$\textbf{proc}()\ 0\ \textbf{end proc}\ \#\ (0) = 1$$

```
> seq(DELTA(x), x = [-1,0,1]);
```

$$0, 1, 0$$

```
> limit(DELTA(x), x = 0);
```

$$0$$

8.10 More about variable scope

This section pursues some further ramifications of variable scoping in Maple.

8.10.1 Lexical scoping

This section applies to both procedures and modules, although references to the `export` declaration apply only to modules, which are discussed in detail in Chapter 15.

Suppose a procedure or module body refers to a variable that is neither a procedure argument nor is declared `local`, `export`, or `global` in that procedure or module. Now suppose that this procedure or module is actually defined within another procedure or module, in which the variable in question is either a procedure argument or is declared `local` or `export`. Then, the variable referenced in the inner context is the same variable as defined in the outer context rather than a global variable. This is called lexical scoping. It means that procedure arguments and variables that are declared `local` or export define a scope for variables that are used without declaration within them.

In short, lexical scoping means that variables behave in the way that you would almost certainly expect. Hence, for most purposes, you can forget about it. You might think it unlikely that you would define one procedure inside another, but in fact it can be useful in larger procedures, and it is especially useful to define modules and procedures within one another, as we will see in Chapter 15. As a simple example, in which it may not be immediately obvious that a procedure is defined within a procedure, consider this procedure to raise every element of a data structure to a specified power:

```
> R2P := proc(struct, p)
     # Raise every element of struct to power p
     map(x->x^p, struct)
  end proc:
```

```
> R2P([a,b,c], 3);
```

$$[a^3, b^3, c^3]$$

Procedure R2P relies on lexical scoping to associate the variable p in the anonymous inner procedure x->x^p with the second argument p of the outer procedure R2P. Alternatively, if p had been defined to be a local variable in procedure R2P then the variable p in the anonymous inner procedure body x->x^p would refer to that local variable and not to a global variable p.

8.10.2 Returning local variables

Normally, local variables should not be allowed to "escape" from their local context. It is entirely possible for a procedure to return a local variable, but any reference to that local variable by name outside of the procedure in which it was defined would refer to a *different variable that happens to have the same name.* This is precisely the reason for using local variables, but it is potentially very confusing in the wrong context. Here is a simple example:

```
> proc() local x; x end proc();
```

$$x$$

```
> evalb(% = x);
```

$$\mathit{false}$$

Usually, this happens by mistake, and the way to avoid it is to ensure that local variables that are returned have values and that the values are returned rather than the variables themselves. In most situations, a local variable will evaluate to its value by default unless that value has last name evaluation semantics (i.e., it is an array, table, procedure, or module), in which case the `eval` function must be explicitly applied to the local variable, as will be illustrated in Chapter 11. However, for special purposes it is possible to take advantage of this behaviour of local variables to create a "protected" symbolic variable that cannot (easily) be accessed. An example might be a symbolic sum in which it is desired to make the summation variable explicitly local to the sum data structure, which could be done like this:

```
> PSeries := proc(cfn, var, maxdeg, mindeg)
     local r;
     sum(cfn(r)*var^r, r=mindeg..maxdeg)
  end proc:
```

```
> S := PSeries(C, x, n, m);
```

$$S := \sum_{r=m}^{n} \mathrm{C}(r)\, x^r$$

Now we can easily change the global variables `C`, `x`, `m`, and `n`, e.g.,

```
> eval(S, {C=(x->1/x), x=y, m=1, n=5});
```

$$y + \frac{1}{2}\, y^2 + \frac{1}{3}\, y^3 + \frac{1}{4}\, y^4 + \frac{1}{5}\, y^5$$

but it is not so easy to change the local variable `r`:

```
> eval(S, r=s);
```

$$\sum_{r=m}^{n} \mathrm{C}(r)\, x^r$$

The only way to access the local variable `r` is by location. By inspecting the operands of the sum:

```
> op(S);
```

$$\mathrm{C}(r)\, x^r,\ r = m..n$$

we see that operand 2 is an equation, of which the local variable `r` is operand 1, and so can be accessed as:

```
> op([2,1],S);
```

$$r$$

We can also substitute for this expression:

```
> subsop([2,1]=s,S);
```

$$\sum_{s=m}^{n} \mathrm{C}(r)\, x^r$$

8.11 Recursion

A procedure is called *recursive* if it calls itself, either directly or indirectly. Many problems naturally have a recursive structure and so are most easily solved by writing a recursive procedure that reflects the structure of the problem. Any recursive definition must always consist of two parts:

- a non-recursive *base case* that will terminate the recursion, and
- the recursive definition itself.

A recursively defined procedure *must* have a name by which it can call itself. The most reliable way for a procedure to call itself is via the special identifier `procname`, but it is still necessary for the procedure to have a name because in all anonymous procedures `procname` is "*unknown*", which *cannot* be used to make a recursive call (otherwise the following anonymous procedure call would lead to infinite recursion, which it does not):

```
> proc() procname() end proc();
```

$$\mathrm{unknown}()$$

It is important to distinguish between the *definition* and the *execution* of a procedure. A procedure must be completely defined before it can be executed, and storage for local variables is set up when a procedure is executed, not when it is defined. (This may not be true in languages that do not allow recursion.) Two executions of the same procedure have completely separate local variables (even though those variables have the same names, which is the whole point of local variables). Hence, one execution of a procedure can stop in the middle and wait while it starts another execution of the same procedure, and this process can be repeated while there is sufficient memory for new sets of local variables (and other local information used by a procedure execution). The storage used temporarily for procedure execution is called a *stack* (see below), which grows and shrinks as necessary (within certain limits).

Repetitive processes can, in principle, be implemented either by using recursive functions or by using explicit iteration or "loops", whichever is more convenient. If a problem is naturally recursive then it is usually easier to program it using recursion, although iteration usually executes faster and requires less memory. Recursive procedures can often be converted into iterative ones quite easily if necessary.

8.11.1 The factorial function

As an example, the factorial function on the non-negative integers can be defined recursively as:

- $0! = 1$ (the base case — no recursion).
- $n! = n(n-1)!$ if $n > 0$.

In Maple, this definition can be implemented as follows:

```
> Factorial := proc(n)
     if n = 0 then 1 else n*Factorial(n-1) end if
  end proc:
```

```
> Factorial(3);
```

$$6$$

(Note that the function `factorial` and the postfix operator `!`, which have identical meanings, are already implemented in Maple.)

8.11.2 Recursion and stacks

A recursive procedure works by building a *stack* of procedure calls until the base case is reached, when the stacked calls are executed from the last to the first. (A stack is a first-in, last-out queue; hence, recursion can be useful for reversing things.) To see the stacking, set the Maple global variable `printlevel` to a large value and then execute a recursive procedure, e.g.,

```
> printlevel := 100:
```

```
> Factorial(3);

  {--> enter Factorial, args = 3

  {--> enter Factorial, args = 2

  {--> enter Factorial, args = 1

  {--> enter Factorial, args = 0
                                    1

  <-- exit Factorial (now in Factorial) = 1}
                                    1

  <-- exit Factorial (now in Factorial) = 1}
                                    2

  <-- exit Factorial (now in Factorial) = 2}
                                    6

  <-- exit Factorial (now at top level) = 6}
```

$$6$$

Do not try this with a large argument to `Factorial`, and reset `printlevel` to its default value of 1 to avoid subsequent diagnostic output that is not required:

```
> printlevel := 1:
```

With a sufficiently large argument to `Factorial`, Maple will run out of memory for the recursion stack, e.g.,

```
> Factorial(2000):

  Error, (in Factorial) too many levels of recursion
```

Such an error is often called a *stack overflow.* The size of the stack depends on the version of Maple.

8.11.3 Avoiding infinite recursion

If the above implementation of `Factorial` is applied to any value that is not a non-negative integer then it will recurse indefinitely (in principle; in practice, until the stack overflows) because the base case condition will never be satisfied, e.g.,

```
> Factorial(-1);

  Error, (in Factorial) too many levels of recursion
```

```
> Factorial(n);

  Error, (in Factorial) too many levels of recursion
```

The neatest solution is to declare the argument to be a non-negative integer as follows:

```
> Factorial := proc(n::nonnegint)
     if n = 0 then 1 else n*Factorial(n-1) end if
  end proc:
```

```
> Factorial(3);
```

$$6$$

```
> Factorial(-1);

  Error, Factorial expects its 1st argument, n, to be of type
  nonnegint, but received -1
```

```
> Factorial(n);
```

```
Error, Factorial expects its 1st argument, n, to be of type
nonnegint, but received n
```

We will consider procedure argument type declarations in detail in Chapter 10, but I will begin to illustrate their use from now on.

8.11.4 Recursive summation

The secret of successfully implementing recursive procedures is to express the problem in a clearly recursive way, which is best illustrated by an example. I will take a very familiar problem so that we can focus on expressing it recursively. The addition of a sequence of values $\sum_{i=1}^{n} v_i$ has a natural recursive structure, namely:

- $\sum_{i=1}^{0} v_i = 0$ — the base case, with no summands, and
- $\sum_{i=1}^{n} v_i = \left(\sum_{i=1}^{n-1} v_i\right) + v_n \quad \text{for} \quad n > 0.$

The recursion can be restricted to the summation itself, as in the following *iterative* implementation of a procedure to add its arguments:

```
> ADD := proc()
     local S, v;
     S := 0;  # base case
     for v in args do S := S + v end do
  end proc:
```

```
> ADD(a, b, c);
```

$$a + b + c$$

Alternatively, the entire procedure can be recursive, as in the following implementation:

```
> ADD := proc()
     if nargs = 0 then 0  # base case
     else ADD(args[1..nargs-1]) + args[nargs] end if
  end proc:
```

```
> ADD(a, b, c);
```

$$a + b + c$$

Note that a procedure definition is recursive only if it calls itself; hence, only the second procedure definition above is recursive in this sense. A recursive procedure definition is frequently simpler and more elegant than its iterative counterpart. For example, a recursive procedure usually requires fewer local variables because it makes better use of the procedure arguments than does an iterative procedure. In the above example, the recursive procedure definition requires no local variables at all, whereas the non-recursive definition requires two.

Note the use of the special identifiers `nargs` and `args`: `nargs` gives the length of the sequence `args`. Note also the use of a range within the selection operation, `[ ]`, to select a *sub-sequence*, which in this case consists of all except the last element. (I could have used negative indexing instead of `nargs` in the selection operations.)

Since addition is symmetrical, we could as easily reverse the order of performing the addition, which might be computationally slightly more efficient, as follows:

```
> ADD := proc()
     if nargs = 0 then 0  # base case
     else args[1] + ADD(args[2..nargs]) end if
  end proc:
```

```
> ADD(a, b, c);
```

$$a + b + c$$

8.11.5 Recursive power computation

The expression X^n, for an integer power n and any X, can be computed using the following formulae, which represent a negative power in terms of a positive power $(-n)$, and a positive power in terms of a smaller non-negative power (either $n/2$ or $n-1$), and use only multiplication and division:

$$X^n = \begin{cases} 1/X^{-n} & n < 0 \\ 1 & n = 0 \text{ (base case)} \\ X^{n/2}X^{n/2} & n \text{ even} \\ X\,X^{n-1} & n \text{ odd} \end{cases}$$

These formulae lead to an efficient recursive algorithm for computing integer powers using the minimal number of multiplications. This algorithm is appropriate for computing powers of many data types, such as numbers, polynomials, and square matrices. In the following implementation I have used the dot operator for multiplication so that it works for both numbers and (rtable-based) matrices. An alternative to using an anonymous squaring mapping (`x->x.x`) would be to assign to a local variable and then explicitly multiply that by itself; it is of course important to compute the value to be squared only once. I have used the special identifier `procname` to make the recursive calls, which allows the name of the procedure to be changed without changing the procedure body.

```
> Pow := proc(X, n::integer)
    if n < 0 then 1/procname(X, -n)
    elif n = 0 then 1
    elif irem(n, 2) = 0 then (x->x.x)(procname(X, n/2))
    else X.procname(X, n-1) end if
  end:
```

```
> Pow(123, 13) = 123^13;
```

$$1474913153392179474539944683 =
1474913153392179474539944683$$

```
> Pow(123, -13) = 123^(-13);
```

$$\frac{1}{1474913153392179474539944683} = \frac{1}{1474913153392179474539944683}$$

```
> Pow(<<1|2>,<3|4>>, 13) = <<1|2>,<3|4>>^13;
```

$$\begin{bmatrix} 741736909 & 1081027478 \\ 1621541217 & 2363278126 \end{bmatrix} = \begin{bmatrix} 741736909 & 1081027478 \\ 1621541217 & 2363278126 \end{bmatrix}$$

```
> Pow(<<1|2>,<3|4>>, -13) = <<1|2>,<3|4>>^(-13);
```

$$\begin{bmatrix} \frac{-1181639063}{4096} & \frac{540513739}{4096} \\ \frac{1621541217}{8192} & \frac{-741736909}{8192} \end{bmatrix} = \begin{bmatrix} \frac{-1181639063}{4096} & \frac{540513739}{4096} \\ \frac{1621541217}{8192} & \frac{-741736909}{8192} \end{bmatrix}$$

To make the algorithm work with polynomials the implementation needs to use the normal commutative multiplication operator * and it needs to expand each product of polynomials in order to be effective:

```
> PolyPow := proc(X, n::integer)
    if n < 0 then return 1/procname(X, -n)
    elif n = 0 then return 1 end if;
    if irem(n, 2) = 0 then (x->x*x)(procname(X, n/2))
    else X*procname(X, n-1) end if;
    expand(%)
  end:
```

```
> PolyPow(x^2+1, 13) = expand((x^2+1)^13);
```

$$1 + 13\,x^2 + x^{26} + 78\,x^4 + 286\,x^6 + 1716\,x^{12} + 1287\,x^{10} + 715\,x^8 + 13\,x^{24} + 78\,x^{22} + 286\,x^{20} + 715\,x^{18} + 1287\,x^{16} + 1716\,x^{14} = 1 + 13\,x^2 + x^{26} + 78\,x^4 + 286\,x^6 + 1716\,x^{12} + 1287\,x^{10} + 715\,x^8 + 13\,x^{24} + 78\,x^{22} + 286\,x^{20} + 715\,x^{18} + 1287\,x^{16} + 1716\,x^{14}$$

```
> PolyPow(x^2+1, -13) = normal((x^2+1)^(-13), expanded);
```

$$1/(1 + 13\,x^2 + x^{26} + 78\,x^4 + 286\,x^6 + 1716\,x^{12} + 1287\,x^{10} + 715\,x^8 + 13\,x^{24} + 78\,x^{22} + 286\,x^{20} + 715\,x^{18} + 1287\,x^{16} + 1716\,x^{14}) = 1/(1 + 13\,x^2 + x^{26} + 78\,x^4 + 286\,x^6 + 1716\,x^{12} + 1287\,x^{10} + 715\,x^8 + 13\,x^{24} + 78\,x^{22} + 286\,x^{20} + 715\,x^{18} + 1287\,x^{16} + 1716\,x^{14})$$

Since a number is a special case of a polynomial, this implementation still works for numbers:

```
> PolyPow(123, 13) = 123^13;
```

$$14749131533921794745399446 83 = 1474913153392179474539944683$$

```
> PolyPow(123, -13) = 123^(-13);
```

$$\frac{1}{1474913153392179474539944683} = \frac{1}{1474913153392179474539944683}$$

8.12 Procedures that output plots

When writing procedures that are intended to produce plots, it is important to remember that plots get displayed only by printing them. (The display of plots is independent of the setting of *Output Display* in the Options menu). Usually, this happens automatically when a top-level call of a plotting function (such as `plot`, `plot3d`, or `display`) that is terminated with a semicolon is evaluated, because its return value is automatically displayed by printing it (to the screen by default). However, a plot can also be displayed by explicitly printing it, in which case the display is a side-effect of the `print` function, which actually *returns* nothing (so the terminator makes no difference), e.g.,

```
> print(plot(sin)):
```

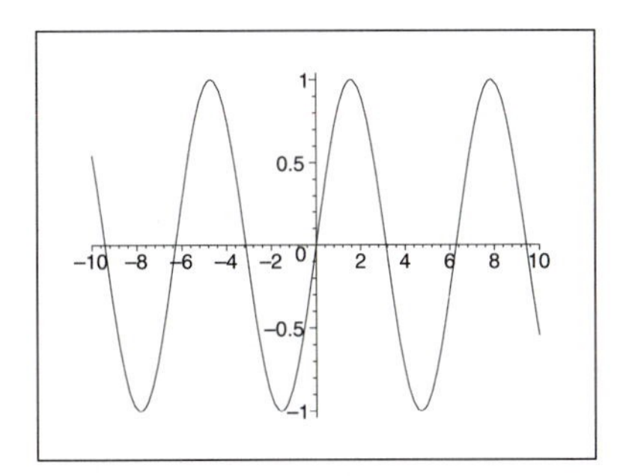

When a procedure is written to output plots, they must be printed either implicitly or explicitly in order to be output. If the purpose of the procedure is to produce a single plot structure then it is probably best to return it as the value of the procedure. If the purpose of the procedure is to produce more than one plot structure then it is probably best to print them explicitly as side-effects of the procedure, and probably return nothing. It is important to remember that simply executing plotting functions at various points within a procedure will generally not cause them to be output. In fact, the plot structures will probably be constructed and then simply discarded, as would the value of any other expression that is computed but then not actually used within a procedure.

Here is an example of a procedure that constructs a *single* plot structure representing the *superposition* of an arbitrary number of power functions, where the powers are specified as its arguments. (The argument type testing is not important for this illustration and could have been omitted; it will be explained in Chapter 10.) In this example, the plot structure is best output by returning it:

```
> PowersPlot := proc()
     # Return one plot of several power graphs superposed
     local x;
     if not {args}::set(numeric) then
        error "arguments must be numeric."
     end if;
     map(p->x^p, [args]);
     plot(%, x=0..2, linestyle=[1,2,3,4],
        legend=map(convert,%,string),
        title="Graph of powers of x")
  end proc:
```

```
> PowersPlot(1/2, 2);
```

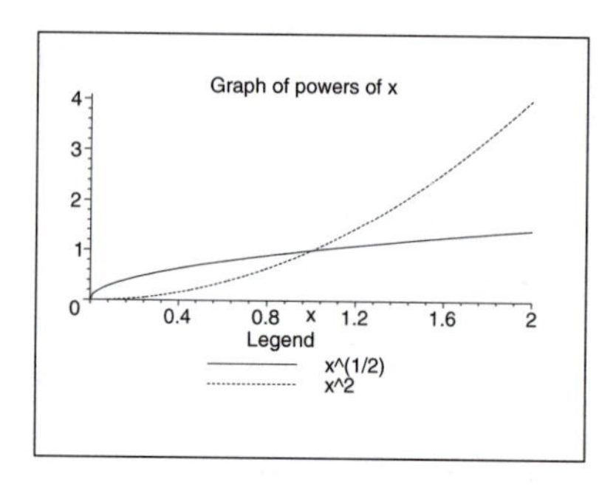

Here is an example of a procedure that constructs an *arbitrary number of plot structures*, each representing one particular power function, where the powers are specified as its arguments. In this example, the plot structures are best output by printing them:

```
> PowerPlots := proc()
    # Output several plots of several power graphs
    local x, p;
    if not {args}::set(numeric) then
       error "arguments must be numeric."
    end if;
    for p in args do
       print(plot(x^p, x=0..2,
          title="Graph of "||(convert(x^p,string))))
    end do
  end proc:
```

```
> PowerPlots(1/2, 2):
```

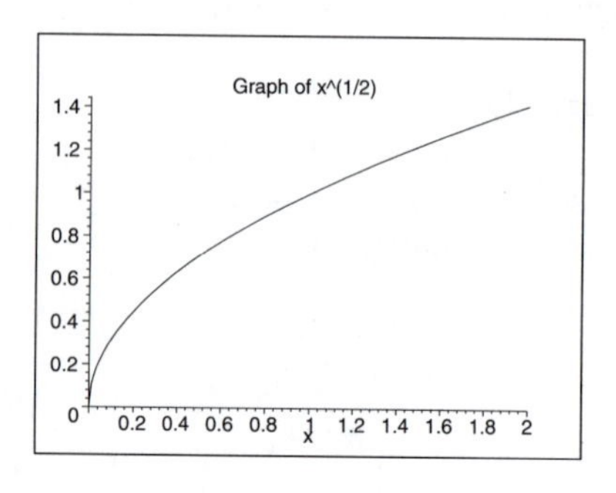

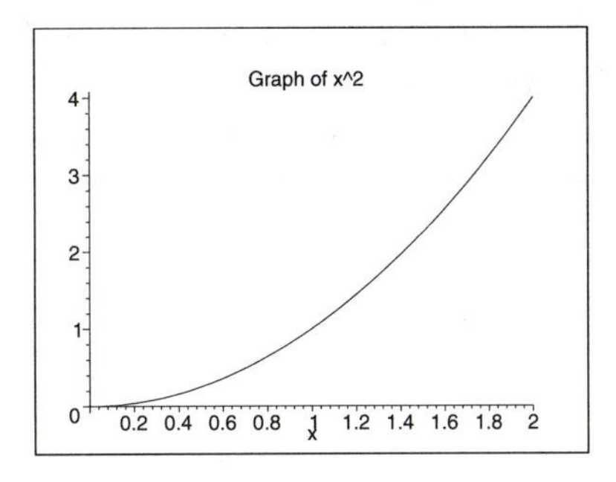

8.13 Recursive plotting

A recursive procedure can be a very good way to generate a plot that has some naturally recursive structure. There are two ways to do it: either recursively generate a plot structure and then output it, or recursively generate a data structure that describes what is to be plotted and then plot it. I will illustrate the former approach in this section and the latter approach, which is likely to be faster, in the next section. To generate a plot structure recursively, the recursive procedure must *return* (not print) it. In general it must explicitly contribute a small piece to the plot structure and combine that piece with the plot structures returned by the recursive calls it makes, so as to build up the entire plot.

As a simple example, consider a binary tree of the form shown in Figure 8.1, in which the *nodes* and *leaves* are indicated by small circles. This

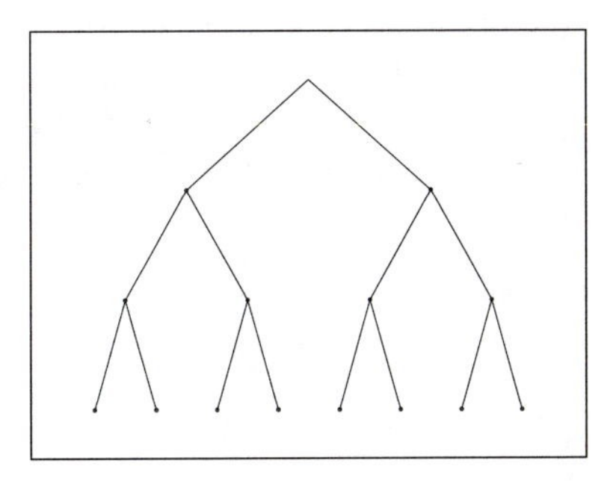

Figure 8.1: A binary tree of depth 3.

particular tree has three levels of *branches* and hence nodes (other than the *root*, which is the top vertex), but it should be clear how it generalizes to any number of levels, which I will call the *depth* of the tree. Hence, the above tree has depth 3. Our aim is to write a procedure called `Tree` that accepts one argument N and returns a two-dimensional plot structure representing a binary tree of depth N with small circles at its nodes (other than the root), so that the call `Tree(3)` would return the above plot structure.

The tree consists of a number of *branch pairs* having the structure shown in Figure 8.2. To be precise, let us specify each branch pair as follows. Suppose

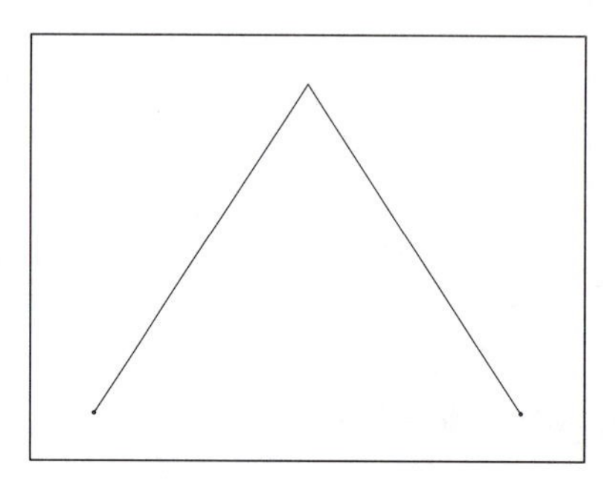

Figure 8.2: A binary tree branch pair.

the top vertex is at Cartesian coordinates (x, y). Then the small circles are at coordinates $(x \pm W, y - 1)$, which defines a *width* parameter W. Let us define the width of the tree to be the width of its top branch. The structure of the tree is recursive because from every node that is not a leaf there grows (downwards!) a similar tree with a smaller depth.

Let us first write a procedure called `SubTree` that takes arguments x, y, W, N and returns a single plot structure representing a tree of width W and depth $N > 0$ with its root at (x, y). It does this by plotting two small circles centred at coordinates $(x \pm W, y - 1)$ and two straight lines from (x, y) to the centres of the two small circles, and then calling itself recursively twice in order to include a tree of width $W/2$ and depth $N - 1$ at each of the two small circles. The four plot structures (two returned by explicit calls of `plot` and two returned by recursive calls of `SubTree`) are combined into one plot structure by `plots[display]`:

```
> SubTree := proc(x,y,W,N)
     local X1, X2, Y;
     if N > 0 then
        X1 := x + W;  X2 := x - W;  Y := y - 1;
        plots[display]({
           plot({[[X1,Y]], [[X2,Y]]},
              style=POINT, symbol=CIRCLE, colour=BLACK),
           plot({[[x,y],[X1,Y]], [[x,y],[X2,Y]]}, colour=BLACK),
           SubTree(X1,Y,W/2,N-1),
           SubTree(X2,Y,W/2,N-1)})
     end if
  end proc:
```

Now we can write procedure `Tree` to accept one argument N, which must be a positive integer, and return a subtree *at the origin* of width 1 and depth N. It is nothing more than a wrapper for the recursive procedure `SubTree`, which checks that its argument N is valid and calls `SubTree` with appropriate initial arguments. This is an example of a very common programming model, in which a simple non-recursive procedure initializes and calls a recursive procedure that does the real computation:

```
> Tree := proc(N::posint)
      plots[display](SubTree(0,0,1,N), axes=NONE)
  end proc:
```

Note that it is good practice not to call the function `with` in a procedure, but rather to access package functions via their full names. It is also good practice not to change the Maple environment within a procedure; hence, I have not called `setoptions` (which I probably would if I were plotting these figures interactively) but rather I have included the options explicitly in the plotting functions. Specifically, I have set the plotting colour to be black and selected no axes.

Finally, we can test procedure `Tree`:

```
> Tree(5);
```

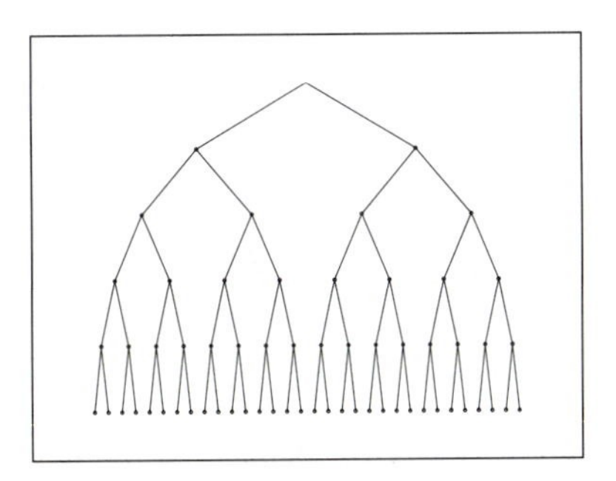

Beware that this procedure is slow when plotting a tree of depth greater than about 5. It could probably be implemented more efficiently. (Figures 8.1 and 8.2 above were generated by executing `Tree(1)` and `Tree(3)`.)

8.14 Animating the Koch snowflake fractal

This section is a case study that brings together a number of the topics that we have studied so far: two-dimensional plotting and animation, elementary numerical computation and linear algebra, and recursive generation of a structure to be plotted. A fractal is an object having a self-similar structure, meaning that it consists of a collection of smaller copies of itself. In other words, it is naturally recursive. One particular fractal is called the Koch snowflake because it was first defined by the mathematician H. von Koch in 1904 and (in the closed form that I will present it here) it looks a bit like a snowflake. A fractal has infinitely fine structure and it is possible to construct explicitly only a finite approximation. The structure in a fractal can normally be labelled by a depth parameter that tends to infinity but is limited to some finite maximum in a concrete representation. The growth of a fractal can be animated if successive frames of the animation represent successively greater maximum depths. (I first saw this done for the Koch snowflake in a posting to the Maple User Group e-mail list by William C. Bauldry of Appalachian State University.)

8.14.1 The snowflake construction

The snowflake can be defined recursively as follows. Start with an equilateral triangle. For one of the sides AB, replace the middle third CD with an equilateral triangle (outwards) with vertex E and delete the base CD to leave the open polygon ACEDB shown in Figure 8.3. Repeat this replacement on each

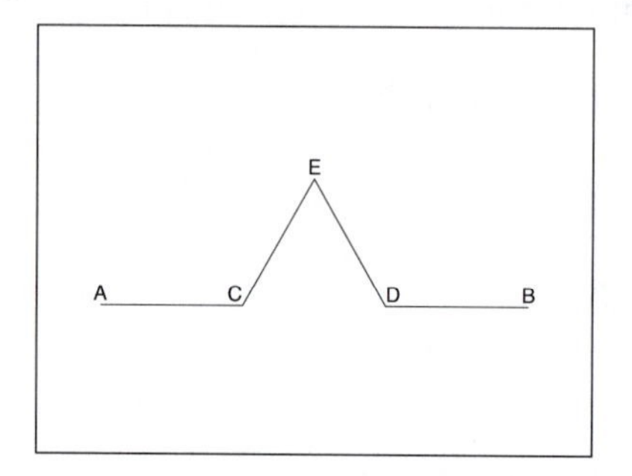

Figure 8.3: Koch snowflake construction.

straight side segment infinitely many times (or in practice until no further replacements are visible).

8.14.2 Basic geometry

First, let us use Maple to help with the coordinate geometry. Let us represent the points A and B as general two-dimensional vectors, and then determine the vectors that represent the points C, D, and E in terms of A and B. (Note that rtable-based arrays cannot remain symbolic in the way that table-based arrays can, so it is more convenient to use table-based linear algebra here. Also, beware that assigning to `D` as a variable prevents use of `D` as the differentiation operator, and so is generally not a good idea!)

```
> A := vector(2):  B := vector(2):
```

```
> C := evalm(A + 1/3*(B-A));
```

$$C := \left[\frac{2}{3}A_1 + \frac{1}{3}B_1, \frac{2}{3}A_2 + \frac{1}{3}B_2\right]$$

```
> unprotect(D):  D := evalm(A + 2/3*(B-A));
```

$$\mathrm{D} := \left[\frac{1}{3}A_1 + \frac{2}{3}B_1, \frac{1}{3}A_2 + \frac{2}{3}B_2\right]$$

Now we need a matrix **R** to rotate a vector through an angle of $\pi/3$:

```
> Pi/3:
  R := matrix([[cos(%), -sin(%)], [sin(%), cos(%)]]);
```

$$R := \begin{bmatrix} \frac{1}{2} & -\frac{1}{2}\sqrt{3} \\ \frac{1}{2}\sqrt{3} & \frac{1}{2} \end{bmatrix}$$

Then, the point E is the vector C plus 1/3 of the vector AB rotated through an angle of $\pi/3$:

```
> E := evalm(C + 1/3*R*(B-A));
```

$$E := \left[\frac{1}{2}A_1 + \frac{1}{2}B_1 - \frac{1}{6}\sqrt{3}\,(B_2 - A_2), \frac{1}{2}A_2 + \frac{1}{2}B_2 + \frac{1}{6}\sqrt{3}\,(B_1 - A_1)\right]$$

We use these formulae for C, D, E below.

8.14.3 Constructing a single side

We now have all the vertex vectors in terms of the vertices A and B. For the snowflake computation it is slightly more convenient to use lists rather than vectors to represent the coordinates of the points. We will use floating-point computation throughout, which is appropriate for graphics, and pre-compute this constant to speed up the computation:

```
> `1/sqrt(3)` := 1.0/sqrt(3.0):
```

Then the following recursive procedure constructs a single snowflake side by using the above replacement algorithm to depth n. What it actually constructs is a sequence of vertices, where each vertex is represented as a list containing the x and y coordinates. Unlike in the previous section, this procedure does not construct a plot structure at all — it constructs recursively the data structure to be plotted.

```
> Side := proc(A, B, n)
    # Return the side AB decorated to depth n
    # as a sequence of vertex vectors, excluding A:
    local n1, A1, A2, B1, B2, C, D, E;
    if n <= 0 then return B end if;
    # These intermediate variables are used for efficiency:
    n1 := n-1;
    A1 := A[1];  A2 := A[2];  B1 := B[1];  B2 := B[2];
    C := [(2.0*A1+B1)/3.0, (2.0*A2+B2)/3.0];
    D := [(A1+2.0*B1)/3.0, (A2+2.0*B2)/3.0];
    E := [((A1+B1)-`1/sqrt(3)`*(B2-A2))/2.0,
          ((A2+B2)+`1/sqrt(3)`*(B1-A1))/2.0];
    Side(A, C, n1), Side(C, E, n1),
    Side(E, D, n1), Side(D, B, n1)
  end proc:
```

8.14.4 Animating the "growth" of a snowflake

The following procedure constructs and animates a sequence of "growing" snowflake fractals. First it defines the points A, B, C of an equilateral triangle, then it constructs a sequence of plot structures representing finite approximations to snowflake fractals to successively greater depths, and finally it displays the sequence of plot structures as the frames of an animation:

```
> Snowflake := proc(nmax)
    local A, B, C, n;
    A := [0.0,0.0];
    B := [0.5, sqrt(3.0)/2.0];
    C := [1.0,0.0];
    # Construct the sequence of snowflake plots:
    seq(plot([A, Side(A, B, n), Side(B, C, n),
       Side(C, A, n)]), n = 0..nmax);
    # Animate the fractal growth:
    plots[display]([%], insequence = true,
       axes = NONE, scaling = CONSTRAINED,
       title = "Koch Snowflake")
  end proc:
```

Now we can display a snowflake animation. I will use a maximum depth of 4, since normally little or no finer structure is visible in practice:

```
> Snowflake(4);
```

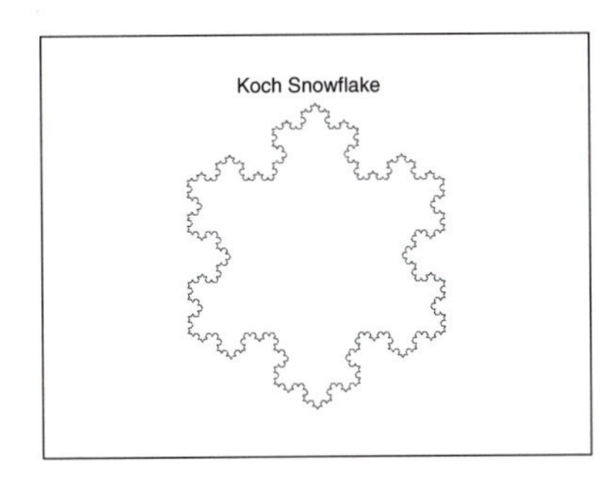

8.14.5 Code to plot the snowflake construction

These are the plotting instructions for Figure 8.3; they call procedure `Side` defined above in the text.

```
> plots[setoptions](colour = BLACK);
  plots[display]({
     plot([[0,0], Side([0,0], [1.,0], 1)]),
     plots[textplot]({[0,0.01,"A"], [1,0.01,"B"],
        [1/2, 1/6*sqrt(3)+0.01, "E"]}, align = {ABOVE}),
     plots[textplot]([1/3, 0.01, "C"], align = {ABOVE,LEFT}),
     plots[textplot]([2/3, 0.01, "D"], align = {ABOVE,RIGHT})
  }, axes = NONE, scaling = CONSTRAINED);
```

8.15 Program design

There are several steps to writing a computer program.

1. ***Understand the problem.*** Usually, it is helpful to experiment with some simple examples.

2. ***Design the algorithm.*** This is the sequence of operations that needs to be performed in order to solve the problem. They can be written in an informal way, without necessarily following the precise syntax of a formal language. The algorithm will be a generalization of the simple examples considered first.

3. ***Write the program.*** This is the implementation of the algorithm, which requires decisions about data and control structures that may have been left a little vague in the algorithm. It usually involves looking up a few details of the programming language to be used (in printed or online documentation), checking what is directly supported by the language, what is easy, what is possible, etc.

4. ***Test the program.*** Always check extreme examples of the input, such as collections that are empty or contain only one element. Then check a few simple but non-special examples. Provided these tests all work correctly and the program is designed and implemented in a general way (e.g., not by looking up special cases in a table), it should work in all cases.

For example, consider the problem of finding all the words that can be constructed from some collection of letters (which is a common "word game"). For simplicity, I will ignore spelling and consider the problem of generating all possible strings of letters, ignoring case. (If you use a language such as Visual Basic for Applications within Microsoft Word then you can use the spelling checker to select only correctly spelt words. This is quite easy to do, although other aspects of the implementation, such as sorting, are harder.)

8.15.1 Understand the problem

It is essential to be systematic. Suppose we start with the letters CAT. Then the "subwords" (ignoring spelling) are all those that start with C, namely:

$$C, CA, CT, CAT, CTA;$$

all those that start with A, namely:

$$A, AC, AT, ACT, ATC;$$

and all those that start with T, namely:

$$T, TA, TC, TAC, TCA.$$

Now we need to start generalizing. The first set of subwords can be expressed as:

$$C \text{ and } C \text{ prepended to all subwords of } AT, \text{ namely } A,\ T,\ AT,\ TA.$$

But what about repeated letter? Suppose we start with the letters $ABBA$. Then the subwords are all those that start with A, namely:

$$A, AA, AB, AAB, ABA, ABB, AABB, ABAB, ABBA;$$

and all those that start with B, namely:

$$B, BA, BB, BAA, BAB, BBA, BAAB, BABA, BBAA.$$

Clearly, we must ignore repeats when choosing the first letter.

8.15.2 Design the algorithm

The generalization of the above examples is this:

> For each *distinct* letter L, the subwords are L and L prepended to all subwords of the remaining letters.

Note that this algorithm is naturally recursive. The easiest way to spot repeated letters is first to sort them, so that repeated letters occur together. This also has the advantage that if the letters are chosen in order then the resulting subwords are automatically in dictionary (lexicographic) order. This gives the following refined algorithm:

> To find all subwords of a collection C of letters, first sort C.
>
> Then, for each letter L in C, if L is different from the previous letter, take L and L prepended to all subwords of the remaining letters.

8.15.3 Write the program

Having decided to use Maple, the sorting is easy after we have looked up the details of the built-in `sort` function. Probably the best data structure to use to represent the collection of letters is a Maple list. (A Maple set is not appropriate because it will not stay sorted and it will discard any repeated letters!) So, the first step of the algorithm could be implemented like this:

```
> Words := proc(Letters)
     SubWords(sort(Letters))
  end proc:
```

Now, let us test this before we proceed:

```
> Words(["C", "A", "T"]);
```

SubWords([“A”, “C”, “T”])

```
> Words(["A", "B", "B", "A"]);
```

SubWords([“A”, “A”, “B”, “B”])

Next, we must implement the recursive sub-procedure that does the real work, to do the following: for each letter L in *Letters*, if L is different from the previous letter, take L and L prepended to all subwords of the remaining letters:

```
> SubWords := proc(Letters)
    # Input: Letters, a list of letters
    # Return: result, a sequence of strings
    local result, subwords, L, lastL, SW, i;
    i := 0;  result := NULL;
    for L in Letters do
       i := i + 1;
       if L <> lastL then
          lastL := L;
          subwords := SubWords(subsop(i=NULL,Letters));
          result := result,
             L, seq(cat(L, SW), SW = subwords)
       end if
    end do;
    result
  end proc:
```

This procedure uses most of the facilities discussed in Chapter 2 for manipulating sequences and lists, and it uses the standard Maple function `cat` to *catenate* or join two strings.

This implementation is actually very inefficient, mainly because it stores the entire sequence of subwords before outputting it. It can be re-implemented so that it stores very little intermediate information and prints each subword as it finds it, although the logic of the recursion is a little more subtle. However, this simple implementation will suffice for the current illustration of program development.

The top-level function `Words` could also be improved to accept a single string rather than requiring a list of separate letters (which is tedious to type), by using `seq` to expand the string into a sequence of characters (see Chapter 2):

```
> Words := proc(Letters::string)
    local c;
    SubWords(sort([seq(c,c=Letters)]))
  end proc:
```

8.15.4 Test the program

First, we test some extreme examples, which for this problem are an empty input string:

```
> Words("");
```

and a string of one letter:

```
> Words("A");
```

"A"

Then we test some simple non-special cases; first, one with no repeated letters:

```
> Words("CAT");
```

"A", "AC", "ACT", "AT", "ATC", "C", "CA", "CAT", "CT",
"CTA", "T", "TA", "TAC", "TC", "TCA"

then, one with one repeat:

```
> Words("ABA");
```

"A", "AA", "AAB", "AB", "ABA", "B", "BA", "BAA"

and ones with two and three repeats:

```
> Words("ABBA");
```

"A", "AA", "AAB", "AABB" , "AB", "ABA", "ABAB", "ABB",
"ABBA", "B", "BA", "BAA" , "BAAB", "BAB", "BABA",
"BB", "BBA", "BBAA"

```
> Words("AAA");
```

"A", "AA", "AAA"

These examples are all simple enough to check by hand. They all appear to be correct, which gives some confidence that the program is correct.

8.16 Exercises

When writing procedures, use local variables as appropriate to avoid computing the same quantity more than once and explicitly declare all local variables as appropriate — do not rely on Maple's default declarations. Test all your procedures carefully. To write a *non-recursive* procedure to solve a general problem, it is often helpful first to solve a specific example interactively by assigning simple/small values to any variables that will later be procedure arguments. Once the interactive code works, put a `proc ... end proc` "wrapper" around it and add a `local` declaration if appropriate.

Non-recursive procedures

Write and test the following non-recursive procedures to perform the specified tasks:

1. `PowSeq(p, n)` to return an expression sequence consisting of the pth power of each of the n smallest positive integers, assuming n is any non-negative integer, including 0. For example:

```
> PowSeq(3, 10);
```

$$1, 8, 27, 64, 125, 216, 343, 512, 729, 1000$$

```
> PowSeq(3, 0);
```

2. `FirstPrimes(n)` to return an expression sequence consisting of the n smallest positive prime numbers, assuming n is any non-negative integer, including 0. (You may use the Maple library function `nextprime` but *not* the library function `ithprime`.) For example:

```
> FirstPrimes(10);
```

$$2, 3, 5, 7, 11, 13, 17, 19, 23, 29$$

```
> FirstPrimes(0);
```

3. `int_n(y, x, n)` to return the value of the nth-order repeated indefinite integral of y with respect to x, obtained by re-integrating y repeatedly n times, assuming n is any non-negative integer, including 0. For example:

```
> int_n(f(x),x,0), int_n(f(x),x,1), int_n(f(x),x,2);
```

$$\mathrm{f}(x),\ \int \mathrm{f}(x)\,dx,\ \int\int \mathrm{f}(x)\,dx\,dx$$

```
> int_n(x,x,0), int_n(x,x,1), int_n(x,x,2);
```

$$x,\ \frac{1}{2}\,x^2,\ \frac{1}{6}\,x^3$$

4. `RangePartition(R, n)` to return an expression sequence consisting of the range R partitioned into n sub-ranges of equal length which together span R, assuming n is any positive integer. (Hint: The range operator `..` is a binary operator, the operands of which may be extracted using `lhs` and `rhs` [or `op`].) For example:

```
> RangePartition(a..b, 1);
```

$$a..b$$

```
> RangePartition(a..b, 3);
```

$$a..\frac{2}{3}\,a+\frac{1}{3}\,b,\ \frac{2}{3}\,a+\frac{1}{3}\,b..\frac{1}{3}\,a+\frac{2}{3}\,b,\ \frac{1}{3}\,a+\frac{2}{3}\,b..b$$

```
> RangePartition(1..2, 2);
```

$$1..\frac{3}{2},\ \frac{3}{2}..2$$

```
> RangePartition(1..2, 3);
```

$$1..\frac{4}{3},\ \frac{4}{3}..\frac{5}{3},\ \frac{5}{3}..2$$

5. `expandifac` to takes a "sequence of lists" data structure of the form returned by the last version of `ifac` developed in the text and convert it to a product, thereby acting as the inverse of the function `ifac`.

Recursive procedures

Write and test both non-recursive and recursive versions of the following procedures to perform the specified tasks. Note that a non-recursive procedure does not call itself and for this exercise should use an explicit loop (`do ... end do` in Maple) instead, whereas ***a recursive procedure is one that calls itself*** (to solve a simpler version of the same problem).

1. `AddList(L)` to return the sum of the elements of the list L. (You should *not* use `+` as a function, only as a binary operator, and you should *not* use the Maple library functions `add` or `sum`. The sum of the elements of an empty list should be returned as 0.) For example:

```
> AddList([a,b,c]);
```

$$a+b+c$$

```
> AddList([]);
```

$$0$$

2. `ReverseArgs()` to return its argument sequence in reverse order. For example:

```
> ReverseArgs(a,b,c);
```

$$c,\ b,\ a$$

```
> ReverseArgs();
```

Chapter 9

Operators and Functions

There is no fundamental difference between operators and functions; it is merely conventional to implement some facilities as operators and some as functions. In most cases, active operators and functions both rely on procedures, which were introduced in the previous chapter primarily from a programming perspective. This chapter takes a more mathematical view, although recursion is still an important concept.

The first section explains operator syntax and introduces inert and active operators. Similar to the equivalence between operators and functions, there is an equivalence between expressions and mappings, which is explained briefly in Section 9.2. The next two sections explore this equivalence further: an expression consisting of a "function of a function" is equivalent to a composition of mappings, and the `D` operator is to mappings what the `diff` function is to expressions. (Terminology is further blurred here because in mathematics functions that act on functions are often called operators even when they use functional syntax!)

Section 9.5 heads off on a different tack and discusses the operators and functions on sets that are built into Maple and shows how to use them to construct the power set of a set recursively. The last section takes a different tack again and briefly explores the analogy between the sets of integers and univariate polynomials over a field, showing how very similar recursively defined functions can perform analogous ring operations. Some of the ideas introduced in this section are pursued further in Chapter 12.

9.1 Operators

9.1.1 Syntax

Operators are really the same as functions except that they use a syntax that does not require parentheses; hence, they can be only nullary, unary, or binary (i.e., they can take at most two operands). A unary operator is usually *prefix*, meaning that it is placed *before* its operand (e.g., "$-a$"), but there is at least one important *postfix* operator, meaning that it is placed *after* its operand, namely the factorial operator "$a!$" (which is indeed defined as a postfix operator in Maple). Binary operators are *infix*, meaning that they are placed *between* their two operands (e.g., "$a + b$"). Denoting an operator by the symbol $\bullet$, the possibilities are these:

unary prefix	binary infix	unary postfix
$\bullet a$	$a \bullet b$	$a\bullet$

The "arity" of an operator or function is the number of operands or arguments it takes. If an operator is to be n-ary with $n > 2$, then functional syntax must be used, i.e., $\bullet(a, b, c, \ldots)$.

Occasionally, the concept of a "nullary" operator is useful; it is an operator that takes no operands. Syntactically, it behaves like a variable, but it has different semantics and cannot be assigned a value. In Maple, the three ditto operators `%`, `%%`, and `%%%` are nullary operators, although they are also environment variables!

9.1.2 Inert operators

Maple provides a rather rudimentary mechanism for introducing *new* operators. (Other languages, such as C++, are much better in this area.) New operator names *must* begin with the special character `&`, which is called "ampersand". Like functions, such operators are inert or neutral by default and just build data structures to be interpreted by other parts of Maple, e.g.,

```
> a &oper b;
```

$$a \,\&\text{oper}\, b$$

The `&` character is recognized by the Maple parser (the part that reads and interprets input) and by the pretty-printer, but internally & operators are treated as functions, e.g.,

```
> lprint(%);
  `&oper`(a,b)
```

All & operators can be used with functional syntax (and backquotes here are optional), e.g.,

```
> &oper(a,b);
```

$$a \,\&\text{oper}\, b$$

```
> `&oper`(a,b);
```

$$a \,\&\mathrm{oper}\, b$$

```
> &*(a,b);
```

$$a \,\&{*}\, b$$

Many standard non-& operators can similarly be used with functional syntax *provided* the operator name is "escaped" using backquotes, e.g.,

```
> *(a,b);
```

```
Error, `*` unexpected
```

```
> `*`(a,b);
```

$$a\, b$$

This is necessary even when the operator name is an identifier, because it is still necessary to escape the operator syntax — backquotes escape both special characters and special syntax, e.g.,

```
> union(A,B);
```

```
Error, reserved words `union` or `minus` unexpected
```

```
> `union`(A,B);
```

$$A \,\mathrm{union}\, B$$

Some & operator names are reserved and already have special meanings within Maple; e.g., `&*` is interpreted by `evalm` as non-commuting multiplication, `&^` is interpreted by `mod` as exponentiation, and `&+` is interpreted by `type` as addition, e.g.,

```
> 2 &* 3 <> 2 * 3;
```

$$2 \,\&{*}\, 3 \neq 6$$

```
> 2&^3 <> 2&^3 mod 3;
```

$$2 \,\&\hat{}\, 3 \neq 2$$

```
> name &+ name <> name + name;
```

$$\mathit{name} \,\&{+}\, \mathit{name} \neq 2\, \mathit{name}$$

An & operator name consists of the character `&` followed by *either* any Maple name that does not require backquotes *or* one or more special characters, but the two forms cannot be mixed, and white space and the following characters are not allowed at all:

```
& | ( ) [ ] { } ; : ' ` # % \
```

An & operator may be used as a unary prefix or binary infix operator, or as a function, in which case it can be n-ary.

The *precedence* of an operator determines which operands are bound to which operators by default in an expression; e.g., the expression `a*b+c` is parsed as `(a*b)+c` because `*` has higher precedence than `+` and so its operands are bound first. All & operators have the same precedence, which is higher than any other operator except "`.`"; parentheses should be used if necessary to ensure the correct interpretation.

9.1.3 Active operators

Operators can be made *active* by assigning procedures to them, exactly as functions can be made active. For example, by default the operator `&div` is inert:

```
> 7 &div 3;
```

$$7 \,\&\text{div}\, 3$$

We can interpret it to mean Euclidean integer division with remainder and make it active by assigning a suitable procedure definition to it (using either `proc` or arrow syntax). Remember that `&` is not an identifier character, so operator names beginning with `&` must be enclosed in backquotes when they appear on the left of an assignment. The following all have effectively the same result:

```
> `&div` := (a,b) -> (iquo(a,b), irem(a,b)):
```

```
> 7 &div 3;
```

$$2,\ 1$$

```
> `&div` := () -> (iquo(args), irem(args)):
```

```
> 7 &div 3;
```

$$2,\ 1$$

```
> `&div` := iquo, irem:
```

```
> 7 &div 3;
```

$$2,\ 1$$

But this operator does not handle unexpected operands very well, e.g.,

```
> (2/7) &div 3;
```

```
Error, (in unknown) wrong number (or type) of parameters in
function iquo
```

An alternative allowed name would be `&/`, and a better definition would check its operand types, e.g.,

```
> `&/` := (a::integer,b::integer) -> (iquo(a,b), irem(a,b)):
```

```
> 7 &/ 3;
```

$$2, 1$$

```
> (2/7) &/ 3;
```

```
Error, &/ expects its 1st argument, a, to be of type integer, but
received 2/7
```

We can similarly define an active unary operator, e.g.

```
> `&!` := a -> a!;
```

$$\&! := \text{factorial}$$

```
> &!5 = 5!;
```

$$120 = 120$$

We can also allow an active operator to be either unary or binary; e.g., this version optionally implements an "n-step factorial" if given a second operand:

```
> `&!` := proc(a::posint, n::posint)
     local result, i;
     if nargs = 1 then a!
     elif nargs = 2 then
        result := 1;
        for i from a by -n to 2 do
           result := result*i
        end do;
        result
     else error "requires 1 or 2 posint arguments."
     end if
  end proc:
```

```
> &! 5;
```

$$120$$

```
> 5 &! 2;
```

$$15$$

```
> 5 &! (-2);
```

```
Error, &! expects its 2nd argument, n, to be of type posint, but
received -2
```

```
> `&!`();
   Error, (in &!) requires 1 or 2 posint arguments.
```

An active operator to append or join two lists might be defined in one of these ways:

```
> `&!` := (a::list,b::list) -> [op(a),op(b)]:

> [a,b] &! [c,d];
```

$$[a, b, c, d]$$

```
> `&!` := (a::list,b::list) -> map(op, [args]):

> [a,b] &! [c,d];
```

$$[a, b, c, d]$$

```
> `&!` := proc(a::list,b::list)
     local arg;
     [seq(op(arg),arg=args)]
  end proc:

> [a,b] &! [c,d];
```

$$[a, b, c, d]$$

The second and third versions have the advantage that they also work as n-ary operators if they are used with functional syntax.

An active operator to add two lists like vectors could be defined like this (note that `&+` is a reserved identifier):

```
> `&++` := proc(a::list, b::list)
     local n, i;
     n := nops(a);
     if n <> nops(b) then
        error "lists must be the same length." end if;
     [seq(a[i]+b[i], i=1..n)]
  end proc:

> [a,b] &++ [c,d];
```

$$[a + c, b + d]$$

```
> [a,b] &++ [c];
   Error, (in &++) lists must be the same length.
```

This functionality could alternatively be implemented by using the built-in function `zip` but, in fact, it is already provided by default by *overloading* the normal + operator:

```
> [a,b] + [c,d];
```

$$[a+c, b+d]$$

9.2 Converting between expressions and mappings

Some ambiguity of nomenclature arises when discussing expressions, functions, and mappings. Consider $f(x)$. According to the Maple type system this is an *expression* of type *function*, and an applied mathematician, scientist, or engineer would probably agree, whereas a pure mathematician or computer programmer would probably regard f as the function and $f(x)$ as the value returned by applying it to x. I will generally take the latter view, although in mathematical contexts I try to avoid the ambiguity by calling f a *mapping*.

If f is a Maple *mapping* then *applying* it to some argument sequence results in an *expression*, regardless of whether f is inert or active, e.g.,

```
> f := 'f': f(x,y);  # gives an expression:
```

$$\mathrm{f}(x, y)$$

```
> f := (a,b)->a^2+b^2: f(x,y);  # gives an expression:
```

$$x^2 + y^2$$

The reverse process is effected by the standard Maple function `unapply`, which must be given both the expression and the variables with respect to which the mapping is to be constructed, e.g.,

```
> unapply(%, x, y);
```

$$(x, y) \to x^2 + y^2$$

The result of using `unapply` is normally assigned to a variable in order to produce a named mapping for later use. This provides an easy way to convert the result of some computation into a mapping, or effectively to have a mapping body evaluated when it is defined. The names of the formal arguments are of no significance — they are just place-holders — and the original definition of the mapping f is the same as that produced by `unapply` (as it should be, although using different formal argument names):

```
> evalb(eval(f) = %);  f := 'f':
```

$$\mathit{true}$$

9.3 Composition of mappings and functions

When a mapping g is applied to the *value* of another mapping f, the two mappings g and f are said to be *composed.* Composition of mappings is normally denoted in mathematics by a small circle, $\circ$. Thus,

$$g(f(x)) = (g \circ f)(x).$$

For this composition to be defined, the domain of g must contain the range of f, e.g., if

$$f : U \to V \text{ and } g : V \to W$$

then $g \circ f$ is defined and

$$(g \circ f) : U \to W$$

maps U to W thus:

$$U \xrightarrow{f} V \xrightarrow{g} W.$$

If

$$f : U \to U$$

then $f \circ f$ is defined, and so is any order of repeated self-composition.

Maple provides the binary infix operators `@` to denote function composition and `@@` to denote repeated self-composition. These operators are analogous to the operators `*` and `**` denoting multiplication and repeated self-multiplication or exponentiation, the latter also (more usually) being denoted by `^`. Maple does not pay much attention to the domains and ranges of functions (although their dimensions must be correct). Here are some examples:

```
> h := g@f;  h(x);  unapply(%, x);
```

$$h := g@f$$
$$\mathrm{g}(\mathrm{f}(x))$$
$$g@f$$

```
> (g@f)(x) = g(f(x));
```

$$\mathrm{g}(\mathrm{f}(x)) = \mathrm{g}(\mathrm{f}(x))$$

Note that repeated self-composition is pretty-printed as a superscript in parentheses:

```
> f@@2 = f@f;
```

$$f^{(2)} = f^{(2)}$$

```
> f@@3 = f@f@f;
```

$$f^{(3)} = f^{(3)}$$

```
> f@@n;  # no explicit form for symbolic n
```

$$f^{(n)}$$

The input syntax and output notation for repeated self-composition are both analogous to that for powers:

```
> f**2 = f*f;  # same as f^2
```

$$f^2 = f^2$$

```
> f**3 = f*f*f;  # same as f^3
```

$$f^3 = f^3$$

```
> f**n;  # same as f^n -- no explicit form for symbolic n
```

$$f^n$$

9.4 The derivative operator D

The derivative of an expression is an expression, e.g.,

```
> diff(sin(x),x);
```

$$\cos(x)$$

but the derivative of a mapping is a mapping, e.g.,

```
> D(sin);
```

$$\cos$$

This new derivative mapping can be applied to an argument (which can be an arbitrary expression) using the normal syntax for function application to give an expression, e.g.,

```
> D(sin)(x);
```

$$\cos(x)$$

```
> D(sin)(0);
```

$$1$$

Hence, `diff(f(x),x)` is the same as `D(f)(x)` and, more generally, `diff(f(x),x$n)` is the same as `(D@@n)(f)(x)`. Although the two forms remain distinct by default, they can be inter-converted:

```
> diff(f(x),x$n):  % = convert(%,D);
```

$$\mathrm{diff}(\mathrm{f}(x),\, x\,\$\,n) = (\mathrm{D}^{(n)})(f)(x)$$

```
> (D@@n)(f)(x):  % = convert(%,diff);
```

$$(\mathrm{D}^{(n)})(f)(x) = \mathrm{diff}(\mathrm{f}(x),\, x\,\$\,n)$$

```
> diff(f(x),x);  unapply(%,x) = D(f);
```

$$\frac{\partial}{\partial x}\,\mathrm{f}(x)$$

$$\mathrm{D}(f) = \mathrm{D}(f)$$

Maple understands the "chain" or "function of a function" rule for differentiation, e.g.,

```
> D(g@f)(x);
```

$$\mathrm{D}(g)(\mathrm{f}(x))\,\mathrm{D}(f)(x)$$

```
> eval(%, {f=sin, g=cos});  # for example
```

$$-\sin(\sin(x))\cos(x)$$

```
> diff(g(f(x)),x);
```

$$\mathrm{D}(g)(\mathrm{f}(x))\,(\frac{\partial}{\partial x}\,\mathrm{f}(x))$$

```
> eval(%, {f=sin, g=cos});  # for example
```

$$-\sin(\sin(x))\cos(x)$$

In fact, this example illustrates a situation where the conventional d/dx notation does not work well. Using d/dx notation and its Maple analogue `diff`, the chain rule must be interpreted in the following rather tortuous manner:

```
> diff(g(f(x)),x);
```

$$\mathrm{D}(g)(\mathrm{f}(x))\,(\tfrac{\partial}{\partial x}\,\mathrm{f}(x))$$

```
> convert(%, diff);
```

$$(\tfrac{\partial}{\partial t1}\,\mathrm{g}(t1))\,(\tfrac{\partial}{\partial x}\,\mathrm{f}(x))\,\&\mathrm{where}\,\{t1=\mathrm{f}(x)\}$$

This is ugly compared to the form we started with, which used `D` where appropriate. If we now evaluate it for specific functions it does not automatically simplify, even though it could, and there seems to be no simple way to force it to simplify:

```
> eval(%, {f=sin, g=cos});   # for example
```

$$(-\sin(t1)\cos(x))\,\&\mathrm{where}\,\{t1=\sin(x)\}$$

```
> subs(`&where`=eval, %);
```

$$(-\sin(t1)\cos(x))\Big|_{t1\,=\,\sin(x)}$$

```
> %;
```

$$-\sin(\sin(x))\cos(x)$$

The last two steps above could have been compressed as follows, but it does not seem possible to use just `eval` here:

```
> eval(subs(`&where`=eval, %%%));
```

$$-\sin(\sin(x))\cos(x)$$

An alternative to using the `&where` operator that is generated automatically by `convert` is to use an unevaluated `subs` explicitly. The resulting expression has the advantage that it is somewhat easier to evaluate, but the disadvantage that the need for careful use of uneval quotes makes it somewhat unstable:

```
> 'subs'(y=f(x), diff(g(y),y))*diff(f(x),x);
```

$$\mathrm{subs}(y=\mathrm{f}(x),\,\tfrac{\partial}{\partial y}\,\mathrm{g}(y))\,(\tfrac{\partial}{\partial x}\,\mathrm{f}(x))$$

```
> eval('%', {f=sin, g=cos});   # for example
```

$$-\sin(\sin(x))\cos(x)$$

9.4.1 Partial derivatives

The Maple `diff` function is actually a *partial* differentiator, which is why derivatives are pretty-printed using the curly partial derivative notation. A partial derivative corresponds to a derivative of a *mapping* with respect to the argument in a particular location (i.e., first, second, etc.) because the actual argument names in a mapping definition have no significance. Hence, the `D` operator uses an index or subscript when it is necessary to specify partial differentiation with respect to a particular variable, like this:

```
> diff(f(x,y), x);
```

$$\frac{\partial}{\partial x}\,\mathrm{f}(x,\,y)$$

```
> convert(%, D) = D[1](f)(x,y);
```

$$\mathrm{D}_1(f)(x,\,y) = \mathrm{D}_1(f)(x,\,y)$$

Unfortunately, `unapply` does not produce the expression using the `D` operator that one might expect:

```
> unapply(%%, x, y);
```

$$(x,\,y) \rightarrow \mathrm{diff}(\mathrm{f}(x,\,y),\,x)$$

Repeated partial differentiations, which are *assumed* by Maple to commute, can be generated like this:

```
> diff(f(x,y), x, y) = diff(f(x,y), y, x);
```

$$\frac{\partial^2}{\partial y\,\partial x}\,\mathrm{f}(x,\,y) = \frac{\partial^2}{\partial y\,\partial x}\,\mathrm{f}(x,\,y)$$

Using `D` operator notation they are represented by repeated subscripts or indices:

```
> convert(diff(f(x,y), x, y), D);
```

$$\mathrm{D}_{1,\,2}(f)(x,\,y)$$

```
> D[1,2](f)(x,y) = D[2,1](f)(x,y);
```

$$\mathrm{D}_{1,\,2}(f)(x,\,y) = \mathrm{D}_{1,\,2}(f)(x,\,y)$$

9.4.2 Specifying conditions on differential equations

When using dsolve to solve differential equations, it is necessary to use the D operator to specify conditions that involve derivatives (because there is no other notation for a derivative evaluated at a specific point); e.g., the solution of the ODE

$$\frac{d^2y(x)}{dx^2} + y(x) = 0$$

such that $y(0) = 0$ and $\frac{dy(x)}{dx} = 1$ at $x = 0$ (i.e., $\mathrm{D}(y)(0) = 1$) is found like this:

```
> dsolve({diff(y(x), x$2) + y(x) = 0,
          y(0) = 0, D(y)(0) = 1}, y(x));
```

$$\mathrm{y}(x) = \sin(x)$$

9.5 Set operations

Maple supports sets as a fundamental data type, and they are used in many parts of the system. It also provides some basic operations on sets, some of which allow their operands to be either symbolic variables or explicit sets.

9.5.1 Union

The *union* of two sets A and B is the set of all elements that are in either A or B or both. It may be defined as follows:

$$A \cup B = \{x \mid x \in A \text{ or } x \in B\}.$$

It is represented in Maple by the binary infix operator union:

```
> A union B;
```

$$A \,\mathrm{union}\, B$$

```
> {1,2,3} union {2,3,4};
```

$$\{1, 2, 3, 4\}$$

The union of many sets can be defined as

$$\bigcup_i A_i = \{x \mid \exists i, x \in A_i\}.$$

This is represented in Maple by the *function* union, which is just the *operator* union used as a function (which requires backquotes to escape the default operator syntax), but it can be input either by repeated use of union as an operator or by using union as a function:

```
> A union B union C = 'union'(A,B,C);
```

$$\mathrm{union}(C,\ A,\ B) = \mathrm{union}(C,\ A,\ B)$$

```
> {1,2,3} union {2,3,4} union {3,4,5}
      = 'union'({1,2,3}, {2,3,4}, {3,4,5});
```

$$\{1,\ 2,\ 3,\ 4,\ 5\} = \{1,\ 2,\ 3,\ 4,\ 5\}$$

(Note that the first [symbolic] example works only because `union` is an active operator that simplifies repeated binary instances of itself to a single functional form; this operator simplification is not performed automatically for all operators so, in particular, it does not happen for inert operators.)

9.5.2 Intersection

The *intersection* of two sets A and B is the set of all elements that are in both A and B. It may be defined as follows:

$$A \cap B = \{x \mid x \in A \text{ and } x \in B\}.$$

It is represented in Maple by the binary infix operator `intersect` (note the shortened spelling!):

```
> A intersect B;
```

$$A \,\mathrm{intersect}\, B$$

```
> {1,2,3} intersect {2,3,4};
```

$$\{2,\ 3\}$$

The intersection of many sets can be defined as

$$\bigcap_i A_i = \{x \mid \forall i, x \in A_i\}.$$

This is represented in Maple by the *function* `intersect`, which is just the *operator* `intersect` used as a function, but it can be input either by repeated use of `intersect` as an operator or by using `intersect` as a function:

```
> A intersect B intersect C = 'intersect'(A,B,C);
```

$$\mathrm{intersect}(C,\ A,\ B) = \mathrm{intersect}(C,\ A,\ B)$$

```
> {1,2,3} intersect {2,3,4} intersect {3,4,5}
      = 'intersect'({1,2,3}, {2,3,4}, {3,4,5});
```

$$\{3\} = \{3\}$$

9.5.3 Difference

The *difference* of two sets A and B is the set of all elements that are in A but not in B. It is denoted by either $A - B$ or $A \backslash B$, and may be defined as follows:

$$A \backslash B = \{x \mid x \in A \text{ and } x \notin B\}.$$

It is represented in Maple by the binary infix operator `minus`:

```
> A minus B;
```

$$A \operatorname{minus} B$$

```
> {1,2,3} minus {2,3,4};
```

$$\{1\}$$

Unlike union and intersection, set difference is not symmetrical; hence, it is strictly a binary operation.

9.5.4 Symmetric difference

The *symmetric difference* of two sets A and B is the set of all elements that are in either A or B but *not* in both and it is provided in Maple as the *function* `symmdiff`. However, this set operation provides a nice illustration of the use of union, intersection, and normal set difference, so let us consider its implementation, initially as an *operator*. Perhaps the most obvious definition is this:

```
> `&-` := (A,B) -> (A union B) minus (A intersect B):
```

```
> {1,2,3} &- {2,3,4};
```

$$\{1, 4\}$$

```
> symmdiff({1,2,3}, {2,3,4});
```

$$\{1, 4\}$$

An alternative definition is this:

```
> `&-` := (A,B) -> (A minus B) union (B minus A):
```

```
> {1,2,3} &- {2,3,4};
```

$$\{1, 4\}$$

There are at least two well-defined generalizations of symmetric difference to an arbitrary number of sets. The first definition above can be generalized very easily by using the n-ary union and intersection and gives the set of all elements that are in *some but not all* of the sets:

```
> `&-` := () -> `union`(args) minus `intersect`(args):
```

```
> {1,2,3} &- {2,3,4};
```

$$\{1, 4\}$$

```
> &-({1,2,3,4}, {2,3,4,5}, {3,4,5,6});
```

$$\{1, 2, 5, 6\}$$

The Maple `symmdiff` function implements a less symmetric generalization that gives the set of all elements that are in *an odd number* of sets:

```
> symmdiff({1,2,3,4}, {2,3,4,5}, {3,4,5,6});
```

$$\{1, 3, 4, 6\}$$

This definition is equivalent to repeatedly taking the pairwise symmetric difference:

```
> {1,2,3,4} &- {2,3,4,5} &- {3,4,5,6};
```

$$\{1, 3, 4, 6\}$$

9.5.5 Membership

Set *membership* can be viewed as a *relation*, akin to a generalization of equality. Maple supports set membership by providing a function called `member` that takes two (or three) arguments, the second of which must be a set or list (or module, which I will not consider here), in which case it returns a Boolean value indicating whether or not its first argument is a member of its second, e.g.,

```
> member(a, {a,b,c});
```

true

```
> member(a, A);
```

```
Error, wrong number (or type) of parameters in function member
```

The built-in `member` function cannot be used as an infix operator, nor is it able to return itself unevaluated. However, the facilities for implementing operators can be used to provide a membership infix operator very easily like this:

```
> `&member` := member;
```

$$\&member := member$$

or slightly better like this:

```
> `&member` := eval(member);
```

$$\&member := \textbf{proc}()\ \textbf{option}\ builtin;\ 193\ \textbf{end proc}$$

```
> a &member {a,b,c};
```

$$true$$

```
> a &member A;
```

```
Error, wrong number (or type) of parameters in function &member
```

Alternatively, a flexible infix member operator could be implemented independently of the built-in `member` function like this (the Maple `type` function will be introduced in the next chapter):

```
> `&member` := proc(elem, SL)
     local el;
     if not type(SL, {set,list}) then 'procname(args)'
     else
        for el in SL do
           if el = elem then return true end if
        end do;
        false
     end if
  end proc:
```

```
> a &member {a,b,c};
```

$$true$$

```
> a &member A;
```

$$a\ \&\text{member}\ A$$

The closely related set relations of subset and superset are not directly supported at all in Maple, but again they are easy to implement as operators (see the exercises).

9.5.6 Set algebra

Set algebra is an example of a general Boolean algebra. The *empty* (or null) set is denoted in Maple simply as `{}` — there is no special notation for it. Maple understands some of the simplest rules of set algebra, e.g.,

```
> A union {};
```

$$A$$

```
> A intersect {};
```

$$\{\}$$

```
> A minus {};
```

$$A$$

But it does not understand less simple rules; e.g., it is clear from the definitions that the following set expression is equivalent simply to A:

```
> (A minus B) union (A intersect B);   # = A
```

$$(A \operatorname{intersect} B) \operatorname{union} (A \operatorname{minus} B)$$

```
> simplify(%);   # = A
```

$$(A \operatorname{intersect} B) \operatorname{union} (A \operatorname{minus} B)$$

Maple can only produce this simplification for explicit sets:

```
> A := {1,2,3}:  B := {2,3,4}:
  (A minus B) union (A intersect B) = A;
```

$$\{1, 2, 3\} = \{1, 2, 3\}$$

```
> unassign('A', 'B');
```

9.5.7 Power sets

The *power set* $P(S)$ of a set S is the *set of all subsets* of S, defined formally as

$$P(S) = \{X \mid X \subseteq S\}.$$

For example, $P(\{1, 2\}) = \{\{\}, \{1\}, \{2\}, \{1, 2\}\}$.

It can be computed recursively as follows:

- The base case is that the only subset of the empty set $\{\}$ is the empty set itself, so
$$P(\{\}) = \{\{\}\}.$$
- Otherwise, S must contain at least one element a and so can be written in the form
$$S = \{a\} \cup T.$$
Each subset of S either does or does not contain a. The set of subsets that do not contain a must be the power set of T, whereas the set of subsets that do contain a must be the set of unions of $\{a\}$ with each element of the power set of T; i.e.,
$$P(S) = P(T) \cup \{\{a\} \cup t \mid t \in P(T)\}$$
where $T = S \backslash \{a\}$, for any $a \in S$.

Implementing this algorithm in Maple leads to a recursive procedure such as this:

```
> PowerSet := proc(S::set)
    # Return the set of all subsets of S.
    local A, P, AP, t;
    if S = {} then {{}} else
       A := {S[1]};  P := PowerSet(S minus A);
       AP := {};
       for t in P do AP := AP union {A union t} end do;
       P union AP
    end if
  end proc:
```

Testing it for some simple examples gives

```
> PowerSet({});
```

$$\{\{\}\}$$

```
> PowerSet({1});
```

$$\{\{\}, \{1\}\}$$

```
> PowerSet({1,2});
```

$$\{\{\}, \{1, 2\}, \{2\}, \{1\}\}$$

```
> PowerSet({1,2,3,4});
```

$$\{\{\}, \{3, 4\}, \{1, 2, 3, 4\}, \{2, 3, 4\}, \{1, 2\}, \{2\}, \{1, 2, 3\}, \{1, 3\},$$
$$\{1, 2, 4\}, \{1, 3, 4\}, \{2, 4\}, \{2, 3\}, \{3\}, \{1, 4\}, \{4\}, \{1\}\}$$

The last result is slightly easier to check if it is sorted by *cardinality*, i.e., number of elements:

```
> sort([op(%)], (a,b)->evalb(nops(a)<nops(b)));
```

$$[\{\}, \{1\}, \{4\}, \{3\}, \{2\}, \{1, 4\}, \{2, 3\}, \{2, 4\}, \{1, 3\}, \{1, 2\},$$
$$\{3, 4\}, \{1, 3, 4\}, \{1, 2, 4\}, \{1, 2, 3\}, \{2, 3, 4\}, \{1, 2, 3, 4\}]$$

The above procedure is straightforward and matches the mathematics, although it could perhaps be implemented slightly more efficiently by using some lower level sequence operations. To see how it works in practice, and that it is efficient (at least at a mathematical level), try executing it with `printlevel := 1000`.

9.6 Examples of recursive integer and polynomial functions

The set of all integers and various sets of polynomials are examples of mathematical structures called *rings*, and several useful and simple computational algorithms apply to both integers and polynomials with little significant difference.

9.6.1 Euclidean division

The set of integers and the set of all univariate (i.e., one variable) polynomials over a field each constitute a *Euclidean ring*, and hence share almost the same division properties. If f, g are elements of the ring then there exist unique elements q, r such that $f = g\,q + r$ and either $r = 0$ or

- for integers, $0 < r < |g|$,
- for polynomials, $\text{degree}(r) < \text{degree}(g)$.

The unique q, r are called the *quotient* and *remainder*, respectively, in the division of f by g, and they are computed by the Maple functions `iquo` and `irem` for integers, and `quo` and `rem` for polynomials, respectively. (*Note that the polynomial versions require the polynomial variable to be specified as the third argument.*)

Here are a couple of random examples that try to ensure non-trivial quotients. First let us use integers:

```
> f := rand();  g := round(rand()/10);
```

$$f := 427419669081$$
$$g := 32111069327$$

The integer quotient q and remainder r are as follows:

```
> q := iquo(f,g);  r := irem(f,g);
```

$$q := 13$$
$$r := 9975767830$$

Now we can check that $f = g\,q + r$:

```
> f - (g*q+r);
```

$$0$$

Now let us repeat the exercise using polynomials:

```
> f := randpoly(x);  g := randpoly(x, degree=3);
```

$$f := -37\,x^5 - 35\,x^4 + 97\,x^3 + 50\,x^2 + 79\,x + 56$$
$$g := 49\,x^3 + 63\,x^2 + 57\,x - 59$$

```
> q := quo(f,g,x);  r := rem(f,g,x);
```

$$q := -\frac{37}{49}\,x^2 + \frac{88}{343}\,x + \frac{6070}{2401}$$
$$r := \frac{492586}{2401} - \frac{57777}{343}\,x^2 - \frac{119967}{2401}\,x$$

Again, we can now check that $f = g\,q + r$:

```
> f - expand(g*q+r);
```

$$0$$

The next two sections provide applications of Euclidean division.

9.6.2 Greatest common divisors

A *greatest common divisor* (GCD) can be computed in a Euclidean ring by an algorithm due to Euclid that uses the remainder in the above division operation. A recursive version of the algorithm for two elements a and b can be stated very simply as follows:

- $\mathrm{GCD}(a, b) = a$ if $b = 0$ — this is the base case.
- $\mathrm{GCD}(a, b) = \mathrm{GCD}(b, \mathrm{remainder}(a, b))$ otherwise.

The remainder must be computed using the *appropriate* remainder function for the domain of computation as illustrated above (i.e., `irem` for integers and `rem` for univariate polynomials). Here is an implementation of the recursive Euclidean GCD algorithm for integers:

```
> IGCD := (a::integer,b::integer) ->
      if b = 0 then a else IGCD(b, irem(a,b)) end if:
```

```
> IGCD(-27, 15);  IGCD(-15, 27);
```

$$3$$
$$-3$$

These results are clearly correct, but perhaps not quite what you would expect. The general definition of GCD does not make it unique: integer GCDs can differ in sign, and univariate polynomial GCDs can differ by a factor of any element of the coefficient field. A unique integer GCD can be defined by requiring it to be *positive*, which can be achieved by returning the absolute value of the GCD. A unique polynomial GCD can be defined by requiring it to be *monic*. A *monic* polynomial has a leading coefficient of one, where the *leading coefficient* is the coefficient of the term with highest degree, and a polynomial over a field can be made monic by dividing it by its leading coefficient.

Here is an implementation of the recursive Euclidean GCD algorithm for integers that returns a *unique* GCD:

```
> IGCD := (a::integer,b::integer) ->
      if b = 0 then abs(a) else IGCD(b, irem(a,b)) end if:
```

```
> IGCD(-27, 15);  IGCD(-15, 27);
```

$$3$$
$$3$$

Implementations for univariate polynomials are very similar and are left as an exercise.

9.6.3 Digits and coefficients

The number 123 in decimal positional representation means:

$$1 \times 10^2 + 2 \times 10^1 + 3 \times 10^0.$$

More generally, any natural number N can be represented with respect to any integer *base* (or *radix*) $B > 1$ as:

$$N = \sum_{i=0}^{m} D_i B^i, \quad 0 \leq D_i < B, \quad D_m > 0.$$

The previous example corresponds to a base of 10. Important bases are 10, 2, 8, and 16, and the corresponding representations are called decimal (or denary), binary, octal, and hexadecimal, respectively. The *digit sequence*

$$(D_m, D_{m-1}, \ldots, D_1, D_0)_B$$

is often written without the commas and parentheses, and the base B is often also omitted when its value is clear from the context, as in the initial example of the number 123. If the base B is symbolic or indeterminate then N is a *polynomial* of degree m in the variable B and the "digits" are called its "coefficients", which once again highlights the close relationship between integers and univariate polynomials.

The digits of a natural number (or the coefficients of a polynomial) can be found by successive Euclidean division operations: the remainder in the division of N by B gives the last digit (coefficient) D_0 and the quotient gives the rest of the number (polynomial) $(D_m, D_{m-1}, \ldots, D_1)$. Hence, recursive application of this algorithm generates the digits (coefficients) in reverse order. Here is an implementation as a recursive Maple procedure to return the digit sequence with respect to the usual base 10 of a natural number. It requires a little care with the base case for the recursion to make the procedure return the digit sequence of 0 as 0 but not return a leading zero in other cases.

```
> N2D := proc(n::nonnegint)
     # Number to digit sequence (decimal)
     local q, r;
     q := iquo(n, 10);  r := irem(n, 10);
     if q = 0 then r else N2D(q), r end if
  end proc:
```

```
> N2D(1579);
```

$$1, 5, 7, 9$$

```
> N2D(0);
```

$$0$$

The inverse operation of constructing a natural number (polynomial) from its digits (coefficients) and the base (variable) is straightforward. However, in the numerical case (which includes evaluating a polynomial for some numerical value of its variable) it is efficient to use a recursive algorithm (Horner's algorithm) based on successive partial factorization, e.g.,

$$a\,x^3 + b\,x^2 + c\,x + d = (a\,x^2 + b\,x + c)\,x + d = ((a\,x + b)\,x + c)\,x + d.$$

Hence, a number can be computed from its digit sequence by adding the last digit to the number corresponding to the previous digits multiplied by the base. Here is an implementation as a recursive Maple procedure to return the natural number specified by a digit sequence with respect to the usual base 10 (without any error checking). The digit sequence is input as the argument sequence of the procedure:

```
> D2N := proc()
      # Decimal sequence to number (decimal)
      if nargs = 0 then 0 else
         D2N(args[1 .. nargs-1])*10 + args[nargs]
      end if
  end proc:
```

```
> D2N(1,5,7,9);
```

$$1579$$

```
> D2N();
```

$$0$$

This algorithm is probably not the best way to *construct* a polynomial because computer algebra systems normally store polynomials in expanded form, whereas this algorithm constructs the partially factorized Horner form. However, it is a good way to *evaluate* a polynomial at a specified numerical value of its variable.

The above two procedures can easily be generalized to work with any base B specified as another argument, essentially just by replacing 10 by B. They can then be used for base conversion. Given a representation of a number as a digit sequence with respect to the old base, compute the corresponding number (with respect to any convenient base internally) and then extract the digit sequence with respect to the new base. The implementation of these extensions is left as an exercise. (There are more direct ways to perform base conversion.)

9.7 Exercises

Operators and mappings

Operators on sets 1: subsets

Implement a binary infix operator `&subset` such that a `&subset` b determines the truth of the (non-strict) set inclusion relation:

$$a \subseteq b,$$

(i.e., evaluates to *true* if a is any subset of b and *false* otherwise). This can be computed as the Boolean value of the expression on the right of the following logical identity (among others):

$$a \subseteq b \iff (a \cup b) = b.$$

You may *assume* that the operands are explicit sets. Test `&subset` *as an infix operator.*

Implement a binary infix operator `&Subset` (with a capital S) such that a `&Subset` b determines the truth of the *proper* (i.e., strict) set inclusion relation:

$$a \subset b,$$

(i.e., evaluates to *true* if a is a *proper* subset of b and *false* otherwise). This can be computed as the Boolean value of the expression on the right of the following logical identity (among others):

$$a \subset b \iff a \subseteq b and a \neq b.$$

You may *assume* that the operands are explicit sets. Test `&Subset` *as an infix operator.*

Operators on sets 2: Cartesian products

Implement a binary infix operator `&X` to return the *Cartesian product* of its two operands *S1* and *S2*, which may be *assumed* to be sets. The Cartesian product of two sets is the set of *all* pairs (which must be represented in Maple as lists) of elements with the first taken from the set *S1* and the second taken from the set *S2*. For example, it should work like this:

```
> {1,2} &X {3,4};
```

$$\{[1, 3], [1, 4], [2, 3], [2, 4]\}$$

Now generalize this to the Cartesian product of an arbitrary number of sets, assuming that sets with list elements are already Cartesian product sets, allowing either infix or prefix syntax. (You may need to read the next chapter and then return to this exercise!) For example, it should work like this:

```
> &X {1,2};
```

$$\{[1], [2]\}$$

```
> {1,2} &X {3,4};
```

$$\{[1, 3], [1, 4], [2, 3], [2, 4]\}$$

```
> {1,2} &X {3,4} &X {5,6};   # infix
```

$$\{[1, 3, 5], [1, 3, 6], [1, 4, 5], [1, 4, 6], [2, 3, 5], [2, 3, 6], [2, 4, 5], [2, 4, 6]\}$$

```
> &X( {1,2}, {3,4}, {5,6} );   # prefix
```

$$\{[1, 3, 5], [1, 3, 6], [1, 4, 5], [1, 4, 6], [2, 3, 5], [2, 3, 6], [2, 4, 5], [2, 4, 6]\}$$

Function composition and derivatives

The identifiers `f`, `g`, and `x` are all assumed to be unbound (i.e., unassigned) in this exercise.

Use the Maple differential operator `D` to compute the differential function (or derivative) of the *function composition* `f@g` (as a *mapping*) and then *apply* the result to `x`. (This illustrates the "chain" or "function of a function" rule for differentiation. A mapping `f` is *applied* to an expression `x` using the syntax `f(x)`.)

Compute the derivative with respect to `x` of the *expression* `f(g(x))` using the Maple `diff` function and then convert it to `D` notation using `convert`. (You should get the same result as you did by using `D` directly above.)

Use Maple to show that `f@@(-1)` is the functional inverse of the mapping `f`, i.e., that `f @ f@@(-1)` is the identity mapping (which maps anything to itself and so does nothing).

Then, compute the differential of the mapping `f@@(-1)` by solving the equation obtained by applying the `D` operator to the equation `f@g = (x->x)` for `D(g)` and *then* substituting `f@@(-1)` for `g` in the result. Check it by computing the differential of the mapping `f@@(-1)` directly. (Note that `(x->x)` is just an explicit representation for the *one-dimensional* identify mapping.)

Recursive procedures

Univariate polynomial GCD computation

A univariate polynomial is one that contains only one indeterminate or variable. By modifying the *first* version of procedure `IGCD` defined above in the text, write a procedure `GCD(a,b,x)` that computes the GCD of two univariate polynomials a and b (over the rationals, i.e., with rational coefficients) in the variable x. The result of this polynomial GCD function will not be normalized in the conventional way. Test procedure `GCD` using simple univariate polynomials with and without obvious common factors. (You may use the Maple library function `gcd` to check your results. Note that it applies to multivariate polynomials and should be called as just `gcd(a,b)`. ***Do not*** give `gcd` the polynomial variable as its third argument; if you do call `gcd(a,b,x)` by mistake then you will need to unassign `x` before you can continue! See the online help for further details.)

Now modify procedure `GCD` so that the result it returns is *monic*, meaning that the leading coefficient (the coefficient of the term of highest degree) is 1. This is the conventional normalization and defines a unique GCD for polynomials over a field. A polynomial is made monic by dividing it by its leading coefficient, and the leading coefficient of a polynomial p with respect to the variable x is returned by the Maple function `lcoeff(p,x)`. (You may find it useful to look this up in the online help.) Making the polynomial GCD monic replaces discarding the sign of the integer GCD. Ensure that procedure `GCD` gives the following result:

```
> GCD(x*(x^2 - 1), -2*(x + 1), x);
```

$$x + 1$$

Finally, revise procedure `GCD` so that it checks that its arguments are univariate polynomials and make the third argument optional. (You may need to read the next chapter and then return to this exercise!) For example, it should work like this:

```
> GCD(x*(x^2 - 1), -2*(x + 1));
```

$$x + 1$$

```
> GCD(a*x*(x^2 - 1), -2*b*(x + 1), x);
```

$$x + 1$$

```
> GCD(a*x*(x^2 - 1), -2*b*(x + 1));

    Error, (in GCD) polynomials must be univariate
```

Integer base conversion

Modify the procedures `N2D` (number to digits) and `D2N` (digits to number) using base 10 (decimal) representation of integers given in the text to use a base (or radix) that is the value of a new *first* argument to each procedure. Thus, `N2D` will take two arguments, the base B followed by the number n to be converted to a base-B digit sequence, and `D2N` will take *at least one* argument, the base B followed by a sequence of zero or more base-B digits to be converted into a number. The base B generalizes the role played by the constant 10 in the procedures defined in the text. Test the procedures, e.g., by specifying the base to be 10. They should now work like this:

```
> 123;  N2D(10, %);  D2N(10, %);
```

$$123$$
$$1, 2, 3$$
$$123$$

Write procedure

```
ConvertBase(DigitList, OldBase, NewBase)
```

that calls procedures `N2D` and `D2N` as appropriate to convert a *list* of digits representing a number in base *OldBase* to the *list* of digits representing the same number in base *NewBase*. Do this by applying `N2D` to the value returned by `D2N`, and converting between sequences and lists as necessary. Test `ConvertBase` by converting a digit list from one base to another and back again. It should work as follows.

Convert 123 from base 10 (decimal) to base 2 (binary) and back again:

```
> [1,2,3]; ConvertBase(%, 10, 2); ConvertBase(%, 2, 10);
```

$$[1, 2, 3]$$
$$[1, 1, 1, 1, 0, 1, 1]$$
$$[1, 2, 3]$$

Convert 789 from base 10 (decimal) to base 7 and back again:

```
> [7,8,9]; ConvertBase(%, 10, 7); ConvertBase(%, 7, 10);
```

$$[7, 8, 9]$$
$$[2, 2, 0, 5]$$
$$[7, 8, 9]$$

Chapter 10

Data Types

Data have not only values but also *types*, which have a hierarchical classification and in principle can be arbitrarily complicated. In compiled languages, such as C, C++, and Java, variables can normally hold data of only a specific type (which might be a union of simpler types), so the type of every variable must be declared before that variable can be used. In interpreted languages, such as LISP, Maple, and many other computer algebra languages, this is not the case and variables can hold data of any type, which can change during the execution of a program. In such languages, the type is associated with the data and not with the variables. Hence, the use of data types is more subtle and can be introduced much later in a programming course. (Data types are the last fundamental programming concept to be introduced in this book.)

The concept of data types is still important in interpreted languages although their role is different from that in compiled languages: in an interpreted language the type of the data assigned to a variable is tested after assignment whereas in a compiled language it is prescribed before assignment. Data types are used for two purposes in interpreted languages: either to ensure that intended operations are well defined or to decide what operations to perform. A value can have type information at many levels; for example, the value of a variable might be a list, whose elements are sets, whose elements are polynomials, whose coefficients are rational numbers, and whose indeterminate is a particular symbol.

We begin in the first section by considering the simplest "top-level" type information, which is called the *surface type*, and then work our way down to the "lower level" type information contained in structured or nested data types. Section 10.2 introduces the main uses for type information in Maple and Section 10.3 explains the Maple syntax for type unions, which are used in the next section to explain the main predefined Maple data types. Section 10.5 presents the syntax for testing data types. The next two sections consider ways to combine type information to construct more subtle data type specifications. Pseudo-types can be used in procedure definitions to circumvent Maple's default argument evaluation rules. Section 10.9 illustrates the use of data type testing to write functions that can be applied to more than one data type and the last section explains how to specify new data type identifiers based on arbitrary criteria.

10.1 Primitive types: `whattype`

A datum (i.e., a value) in Maple carries a *type* with it, which is automatically fixed by the datum itself. (In lower level compiled languages, such as C, C++, and Java, data types are handled when the program is compiled and not when it is executed.) For purposes of interactive investigation, the primary type of a datum can be found by applying the function `whattype`. (But this function should not normally be used within programs.) Here are some examples of the Maple data types returned by `whattype`:

```
> whattype(13);
```

integer

```
> whattype(13.0);
```

float

```
> whattype(x);
```

symbol

```
> whattype(f(x));
```

function

```
> whattype(x->x);
```

procedure

```
> whattype(x+y);
```

+

```
> whattype(3*x);
```

*

```
> whattype(x^2);
```

^

```
> whattype(x,y);
```

exprseq

```
> whattype([x,y]);
```

list

```
> whattype({x,y});
```

set

```
> whattype(f[x]);
```

indexed

There are a few other types that `whattype` can return (see the online help). The `whattype` function returns only the most important "top-level" type information, namely a subset of what are called *surface types*, which I will call *primitive types*. Note, however, that *exprseq* is not a Maple type as defined within the Maple type system and so is not a type that can be tested for in the normal way. Many more complicated types are defined.

Note that `whattype` performs only default evaluation of its argument(s). Hence, because a variable to which a procedure or array (or table or module) has been assigned evaluates by default to itself, `whattype` will return the type *symbol*, and explicit evaluation is required to see the type of the *value* of the variable, e.g.,

```
> f := x -> x;
```

$$f := x \to x$$

```
> whattype(f);
```

symbol

```
> whattype(eval(f));
```

procedure

```
> a := array(1..2);
```

$$a := \mathrm{array}(1..2, [\,])$$

```
> whattype(a);
```

symbol

```
> whattype(eval(a));
```

array

This remark does not, however, apply to the type testing function `type`, to which we will return shortly, e.g.,

```
> type(f, symbol);
```

true

```
> type(f, procedure);
```

true

10.2 Use of type information: error checking, polymorphism

Types are used almost exclusively within procedures for one or both of the following purposes:

- to ensure that operations are not applied to invalid data types (e.g., it is not possible to add the elements of an identifier);
- to choose the interpretation of an operation that is appropriate for the given data types.

The latter is called *polymorphism* (meaning multiform) or *overloading*, and is fundamental to object oriented programming. (See the last chapter for more about operator overloading in Maple.)

10.3 Alternative types: {*type1*, *type2*, ...}

Often, a classification that is broader than the primitive types of individual data is appropriate — for example, *number* is a rather broad classification that includes integer, fraction, and floating-point representations. Hence, it is common to be interested in a *set of types* rather than one specific type. A set of types in Maple always means any element of a set of *alternative* types, and is specified using the normal set syntax, i.e., {*type1*, *type2*, ...} (which can therefore not be used for other purposes within the type system). Important sets of *disjoint* (i.e., non-overlapping) types are predefined; e.g., the type `name` means either type `symbol` or type `indexed` and is defined as (essentially)

```
name := {symbol, indexed}
```

the type `anything` means *any* type, and corresponds to the *set of all defined types.*

Types can themselves be considered to be sets and this is often a useful paradigm. For example, the type *integer* could be considered to represent the set of all integers, since a datum is of type integer if it is a member of the set of all integers. In this context, a set of alternative types is a *union* of types, and indeed this is the terminology use in languages such as C and C++.

10.4 The main Maple data types

This section attempts to classify the main Maple data types in a helpful rather than a formal way — for the full set see the online help. In terms of a set algebra of types, the primitive types are disjoint, but many non-primitive types that are useful in practice are defined to be unions of simpler types; such equivalences are indicated below by an equal sign (=). Hence, there are natural inclusions among the types, indicated below by a set inclusion symbol (⊂). The convention in this classification is that subtypes appear to the left of supertypes, separated either by an equal sign (=) if the subtypes are contained within a type union or by a set inclusion symbol (⊂) otherwise. Some types have alternative names, usually an operator name composed of special symbols that needs escaping and an equivalent descriptive name composed only of letters that does not need escaping; such equivalences are also indicated below by an equal sign (=). Meanings that may not be obvious are shown in parentheses (after a space) or explained at the end of each class. Descriptions that are not part of the Maple syntax are shown in a different font.

10.4.1 Numerical types

$$\{\texttt{integer, fraction}\} = \texttt{rational}$$

$$\{\texttt{rational, float}\} = \texttt{numeric}$$

$$\texttt{numeric, realcons, complex} \subset \texttt{constant}$$

$$\texttt{posint}\ (>0),\ \texttt{nonnegint}\ (\geq 0),\ \texttt{negint}\ (<0),\ \texttt{nonposint}\ (\leq 0) \subset \texttt{integer}$$

$$\texttt{even, odd, prime} \subset \texttt{integer}$$

$$\texttt{positive}\ (>0),\ \texttt{nonnegative}\ (\geq 0),\ \texttt{negative}\ (<0),\ \texttt{nonpositive}\ (\leq 0) \subset \texttt{numeric}$$

10.4.2 Algebraic types

$$\texttt{`=`} = \texttt{equation}$$

$$\{\texttt{symbol, indexed}\} = \texttt{name}$$

$$\{\texttt{numeric, name, `+`, `*`, `.`, `\^{}`, function,...}\} = \texttt{algebraic}$$

$$\texttt{`**`} = \texttt{`\^{}`}$$

$$\texttt{constant} \subset \texttt{algebraic}$$

$$\texttt{polynom} \subset \texttt{ratpoly} \subset \texttt{algebraic}$$

Here, `polynom` means any polynomial; `ratpoly` means any quotient of polynomials.

10.4.3 Boolean types

$$\{\texttt{`=`, `<>`, `<`, `<=`}\} = \texttt{relation}$$

$$\{\texttt{`and`, `or`, `not`}\} = \texttt{logical}$$

$$\{\texttt{true, false, relation, logical}\} = \texttt{boolean}$$

10.4.4 Structural types

$$\texttt{`..`} = \texttt{range}$$

$$\texttt{list}, \texttt{list(}\textit{type}\texttt{)}, \texttt{set}, \texttt{set(}\textit{type}\texttt{)}$$

$$\texttt{vector}, \texttt{vector(}\textit{type}\texttt{)}$$

$$\texttt{matrix}, \texttt{matrix(}\textit{type}\texttt{)}, \texttt{matrix(}\textit{type}\texttt{, square)}$$

$$\texttt{array}, \texttt{array(}\textit{type}\texttt{)}$$

$$\texttt{vector}, \texttt{matrix} \subset \texttt{array} \subset \texttt{table}$$

$$\texttt{Vector}, \texttt{Matrix}, \texttt{Array} \subset \texttt{rtable}$$

`table, rtable` ⊂ `tabular(`*type*`)`

The parentheses above are part of the Maple syntax and "*type*" indicates a Maple type that specifies the type of the elements. An `array` is a special case of Maple's general hash-table data type `table` that is indexed using only integers, whereas an `rtable` is a rectangular-table data type that can be indexed only by integers anyway. The following are special cases of `array` (respectively, `rtable`) data structures: a `vector` (`Vector`) is a one-dimensional (i.e., one index) `array` (`rtable`) that has lower index bound 1; a `matrix` (`Matrix`) is a two-dimensional (i.e., two index) `array` (`rtable`) that has lower index bound 1.

10.4.5 Programming types

`string` ⊂ `constant`

{`symbol, indexed`} = `name`

`procedure, 'module'`

Note that type identifiers that are special symbols or operator names or keywords must be enclosed in name (backward) quotes, as indicated above, and type identifiers that are names of active (constructor) functions need to be enclosed in unevaluation (forward) quotes if they are qualified using functional syntax.

10.5 Type testing: type and ::

The double-colon operator (`::`) is used to bind a data value on its left to a data type on its right. It can be used in several contexts, two of which are explained below. Another use is in type-based pattern matching using the `typematch` function. Pattern matching is a programming paradigm that is not explored in this book, mainly because it is neither a mainstream programming concept nor a crucial feature of Maple. However, pattern matching is of fundamental importance in some other computer algebra languages, in particular REDUCE and Mathematica.

10.5.1 Error-checking of procedure arguments using ::

The simplest use of type testing is to check that the *actual* arguments to a procedure are of an acceptable type and to stop execution and report an error otherwise. This is effected *automatically* by following a *formal* argument in a procedure definition by the double-colon operator and a data type, as we have already seen, e.g.,

```
> f := proc(a::name) end proc:
  f([]);
```

```
Error, f expects its 1st argument, a, to be of type name, but
received []
```

The :: operator is also allowed in procedures defined using "mapping" syntax (->), although then even a single argument must be enclosed in parentheses, e.g.,

```
> f := (a::{set,list})->NULL:
  f(x+y);
```

```
Error, f expects its 1st argument, a, to be of type {set, list},
but received x+y
```

This type-checking facility need not be applied to all formal arguments and it can be applied to optional arguments that are specified as formal arguments; it does not require that an actual argument be provided.

10.5.2 General type testing: type or ::

Completely general type testing is provided by the Boolean-valued function (or predicate) type, which takes two arguments, the first of which is the data value and the second is the data type for which it is to be tested:

type(*data*, *required_type*)

e.g.,

```
> type(F(x), name);
```

false

```
> type(F(x), function);
```

true

```
> type(F[x], name);
```

true

The type function is nearly always used within a conditional statement. For example, the first procedure in the previous sub-section could be implemented a lot less elegantly like this:

```
> f := proc(a)
     if not type(a, name) then
        error "argument a should be of type name." end if
  end proc:
```

```
> f([]);
```

```
Error, (in f) argument a should be of type name.
```

The `::` operator is inert, rather like the relational operators, and just builds a data structure to be interpreted by some other part of Maple. When used to bind procedure arguments to types it is interpreted when the procedure is executed. When used in a Boolean context, such as in an `if` or `while` statement or when the `evalb` function is applied, then

data :: *required_type*

is equivalent to `type(`*data*`,` *required_type*`)`, e.g.,

```
> F(x) :: name;
```

$$F(x)::name$$

```
> evalb(%);
```

false

10.5.3 Quoting type specifications

Although it is not always *necessary*, it is generally good practice to enclose type specifications in uneval (forward) quotes, *except* when specifying the types of procedure arguments because these type specifications are not evaluated. Note the following examples, which use the integer type for illustration but apply to all types:

```
> type(2, 'integer');  # quoted type advisable
```

true

```
> evalb(2::'integer');  # quoted type advisable
```

true

```
> proc(n::'integer') end proc (2);  # DO NOT quote type
```

```
Error, unknown expects its 1st argument, n, to be of type
'integer', but received 2
```

```
> proc(n::integer) end proc (2);  # OK
```

10.6 Boolean combinations of types: And, Or, Not

There are various ways in Maple to test for combinations of types. One way is to test types at various levels below the surface type using structured or nested types, which is the subject of the next section. The other way is to combine types at the same level, which for purposes of introduction we can consider to be the surface type level, and there are two ways to do this, modelled, respectively, on the set theoretic and Boolean algebra view of types. We have already met the former.

The Boolean type combiners `And`, `Or`, and `Not` *must* be used with functional syntax; `And` and `Or` are n-ary, whereas `Not` is unary. They allow combinations of types to be tested efficiently without the need to explicitly define a new type, which may be particularly useful for argument checking. The types combined by `And` and `Or` are checked in the order specified, which is the only significant difference between using `Or` and set notation for alternative types; i.e., `Or(`*`type1, type2,`* `...)` matches the same types as `{`*`type1, type2,`* `...}`. Here are some simple examples that test for types that are not predefined.

The first example tests for a *negative fraction*, which (after simplification) is a quotient of a negative integer and a positive integer greater than 1:

```
> type(-1/2, And(fraction,negative));
```

true

```
> type(-2, And(fraction,negative));
```

false

```
> type(+1/2, And(fraction,negative));
```

false

Here is a test for types that can be used to represent text:

```
> type(`a b c`, Or(symbol,string));
```

true

```
> type("a b c", Or(symbol,string));
```

true

```
> type(123, Or(symbol,string));
```

false

The final example illustrates a test for an integer that is not prime:

```
> type(1, And(integer,Not(prime)));
```

true

```
> type(2, And(integer,Not(prime)));
```

false

```
> type(4, And(integer,Not(prime)));
```

true

```
> type('x', And(integer,Not(prime)));
```

false

Beware that `Not(`*type1*`)` matches a very large class of types, and so will often need to be constrained by appearing in an `And` data-type expression, as in the above example.

10.7 Structured or nested data types

Most data have a tree structure, in which the branches and leaves have their own structures. For example, [1, 2, 3] is a *list* of *integers*, and the expression $x+y$ is a *sum* of *symbols*. The function `whattype` returns only the primitive *surface* type, e.g.,

```
> whattype( [1,2,3] );
```

list

```
> whattype( x+y );
```

$$+$$

However, type testing can test the type of a data structure in more detail by specifying a *structured (or nested) type*, which consists of a *pattern* that can be specified in two different ways.

10.7.1 Generic structured types

One way to specify a structured type is as a structure of one type, *type1*, whose elements are all structures of another type, *type2*, which is specified using a functional notation thus: *type1*(*type2*). I will call this type syntax a *generic structured type*; it is generic because it does not specify the *number* of elements and all elements must have the *same* type. The inner type can be *any* type; hence, the generic structured type syntax can be nested arbitrarily deeply, e.g., *type1*(*type2*(*type3*(...))). The list of integers discussed above would be represented as the type `list(integer)` (which represents a list of *any number* of integers), e.g.,

```
> L := [1,2,3];
```

$$L := [1, 2, 3]$$

```
> whattype(L);
```

list

```
> type(L, list);
```

true

```
> type(L, list(integer));
```

true

```
> type(L, list(name));
```

false

The sum of variables discussed above would, in principle, be represented as the type `‘+‘(symbol)` (which represents a sum of *any number* of symbols). However, it is necessary to avoid evaluating any active functions, such as `‘+‘` and `‘*‘`, in a generic structured type specification, because

```
> ‘+‘(symbol) <> ’‘+‘’(symbol);
```

$$symbol \neq +(symbol)$$

Therefore, such type specifiers can be used *only* within unevaluation quotes unless they are part of a procedure argument type specification, as explained above.

```
> S := x+y;
```

$$S := x + y$$

```
> whattype(S);
```

$$+$$

```
> type(S, `+`);
```

true

```
> type(x+y, '`+`(symbol)');
```

true

Other examples of generic structured types were shown above in Section 10.4; e.g., the type `set(polynom)` represents a set of any number of polynomials.

10.7.2 Specific structured types

The other way to specify a structured type uses a *pattern* that is a normal Maple expression with *values* (optionally) replaced by *types*, e.g., ***type1 &op type2***. I will call this type syntax a *specific structured type*; it is *specific* because it specifies both the *number* of elements and the *type* of *each* element. The inner types can be *any* types, and the specific structured type syntax can be nested arbitrarily deeply, e.g., ***type1 &opa (type2 &opb type3)***. The list of integers discussed above would be represented as the type "`[integer,integer,integer]`" (which matches only a list of *precisely three* integers), e.g.,

```
> L := [1,2,3];
```

$$L := [1, 2, 3]$$

```
> type(L, [integer,integer,integer]);
```

true

```
> type(L, [integer,integer]);
```

false

However, it is necessary to circumvent the automatic simplification that Maple applies to expressions, e.g.,

```
> 'symbol + symbol' <> symbol &+ symbol;
```

$$2\,symbol \neq symbol \,\&{+}\, symbol$$

Hence, the Maple type system uses the inert operators `&+` and `&*` to match `+` and `*` in specific structured types. The sum of variables discussed above could therefore be represented as the type "`symbol &+ symbol`" (which matches only a sum of *precisely two* symbols):

```
> S := x+y;
```

$$S := x + y$$

```
> type(S, symbol &+ symbol);
```

$$true$$

```
> type(S, symbol &+ symbol &+ symbol);
```

$$false$$

Note that the inert operators `&+` and `&*` cannot be used at all in generic structured types.

Generic and specific type patterns can be mixed freely, and in some cases the two are equivalent. For example, an optional procedure argument might be required to have one of the following types (which all mean the same):

```
symbol = numeric..numeric
symbol = range(numeric)
symbol = `..`(numeric)
```

Remember that a set $\{\dots\}$ cannot be used as a specific type pattern to match a set of data because a set represents alternative types within the type system, so the generic type syntax `set(`*`type`*`)` must be used, although `[...]` can be used as a specific type pattern (and is used a lot). This is not a restriction in practice, because it is not possible to match the types of specific elements in a set anyway and it is rarely necessary to match the number of elements, although this could be done using a procedural type definition (see the final section of this chapter).

10.7.3 Type-testing commuting operators

Maple considers addition and multiplication using + and * to be commutative, which means that it will change the order of the operands to suit itself. To be precise, it commutes them into a canonical order for any given expression, which is determined by the first equivalent expression that has been stored, e.g.,

```
> x+3 = 3+x;
```

$$x + 3 = x + 3$$

```
> 3+y=y+3;
```

$$3 + y = 3 + y$$

Hence, only one of the following type tests will succeed, but there is no reliable way to know which:

```
> type(x + 3, symbol &+ integer);
```

true

```
> type(x + 3, integer &+ symbol);
```

false

Assuming you do not care which way round this expression is stored then it is necessary to allow either type. The best way to do that is probably to allow both operands to be of either type, for which a generic structured type is more elegant and succinct than a specific structured type, except that it does not constrain the number of operands:

```
> type(x + 3, '`+`({integer,symbol})');
```

true

```
> type(x + 3, '`+`({symbol,integer})');
```

true

Although the numerical factor in any product can be guaranteed to be the first operand, commutativity problems involving other data types could similarly arise with the `*` operator. This discussion does not apply to the `.` operator because it never commutes non-numeric operands. But beware that `.` cannot be used in a specific structured type and, unlike `*`, is regarded as a function, e.g.,

```
> a . b <> b . a;
```

$$a \, . \, b \neq b \, . \, a$$

```
> type(a.b, '`.`(symbol)');
```

true

```
> whattype(a.b);
```

$$function$$

```
> type(a.b, function);
```

$$true$$

```
> type(a*b, function);
```

$$false$$

Moreover, with at least one numeric operand it produces a normal commutative product:

```
> 3 . x = x . 3;
```

$$3\,x = 3\,x$$

```
> whattype(3 . x);
```

$$*$$

10.7.4 Type-testing sequences

In general, to test the type of a *sequence* (as a single data type) it must be made into a *list* and the type of the resulting list tested. However, if a generic structured type is used then the sequence can be made into a *set*, as illustrated in Chapter 8. These techniques are commonly used to type-test the actual argument sequence of a procedure. If a specific structured type is used for the test then the number of elements in the sequence (or the number of actual procedure arguments) is tested as well as their types (whereas direct procedure argument type declaration using the `::` operator does not constrain the number of actual arguments), e.g.,

```
> Div := proc()
    if not type([args], [algebraic,algebraic]) then
       error "requires 2 algebraic arguments."
    end if;
    `/`(args)
  end proc:
```

```
> Div(a, b);
```

$$\frac{a}{b}$$

```
> Div("a", "b");
```

```
Error, (in Div) requires 2 algebraic arguments.
```

```
> Div(a, b, c);
```

```
Error, (in Div) requires 2 algebraic arguments.
```

If the length of a sequence is not important, as when a procedure can take any number of arguments, then a generic structured type is appropriate (if all elements or arguments should have the same type), e.g.,

```
> AddArgs := proc()
     if not type([args], list(algebraic)) then
        error "arguments must be algebraic."
     end if;
     `+`(args)
  end proc:
```

```
> AddArgs(a, b, c);
```

$$a + b + c$$

```
> AddArgs("a", "b", "c");
```

```
Error, (in AddArgs) arguments must be algebraic.
```

This technique is much more efficient than testing the type of each element (or argument) separately and should always be used if possible.

Recall that the double-colon operator (`::`) for procedure argument type checking does not require that the actual argument be present, only that it have the correct type if it is present, and some explicit test is always necessary to check the number of actual arguments. This can be either a test of `nargs` or a specific structured type test applied to `[args]`.

10.8 Special procedure-argument types

Two special "pseudo-types" exist for use in procedure argument type declarations with the double-colon operator (`::`) that control how actual arguments to procedures are evaluated: `evaln` causes the actual argument to be evaluated to a *name*, but no further, and `uneval` prevents the actual argument from being evaluated at all. This allows the user to define procedures with the special semantics of built-in functions such as `seq`. If `evaln` is qualified by another type in the form `evaln(`*type*`)` then the value that has been assigned to the variable with the resulting name must be of the specified type.

Control over argument evaluation is an important facility for which some mechanism is usually available in other programming languages; for example, in C it could be achieved by accessing an argument by address rather than by value and in LISP by implementing a function as a macro. In other contexts within Maple, evaln is a standard function that evaluates an expression to a name, whereas the type uneval can be used for normal type testing and matches any expression contained in forward (i.e., uneval) quotes.

As an example, consider a procedure intended to increment by 1 the value of the variable supplied as its argument and return the incremented value. This operation is well defined if the argument evaluates to the name of a variable that has a numeric value assigned to it. (This condition could be generalized at the expense of a little more complexity.) Here is a first implementation using a qualified evaln as the argument type declaration. Note the need to use eval to evaluate the argument explicitly within the procedure body since it was not automatically (fully) evaluated:

```
> inc := proc(x::evaln(numeric))
     x := eval(x) + 1
  end proc:
```

```
> x := 'x';
```

$$x := x$$

```
> inc(x);
```

```
Error, inc expects its 1st argument, x, to be of type
evaln(numeric), but received x := x
```

```
> x := 0;
```

$$x := 0$$

```
> inc(x);
```

$$1$$

```
> x;
```

$$1$$

```
> inc(0);
```

```
Error, illegal use of an object as a name
```

Everything in this example is fine, except that the final error message above is not under our control and would not be very helpful to a user. We can take total control by using `uneval` as the argument type declaration, which allows us to check that the actual argument is a name as well as evaluating appropriately:

```
> inc := proc(x::uneval)
     if x::name and eval(x)::numeric then
        x := eval(x) + 1
     else
        error "arg must be a variable with a numeric value"
     end if
  end proc:
```

```
> x := 'x';
```

$$x := x$$

```
> inc(x);
   Error, (in inc) arg must be a variable with a numeric value
```

```
> x := 0;
```

$$x := 0$$

```
> inc(x);
```

$$1$$

```
> x;
```

$$1$$

```
> inc(0);
   Error, (in inc) arg must be a variable with a numeric value
```

10.9 Examples of polymorphic procedures

Polymorphic procedures behave differently with different *types* of argument and provide a mechanism for overloading functions and operators. One very simple example of polymorphism is when a procedure returns itself unevaluated if, for some reason, it is unable to return any more useful value; it may do this when it does not have enough information to do otherwise, e.g., because its arguments are symbolic. This section gives four examples of polymorphism, all but the first of which illustrate natural developments of procedures that were introduced in Chapters 8 and 9. Implementing polymorphism is one of the main uses for a type system.

10.9.1 Active relational operators

As we have seen, the standard relational operators are inert, e.g.,

```
> 1 < 2;
```

$$1 < 2$$

However, an active version could be defined as follows. If possible, it would evaluate to a Boolean value; otherwise, it would remain symbolic (i.e., return itself unevaluated). The type `constant` includes explicit numerical values, which are of type `numeric`, and symbolic numerical values such as `Pi` and unevaluated functions of constants for which `evalf` should return an explicit numerical value. Hence, if the first type test below fails but the second succeeds then both arguments are constants but at least one of them is symbolic and so needs floating-point evaluation before the arguments can be compared. (Beware that this is not reliable!) The first version uses *generic* structured types:

```
> `&<` := proc(a,b)  # An active "<" relation
     if nargs <> 2 then
        error "requires precisely 2 operands."
     elif type([args], list(numeric)) then
        evalb(a < b)
     elif type([args], list(constant)) then
        # Constants that are not explicitly numeric:
        evalb(evalf(a < b))
     else
        'procname(args)'
     end if
  end proc:
```

The second version uses *specific* structured types:

```
> `&<` := proc(a,b)  # An active "<" relation
     if type([args], [numeric,numeric]) then
        evalb(a < b)
     elif type([args], [constant,constant]) then
        # Constants that are not explicitly numeric:
        evalb(evalf(a < b))
     elif nargs = 2 then
        'procname(args)'
     else
        error "requires precisely 2 operands."
     end if
  end proc:
```

Both implementations work identically:

```
> 1 &< 2;
```

$$true$$

```
> a < b;
```

$$a < b$$

```
> a &< b;
```

$$a \mathbin{\&} < b$$

```
> &< a;
Error, (in &<) requires precisely 2 operands.
```

```
> 3 < Pi;
```

$$3 < \pi$$

```
> 3 &< Pi;
```

$$true$$

```
> Pi &< 3;
```

$$false$$

10.9.2 A more general maximum function

It may well be convenient to have a maximization function that will maximize a *list* or *set* as well as a sequence of values. An elegant way to provide this functionality is to allow any argument to be a list or set and to maximize all the values that are supplied either as arguments or as elements of lists or sets. While it is unlikely that this full generality would be used in practice, it is often easier to implement more generality than may be required and there is certainly no loss in doing so. This version of the maximum function (a simpler version of which was discussed earlier) also allows values to be *constants* that are not explicitly numerical or to be purely *symbolic.* It maximizes symbolic constants but retains their symbolic forms and it retains other purely symbolic values without any further processing.

The easiest way for a procedure to process elements of arguments that are lists or sets in the same way as other arguments is to *flatten* its argument sequence (one level deep) by splicing the elements of any list or set into the argument sequence, e.g., to flatten the sequence

$$a,\ b,\ \{c,\ d\},\ e,\ [f,\ g]$$

into simply

$$a,\ b,\ c,\ d,\ e,\ f,\ g.$$

Here is an example sequence in Maple to flatten:

```
> A := a, b, {c, d}, e, [f, g];
```

$$A := a,\ b,\ \{c,\ d\},\ e,\ [f,\ g]$$

The flattening (one level deep) can be done either by using a type test within the first argument of the `seq` function, for which the functional form of `if` is convenient:

```
> seq(`if`(type(el,{list,set}), op(el), el), el = A);
```

$$a,\ b,\ c,\ d,\ e,\ f,\ g$$

or by mapping a type test function over the sequence using the `map` function. To do this, it is necessary beforehand to make the sequence into a *list* and then afterwards to extract the sequence by applying `op`:

```
> op(map(el -> if type(el, {list,set}) then
      op(el) else el end if, [A]));
```

$$a,\ b,\ c,\ d,\ e,\ f,\ g$$

For the purpose of maximizing, it is best to make the sequence into a set and then flatten that. This automatically discards any repeated elements. Note that flattening can be performed to any depth by applying the technique described above recursively, usually by writing a recursive procedure; see the exercises. (Flattening is a standard operation in the context of list processing languages such as LISP and Prolog.)

Symbolic constants are handled much as described in the previous section. Because comparing floating-point approximations is unreliable it is done only when necessary, and the explicit numerical value of the current maximum is stored together with its exact value, which is what is returned as the explicit maximum. (The two will be the same except for symbolic constants.) Initialization of the explicit maximum `m` is no longer straightforward because we do not know in advance which is the first constant element in the input data. Hence, `m` is initialized to `NULL` as a *sentinel* value that cannot occur in the input data, as introduced in Chapter 7. The test whether `m` is `NULL` then allows the maximization loop to determine how to proceed without duplicating a lot of code. (But this "dynamic initialization" technique should not be used in straightforward cases.) Purely symbolic arguments are simply collected separately from the maximum. Here is the procedure definition:

```
> MAX := proc()  # args can be algebraic, lists or sets
     local m, ARGS, SYM, el, NumMax;
     m := NULL;  # default maximum value
     # Flatten arguments to depth 1 (as a set):
     ARGS := map(el -> if type(el, {list,set}) then
        op(el) else el end if, {args});
     # SYM accumulates non-maximizable arguments
     SYM := NULL;
     # NumMax stores EXPLICIT NUMERIC value of maximum
     for el in ARGS do
        if type(el, numeric) then
           # explicitly numerical
           if m = NULL or el > NumMax then
              m := el; NumMax := el
           end if
        elif type(el, constant) then
           # symbolic constant
           if m = NULL or evalf(el) > NumMax then
              m := el; NumMax := evalf(el)
           end if
        else
           # symbolic non-constant
           SYM := SYM, el
        end if
     end do;
     SYM := SYM, m;
     if nops([SYM]) <= 1 then SYM else 'procname'(SYM) end if
  end proc:
```

When the maximization loop terminates, the explicit maximum `m` is appended to the sequence of symbolic arguments. If the resulting sequence contains no more than one element, then it is returned alone; otherwise, the sequence is returned as the argument sequence of a new invocation of original function — i.e., the function returns a copy of itself with a simplified argument sequence. Here are some simple tests:

```
> MAX();
```

```
> MAX(5);
```

$$5$$

```
> MAX(1, [2,3], {4,5}, 6);
```

$$6$$

```
> MAX(Pi);
```

$$\pi$$

```
> MAX(5, Pi);
```

$$5$$

```
> MAX(5, 2*Pi);
```

$$2\pi$$

```
> MAX(foo, 1, [2,3], {4,5}, Pi);
```

$$\mathrm{MAX}(foo, 5)$$

```
> MAX(foo, 1, [2,3], {4,5}, 2*Pi);
```

$$\mathrm{MAX}(foo, 2\pi)$$

10.9.3 A more general (rational) factorial function

The factorial of any half-integer can be computed by using the usual recursive relation together with the base case:

```
> (1/2)!:  % = simplify(%);
```

$$(\frac{1}{2})! = \frac{1}{2}\sqrt{\pi}$$

This comes from the generalization of the factorial function on the natural numbers to the Gamma function on the real numbers (excluding the negative integers, on which it is undefined). Let us use it to implement a generalization of our previous factorial procedure.

For efficiency, it is best to separate the procedure into two parts: one that just checks that the initial arguments are valid and the other that recursively computes the factorial. This avoids repeatedly making essentially the same type tests. The constant 2 is used as a specific type pattern here, although the test `type(denom(n),2)` could just as well be implemented as `denom(n)=2`. The error message is constructed using the error formatting code `%1`, which is replaced by the value (using `lprint` representation) of the first expression after the message string. Here is the non-recursive top-level procedure:

```
> Factorial := proc(n)  # Argument type testing
     if type(n, nonnegint) or
        ( type(n, fraction) and type(denom(n), 2) )
     then
        Factorial1(n)
     else
        error "%1 invalid as argument.", n
     end if
  end proc:
```

We can immediately test the handling of invalid arguments:

```
> Factorial(-3);
```

```
Error, (in Factorial) -3 invalid as argument.
```

```
> Factorial(2/3);
```

```
Error, (in Factorial) 2/3 invalid as argument.
```

```
> Factorial(0.3);
```

```
Error, (in Factorial) .3 invalid as argument.
```

The following recursive sub-procedure uses the observation that

$$(n+1)! = (n+1)n! \quad \Longrightarrow \quad n! = (n+1)!/(n+1)$$

to recursively increase negative half-integer arguments towards $1/2$, whereas the normal recursion is used to decrease any positive integer or half-integer argument towards either 0 or $1/2$, respectively. It uses `option remember` to improve its efficiency:

```
> Factorial1 := proc(n) # Recursive support procedure
     option remember;
     if n >= 1 then      # downward recursion
        n*procname(n-1)
     elif n = 1/2 then   # half-integer base
        sqrt(Pi)/2
     elif n = 0 then     # integer base
        1
     else                # upward recursion
        procname(n+1)/(n+1)
     end if
  end proc:
```

Here are some simple tests:

```
> Factorial(0),  Factorial(1),  Factorial(3);
```

$$1, 1, 6$$

```
> Factorial(1/2),  Factorial(3/2),  Factorial(-3/2);
```

$$\frac{1}{2}\sqrt{\pi},\ \frac{3}{4}\sqrt{\pi},\ -2\sqrt{\pi}$$

From a pedagogical point of view it may be clearer to have two separate procedures as above. However, from an implementation point of view it would be better to make procedure `Factorial1` local to procedure `Factorial`, since `Factorial1` is intended only ever to be called by `Factorial`. Making `Factorial1` local to `Factorial` would enforce this by making `Factorial1` inaccessible outside `Factorial` and justify `Factorial1` having no argument type checking. I will follow this implementation model in the final example below.

10.9.4 A more general digit-sequence function

The digits of a floating-point number can be computed from its integer mantissa, using the integer exponent to put the decimal point in the correct place if necessary. The following generalization of our previous procedure to convert a non-negative integer to its decimal digit sequence accepts either an integer or a float and also allows it to be negative. This implementation has the same structure as was used in the previous section: the type test in the non-recursive interface procedure (`N2D`) is used not only to exclude invalid arguments but also to initialize correctly the recursive sub-procedure (`N2D1`) that actually computes the digit sequence. However, the sub-procedure is now local to the main procedure and so is inaccessible externally:

```
> N2D := proc(n)
    # Number to decimal digit sequence,
    # including decimal point if necessary
    local N2D1;
    N2D1 := proc(n, e)  # mantissa, exponent
       local q, r;
       q := iquo(n, 10, 'r');  # r = irem(n, 10)
       if e = -1 then r := `.`, r end if;
       if q = 0 then  # Check for leading zeros
          if e < -1 then `.`, 0$(-e-1), r else r end if
       else procname(q, e+1), r end if
    end proc:

    if type(n, integer) then
       N2D1(abs(n), 0)   # mantissa, exponent
```

```
    elif type(n, float) then
       N2D1(op(abs(n)))  # mantissa, exponent
    else
       error "argument %1 invalid as integer or float", n
    end if;
    if n < 0 then `-`, % else % end if
 end proc:
```

```
> N2D(2/3);
```

```
Error, (in N2D) argument 2/3 invalid as integer or float
```

```
> N2D(1234500);
```

$$1, 2, 3, 4, 5, 0, 0$$

```
> N2D(-1234500.);
```

$$-, 1, 2, 3, 4, 5, 0, 0$$

```
> N2D(123.45);
```

$$1, 2, 3, ., 4, 5$$

```
> N2D(0.0012345);
```

$$., 0, 0, 1, 2, 3, 4, 5$$

10.10 Defining new types

A new type called `TYPE` can be defined by assigning to the global variable named `` `type/TYPE` ``, which must *always* be enclosed in backward name quotes because of the special symbol `/`. When the type function is called with second argument `TYPE`, it constructs the (global) variable name `` `type/TYPE` `` and then tries to access its value. If it has no value then an error occurs, e.g.,

```
> type(data,TYPE);
```

```
Error, type `TYPE` does not exist
```

```
> `type/TYPE` := 'anything';
```

$$type/TYPE := anything$$

```
> type(data,TYPE);
```

true

The value assigned to a type identifier must evaluate, when the type is used, to either a defined type expression (as illustrated above) or a Boolean-valued procedure. A types expression is a combination of defined types, whereas a procedural definition can execute arbitrary code in order to test the type. For example, we might define a new type `listset` to match either a list or a set, as follows:

```
> `type/listset` := '{list,set}';
```

$$type/listset := \{set, list\}$$

```
> type(data, 'listset');
```

false

```
> type([data], 'listset');
```

true

```
> type({data}, 'listset');
```

true

To test for a half-integer type is more difficult, because it requires two different tests as illustrated in the previous section, one on the value itself and *then* one on its denominator, and there is no way to combine these into a single type expression. The flexibility of a procedural type definition is required:

```
> `type/halfinteger` :=
     n -> type(n,'fraction') and denom(n)=2:
```

```
> type(1, 'halfinteger');
```

false

```
> type(1/2, 'halfinteger');
```

true

If functional syntax is used to qualify a type then the "arguments" to the type are passed as the second and subsequent arguments when a procedural type definition is called. For example, here is a more general fractional-integer type definition:

```
> `type/fracinteger` :=
    (n,d) -> type(n,'fraction') and denom(n)=d:
```

```
> type(1, 'fracinteger(2)');
```

false

```
> type(1/2, 'fracinteger(2)');
```

true

```
> type(2/3, 'fracinteger(3)');
```

true

10.11 Exercises

Recursively flattening a data structure

Write and test a procedure called `flatten` that converts its arguments, which may be *arbitrarily* deeply nested lists and sets, into a sequence that contains no lists or sets at all, by testing the type of each argument and if it is a list or set then recursively inserting *its elements* into the rest of the argument sequence. One way to do this is recursively to flatten the first argument and append the flattened sequence of remaining arguments. The base case for this recursion is when there are no arguments. It should work like this:

```
> flatten([a, b], [[c,d],[[e,f],g]], h);
```

$$a, b, c, d, e, f, g, h$$

Take care with the various special cases, such as these:

```
> flatten();
```

```
> flatten(a);
```

$$a$$

```
> flatten(a, b);
```

$$a, b$$

Overloading an operator — division

Write and test a binary infix operator called `&/` that performs Euclidean division to return an expression sequence consisting of the quotient and remainder of its two operands. (See Chapter 9 for further details.) A set in which Euclidean division is defined is called a Euclidean domain, of which the principal example is the set of integers. Any field is also trivially a Euclidean domain, and important fields are the set of rational numbers and the set of real numbers (represented computationally as floating-point numbers). The operator `&/` should work correctly with operands of *any* of these types — all elements of a field are divisible by all other elements (except 0) and the remainder is always 0.

Operand types should be "promoted" as necessary. Thus, if both operands are integers, then perform Euclidean division in the ring of integers; otherwise, if both operands are numeric, then perform division in the field of real numbers. The operands are invalid otherwise, which is an error. The operator should work as follows:

```
> 7 &/ 3;  # ring
```

$$2, 1$$

```
> 7. &/ 3;  # field
```

$$2.333333333, 0$$

```
> (1/7) &/ 3;  # field
```

$$\frac{1}{21}, 0$$

```
> [0] &/ 3;  # error
   Error, (in &/) invalid arguments
```

```
> &/(a,b,c);
   Error, (in &/) invalid arguments
```

Overloading an operator — appending

Write and test an infix operator `&!` to either append two lists or return the union of two sets. If one operand is a list and one a set then the list should be treated as a set. Either operand should be allowed to be a variable (i.e., a name), in which case the operator should return either itself or a symbolic union if one operand is an explicit set. The operator should detect invalid operands, and it must not change the order of the elements in any lists. It should work like this:

```
> [1,2] &! [3,4];
```

$$[1, 2, 3, 4]$$

```
> [ ] &! [ ];
```

$$[]$$

```
> [1,2] &! {3,4};
```

$$\{1, 2, 3, 4\}$$

```
> [ ] &! { };
```

$$\{\}$$

```
> a &! b;
```

$$a \mathbin{\&!} b$$

```
> [1,2] &! a;
```

$$[1, 2] \mathbin{\&!} a$$

```
> [ ] &! a;
```

$$a$$

```
> a &! {1,2};
```

$$a \operatorname{union} \{1, 2\}$$

```
> a &! { };
```

$$a$$

```
> [1,2] &! (x+y);
   Error, (in &!) invalid arguments
```

Generalized division

The set of univariate polynomials (polynomials in one variable or indeterminate) over a field is a Euclidean domain, and the set of rational expressions (i.e., quotients of polynomials) in *any number of variables* is a field. Generalize the division operator `&/` defined in the second exercise above to work correctly with mixed operands of any of its previous or these new types. Choose the leftmost set that contains both operands in the following sequence of inclusions:

$$\mathbb{Z} \text{ (integers)} \subset \mathbb{R} \text{ (reals)} \subset \mathbb{K}[x] \text{ (univariate polynomials)}$$
$$\subset \mathbb{K}(x, y, \ldots) \text{ (rational expressions)},$$

where $\mathbb{K}$ denotes any field and x, y denote *any* variables. You may use the Maple function `indets`, which returns a *set* of the indeterminates in any expression, to determine whether a polynomial is univariate and if so what its indeterminate (variable) is (as illustrated in Chapter 9). The coefficients of polynomials and rational expressions are *always* to be regarded as field elements. If both operands are not (subtypes of) one of these types then report an error. The operator should work as follows:

```
> (2*x^3 + 3) &/ (x^2 + 1);  # ring (univariate)
```

$$2\,x,\ 3 - 2\,x$$

```
> (2*x^3 + 3/x) &/ (x^2 + 1);  # field (rational)
```

$$\frac{2\,x^4 + 3}{x\,(x^2 + 1)},\ 0$$

```
> (a*x^3 + b) &/ (x^2 + 1);  # field (multivariate)
```

$$\frac{a\,x^3 + b}{x^2 + 1},\ 0$$

Generalized appending of lists and sets

Generalize the definition of the operator `&!` to allow it also to be used as a function taking an arbitrary number of arguments, as illustrated by the following examples:

```
> &!([1,2], a, [3,4]);
```

$$\&!([1,\ 2],\ a,\ [3,\ 4])$$

```
> &!([1,2], [3,4], a);
```

$$[1,\ 2,\ 3,\ 4]\ \&!\ a$$

```
> &!([1,2], a, {3,4});
```

$$a\ \text{union}\ \{1,\ 2,\ 3,\ 4\}$$

```
> &!([1,2], a, {3,4}, b);
```

$$\text{union}(a,\ b,\ \{1,\ 2,\ 3,\ 4\})$$

Chapter 11

Conventional Programming

This chapter contains further simple examples of programming in Maple, in particular some programming tasks that are more important with "conventional" languages, and illuminates further a few more standard Maple functions. However, the first section is "conventional" only in the context of symbolic programming languages such as LISP; generally, this chapter is about programming tasks that are not specifically algebraic or numerical.

The first section is intended to reinforce the concept of mapping a function over a structure, using `map` and its relatives, by illustrating how such general-purpose mapping functions might be written. Then, as an illustration of the use of arrays, the second section shows how to implement some simple operations in matrix algebra (without the aide of any of the Maple packages). Often, computers are used to process large amounts of data, which is done "offline" rather than interactively, and Section 11.3 provides an introduction to the use of Maple for such "data processing" tasks. Maple is normally used interactively, but not in the same way that a program written in a compiled language might be run interactively, when the program itself would initiate a dialogue with the user. While this mode of interaction is not normally appropriate when using Maple, simple command-line interaction can easily be implemented much as it would in a language such as C, as illustrated in Section 11.4.

The last two sections discuss two further aspects of errors in computer programs. Section 11.5 introduces the Maple mechanism for "catching" anticipated errors within a program and deciding how to handle them rather than always allowing them to stop execution of the program. Section 11.6 introduces the Maple facilities for dealing with unanticipated errors caused by mistakes at some point in the design or implementation of a program.

11.1 Operations on structures

Maple provides a number of high-level data structures: sequences, lists, sets, sums, products, relations, tables, arrays, etc., and it is frequently convenient to be able to perform operations on these structures *without* needing to write explicit loops to access each element of the structure. The most common operations are to apply the same function to every element, or to select or remove some elements. These operations are supported, respectively, by the Maple functions `map` and `map2`, and `select`, `remove`, and `selectremove`. The ideas go back to much earlier languages designed for symbolic programming, such as LISP.

None of these functions directly changes the data structures to which they are applied; they return a *copy* with the appropriate changes made.

11.1.1 Mapping over structures: map

The function `map` applies a mapping to each element of a data structure and returns the result. The syntax is

map(*mapping, structure, additional_arguments*)

where *additional_arguments* represents an optional sequence of additional arguments for *mapping*. Here is a simple example:

```
> map(f, [x,y,z]);
```

$$[f(x), f(y), f(z)]$$

If the mapping takes more than one argument then the second and subsequent arguments must be supplied as the optional additional arguments to `map`; each successive invocation of the mapping takes its first argument as the next element of the data structure but its subsequent arguments are always the *same* set of optional additional arguments, e.g.,

```
> map(f, [x,y,z], a, b);
```

$$[f(x, a, b), f(y, a, b), f(z, a, b)]$$

This is necessary, for example, to differentiate a list of expressions (each element with respect to the same variables):

```
> map(diff, [f(x,y), g(x,y), h(x,y)], x, y);
```

$$[\frac{\partial^2}{\partial y\,\partial x} f(x, y), \frac{\partial^2}{\partial y\,\partial x} g(x, y), \frac{\partial^2}{\partial y\,\partial x} h(x, y)]$$

It is often required to perform the same operation on both sides of an equation (or other relation). Some functions automatically map themselves over data structures, e.g.,

```
> eq := (x*(x+1) = y*(y+1));   expand(eq);
```

$$eq := x\,(x+1) = y\,(y+1)$$
$$x^2 + x = y^2 + y$$

Other functions do not, in which case map must be applied explicitly, e.g.,

```
> eq := (a = b);   sin(eq);
```

$$eq := a = b$$

```
Error, sin expects its 1st argument, x, to be of type algebraic,
but received a = b
```

```
> map(sin, eq);
```

$$\sin(a) = \sin(b)$$

Some operations can be applied to relations in (almost) the obvious way, e.g.,

```
> eq + (2=2);
```

$$a + 2 = b + 2$$

But, for example, the only way to square both sides of a relation is to map a squaring function over it:

```
> eq^2;
```

```
Error, (in simpl/reloprod) invalid terms in product
```

```
> map(x->x^2, eq);
```

$$a^2 = b^2$$

This is a good illustration of the use of an anonymous mapping, which is appropriate when the operation to be performed is simple but not represented by a built-in Maple function.

11.1.2 Mapping on other arguments: map2

Sometimes it is not the first argument of a mapping that should take each element of a data structure. If it is the second, then the function map2 can be used instead of map. The function map2 takes its arguments in the natural order for the operation being performed and so maps over its *third* argument. For example, an expression can be integrated with respect to each variable in a list as follows:

```
> map2(int, f(x,y), [x,y]);
```

$$[\int f(x,\, y)\, dx,\ \int f(x,\, y)\, dy]$$

This can also be accomplished using `map` together with an anonymous mapping like this:

```
> map((b,a)->int(a,b), [x,y], f(x,y));
```

$$[\int f(x,\, y)\, dx,\ \int f(x,\, y)\, dy]$$

Note the reversal of the arguments in the anonymous mapping and in `map` itself. The version using `map` is a little less clear than that using `map2`, but it has the advantage that it generalizes easily to mapping with respect to *any* argument of a function.

11.1.3 How is `map` defined?

Let us take a look:

```
> print(map): print(map2):
```

proc() **option** *builtin*; 189 **end proc**

proc() **option** *builtin*; 190 **end proc**

Since these procedures are both built in we cannot read their source code.

The best way to understand how `map` works is to write our own version. Here is a first attempt:

```
> MAP := proc(f::{name,procedure}, struct)
      local extra_args, res_type;
      extra_args := args[3..nargs];
      res_type := whattype(struct);
      seq(f(el, extra_args), el = struct);
      if type(struct,function) then op(0,struct)(%)
      else convert([%], res_type) end if
  end proc:
```

This procedure collects any optional argument sequence and assigns it to the variable `extra_args`, and determines the primitive type of the data structure `struct` and assigns it to the variable `res_type`. (This is one of the few situations where it is appropriate to use `whattype` within a program.) It then uses `seq` to build a sequence of the mapping `f` applied to each element of the data structure `struct`, passing any optional arguments as second and subsequent arguments to `f`. Finally, this sequence is converted back to the type of the original data structure and returned as the value of `MAP`. The conversion is tricky and needs to be performed differently for different original structure types: if the original structure was a function then it needs to be *applied* to the new sequence as its arguments; otherwise, `convert` can be applied to the new sequence as a list.

It works like this:

```
> MAP(f, [x,y,z]);
```

$$[\mathrm{f}(x),\, \mathrm{f}(y),\, \mathrm{f}(z)]$$

```
> MAP(diff, [f(x,y), g(x,y), h(x,y)], x, y);
```

$$[\frac{\partial^2}{\partial y\,\partial x}\,\mathrm{f}(x,\, y),\, \frac{\partial^2}{\partial y\,\partial x}\,\mathrm{g}(x,\, y),\, \frac{\partial^2}{\partial y\,\partial x}\,\mathrm{h}(x,\, y)]$$

```
> eq := (a = b);  MAP(sin, eq);
```

$$eq := a = b$$

```
Error, (in MAP) unable to convert
```

The neatest way to avoid this error is to extend `convert` so that it can convert a structure containing two operands into an equation, by defining a procedure with the following special name:[1]

```
> `convert/=` := proc(A)
     if nops(A) = 2 then A[1] = A[2]
     else error "two operands required." end if
  end proc:
```

```
> MAP(sin, eq);
```

$$\sin(a) = \sin(b)$$

```
> MAP(x -> x^2, eq);
```

$$a^2 = b^2$$

Similar conversion functions would be required to handle also `<>`, `<`, and `<=`. (Facilities like this for extending built-in Maple functions are supported fairly widely; see the online help for the function you want to extend for details.)

---

[1]By default, `convert` will convert to `=` only from `<`, `<=` or `<>` and it accepts only `equality` rather than `` `=` `` as the target type.

11.1.4 Mapping over arrays and tables

Mappings can also be mapped over arrays and tables (and similarly over rtables, although I will not illustrate that here):

```
> A := array([[a,b],[c,d]]);  # 2D array (matrix)
```

$$A := \begin{bmatrix} a & b \\ c & d \end{bmatrix}$$

```
> map(f, A);
```

$$\begin{bmatrix} \mathrm{f}(a) & \mathrm{f}(b) \\ \mathrm{f}(c) & \mathrm{f}(d) \end{bmatrix}$$

```
> V := array([a,b]);  # 1D array (vector)
```

$$V := [a, b]$$

```
> map(f, V);
```

$$[\mathrm{f}(a), \mathrm{f}(b)]$$

```
> T := table([x = y, u = v]);  # general table
```

$$T := \mathrm{table}([u = v, x = y])$$

```
> map(f, T);
```

$$\mathrm{table}([u = \mathrm{f}(v), x = \mathrm{f}(y)])$$

It requires some more code in `MAP` to handle arrays and tables. Their operand structures are these:

array: *index_function, index_range_seq, data_list*

table: *index_function, data_list*

where *data_list* is a list of equations with indices on the left and values on the right and the *index_function* operand is `NULL` by default. This structure must be preserved, and one way to do so is as follows. An array is a special case of a table, and we must assume that it was the *name* of an array or table that was passed to `MAP`, so `eval` must be explicitly applied in order to obtain the array or table data structure. We can use `subsop` within an `seq` to apply the mapping `f` to every data value in the *data_list* operand (which is always the last) and then call the `array` or `table` constructor function to construct a new data structure of the correct type:

```
> MAP := proc(f::{name,procedure}, struct)
    local xargs, data_list;
    xargs := args[3..nargs];
    if type(struct, table) then
       data_list :=
          [seq(subsop(2=f(op(2,el),xargs), el),
             el = op(-1,eval(struct)))];
       if type(struct, array) then
          array(op(1..2,eval(struct)), data_list)
       else
          table(op(1,eval(struct)), data_list)
       end if
    else
       seq(f(el,xargs), el = struct);
       if type(struct,function) then op(0,struct)(%)
       else convert([%], whattype(struct)) end if
    end if
  end proc:
```

```
> MAP(f, A);
```

$$\begin{bmatrix} \mathrm{f}(a) & \mathrm{f}(b) \\ \mathrm{f}(c) & \mathrm{f}(d) \end{bmatrix}$$

```
> MAP(f, V);
```

$$[\mathrm{f}(a), \mathrm{f}(b)]$$

```
> MAP(f, T);
```

$$\mathrm{table}([u = \mathrm{f}(v), x = \mathrm{f}(y)])$$

This implementation illustrates the essence of `map` although it may not correctly handle all the data structures handled by the builtin `map` function.

11.1.5 Selecting from structures: `select`, `remove`

The function `select` applies a *Boolean-valued* mapping to each element of a data structure and returns a copy of the structure containing *only* those elements for which the mapping evaluates to true. The syntax is

```
select(mapping, structure, additional_arguments)
```

where *additional_arguments* represents an optional sequence of additional arguments for *mapping*.

The function `select` is very similar to `map`, except that the mapping that it applies *must* be explicitly Boolean-valued (and so cannot be inert) and it returns the same structure but with (generally) fewer elements. Hence, it can be applied only to a restricted set of data structures (which excludes relations and tables), namely lists, sets, sums, products, and functions. Here is a simple example:

```
> select(type, [1,2,5,8,11], even);
```

$$[2, 8]$$

```
> select(type, 23*a*x^3, constant);
```

$$23$$

The function `remove` is the inverse of `select`, e.g.,

```
> remove(type, [1,2,5,8,11], even);
```

$$[1, 5, 11]$$

```
> remove(type, 23*a*x^3, constant);
```

$$a\,x^3$$

and the function `selectremove` performs both operations together (which is more efficient if you want a data structure split into two components), e.g.,

```
> selectremove(type, [1,2,5,8,11], even);
```

$$[2, 8], [1, 5, 11]$$

```
> selectremove(type, 23*a*x^3, constant);
```

$$23, a\,x^3$$

Let us take a look at how `select` and `remove` are defined:

```
> print(select):  print(remove):  print(selectremove):
```

proc() **option** *builtin*; 229 **end proc**

proc() **option** *builtin*; 216 **end proc**

proc() **option** *builtin*; 230 **end proc**

Not very informative!

Once again, let us implement our own version of `select`, to make sure we understand it:

```
> SELECT := proc(f::procedure,
        expr::{list,set,`+`,`*`,function})
     local xargs, result, el;
     xargs := args[3..nargs];
     result := NULL;
     for el in expr do
        if f(el, xargs) then result := result, el end if
     end do;
     if type(expr,function) then op(0,expr)(result)
     else convert([result], whattype(expr)) end if
  end proc:
```

```
> SELECT(type, [1,2,5,8,11], even);
```

$$[2, 8]$$

```
> SELECT(type, 23*a*x^3, constant);
```

$$23$$

```
> SELECT(type, F(1,2,5,8,11), even);
```

$$F(2, 8)$$

The code is very similar to that for `MAP` although I have streamlined it slightly. Instead of constructing the result sequence in a loop as done here, I could have used `seq` again; its first argument would need to be a conditional expression that would evaluate to either the required element or `NULL`. Analogues of the other two functions, `remove` and `selectremove`, could be implemented similarly or defined in terms of `SELECT`.

11.2 Implementing vector and matrix algebra

Implementing vector and matrix algebra provides good examples of "conventional" programming with arrays and loops, and provides some insight into the operation of the standard Maple linear algebra packages, `linalg` and `LinearAlgebra`. I will use table-based arrays (as used by the older `linalg` package) in this section and leave it as an exercise for the reader to re-implement the facilities described here using rtable-based arrays (as used by the newer `LinearAlgebra` package). However, I will not use functions from any package in this section, preferring to implement everything using more basic facilities to illustrate how to do it.

11.2.1 Vector addition

A simple task to begin with is to write a procedure to add two vectors with specified dimensions:

```
> VecAdd := proc(a::vector, b::vector, dim::posint)
     # Add two vectors a, b of dimension dim.
     local c, i;
     c := array(1..dim);
     for i to dim do
        c[i] := a[i] + b[i]
     end do;
     eval(c)  # Return the array, not the identifier!
  end proc:
```

A local array `c` is used to represent the sum, and it is very important to return the array data structure using `eval(c)` and not return just the local identifier `c`:

```
> a := array([1,2,3]);  b := array([4,5,6]);
```

$$a := [1, 2, 3]$$
$$b := [4, 5, 6]$$

```
> VecAdd(a, b, 3);
```

$$[5, 7, 9]$$

A better implementation determines the dimensions of the vectors from the array data structure and checks that both vectors have the same dimensions. This task is best delegated to a sub-procedure that I will call `GetDim`. Otherwise, the new procedure is almost identical to the previous version. In the spirit of "top-down" design, I will use `GetDim` now and defer writing it until later:

```
> VecAdd := proc(a::vector, b::vector)
     # Add two vectors a, b.
     local dim, c, i;  dim := GetDim(a, b);
     c := array(1..dim);
     for i to dim do
        c[i] := a[i] + b[i]
     end do;
     eval(c)
  end proc:
```

In order to write `GetDim` we need to recall the array operand structure, which is *index_function, index_range_seq, data_list.*

The Maple data-type `vector` represents an array indexed by a single integer with lower bound 1, so the dimension of the vector is the upper index bound. This is the *right-hand side* of the index bound range, which is stored as the *second operand* of the array data structure. The following procedure checks that two vectors have the same dimension, and if so returns it:

```
> GetDim := proc(a, b)  # assumes a, b vectors
     # Get common dimension of two vectors a, b.
     local dim;
     dim := rhs(op(2, eval(a)));
     if dim <> rhs(op(2, eval(b))) then
        error "incompatible dimensions."
     end if;
     dim
  end proc:
```

```
> VecAdd(a, b);
```

$$[5, 7, 9]$$

```
> c := array([4,5]);
```

$$c := [4, 5]$$

```
> VecAdd(a, c);
```

```
Error, (in GetDim) incompatible dimensions.
```

While the code `rhs(op(2,eval(A)))` is straightforward, it could be written marginally more succinctly and efficiently using recursive operand access as `op([2,2],eval(A))`, where `A` is some array identifier. (Note that there are functions in the `linalg` package called `vectdim`, `rowdim`, and `coldim` that find the dimension of a vector and the row and column dimensions of a matrix, respectively. They could be used here, but they would not illustrate explicitly how to access the array data structure. There are similar functions in the `LinearAlgebra` package.)

11.2.2 Scalar product

The scalar product of two vectors $\boldsymbol{a}$, $\boldsymbol{b}$ with dimensions *dim* is defined to be the scalar quantity

$$\boldsymbol{a} \cdot \boldsymbol{b} = \sum_{i=1}^{dim} a_i \, b_i.$$

It can be computed using a procedure such as the following, which re-uses procedure `GetDim` to find and check the dimensions:

```
> ScaProd := proc(a::vector, b::vector)
     # Scalar product of two vectors a, b.
     local s, dim, i;  dim := GetDim(a, b);
     s := 0;
     for i to dim do
        s := s + a[i]*b[i]
     end do;
     s
  end proc:
```

Note the use above of the local variable s to accumulate the sum.

```
> a := array(1..3):  evalm(a);
```

$$[a_1, a_2, a_3]$$

```
> b := array(1..3):  evalm(b);
```

$$[b_1, b_2, b_3]$$

```
> ScaProd(a, b);
```

$$a_1 b_1 + a_2 b_2 + a_3 b_3$$

```
> a := array([1,2,3]);
```

$$a := [1, 2, 3]$$

```
> b := array([2,-1,0]);
```

$$b := [2, -1, 0]$$

```
> ScaProd(a, b);
```

$$0$$

The above implementation is typical of the code that would be required in most programming languages, although in Maple it could (and probably should) be implemented rather more succinctly like this:

```
> ScaProd := (a::vector, b::vector)->
     # Scalar product of two vectors a, b.
     add(a[i]*b[i], i=1..GetDim(a, b)):
```

11.2.3 Matrix multiplication

The product **C** of two matrices **A** and **B** is defined by

$$c_{ij} = \sum_{k=1}^{k\_max} a_{ik}\, b_{kj},$$

where c_{ij} denotes the $(i{,}j)$-element of the matrix **C**, etc. The conformability (i.e., compatibility) condition is that the number of columns of **A** must be equal to the number of rows of **B**, in which case this number defines the upper bound *k_max* for the summation index k. The Maple data-type `matrix` represents an array indexed by two integers with lower bound 1. Hence, the row and column dimensions of the matrix are the first and second upper index bounds, respectively. These are the right-hand sides of the two elements of the sequence of index bound ranges that is stored as the second operand of the array data structure.

The following procedure implements the above formula for matrix multiplication. A double loop is required to run over the values of i and j to build the product matrix **C**. The value of each element c_{ij} of **C** is accumulated in a simple local variable (s) and then assigned to c_{ij}. This is much more efficient than accumulating it directly into an element of an array, because the indexing required to access an array element makes it much slower than accessing a simple variable:

```
> MatMul := proc(A::matrix, B::matrix)
     # Multiply two matrices A, B.
     local A_dim, B_dim, k_max, i_max, j_max, C, s, i, j, k;
     A_dim := op(2, eval(A));  # 1..i_max, 1..k_max
     B_dim := op(2, eval(B));  # 1..k_max, 1..j_max
     k_max := rhs(A_dim[2]);
     if k_max <> rhs(B_dim[1]) then
        error "incompatible dimensions"
     end if;
     i_max := rhs(A_dim[1]);  j_max := rhs(B_dim[2]);
     C := array(A_dim[1], B_dim[2]);
     for i to i_max do
        for j to j_max do
           s := 0;
           for k to k_max do s := s + A[i,k]*B[k,j] end do;
           C[i,j] := s
        end do
     end do;
     eval(C)
  end proc:
```

```
> A := array([[1,2],[3,4],[5,6]]);
```

$$A := \begin{bmatrix} 1 & 2 \\ 3 & 4 \\ 5 & 6 \end{bmatrix}$$

```
> B := array([[1,2],[3,4]]);
```

$$B := \begin{bmatrix} 1 & 2 \\ 3 & 4 \end{bmatrix}$$

```
> MatMul(A, B);
```

$$\begin{bmatrix} 7 & 10 \\ 15 & 22 \\ 23 & 34 \end{bmatrix}$$

```
> MatMul(B, A);
```

```
Error, (in MatMul) incompatible dimensions
```

Of course, in Maple the summation loop in this procedure would normally be implemented more succinctly by using `add`, as illustrated at the end of the previous section, but this is left as an exercise.

11.3 Data processing in Maple

This usually implies reading data from a file, processing it, and writing the results to another file. It is convenient to introduce the second step first, namely outputting data to a file.

11.3.1 Outputting to a file

There are several ways to do this, but probably the simplest is to redirect normal Maple output from the screen to a file. The function `writeto` does this. Its argument should be a Maple string (enclosed in double quotes) representing a filename or full pathname, with one important exception: the function-call `writeto(terminal)`, where `terminal` is an *identifier*, closes any output file and directs output back to the terminal. (If the string `"terminal"` is used then it refers to a file called `"terminal"`.)

The directory separator in file paths in Maple should be a forward slash on all systems, which is converted by Maple into whatever is appropriate for the actual operating system on which Maple is running. (On Windows the native directory or folder separator is a backward slash, which can be included explicitly in a string provided it is doubled, e.g.,

```
> printf("directory\\file");
```

```
directory\file
```

as explained in Chapter 2, because a backward slash within a string is an escape character exactly as in C, which is used mainly to insert ASCII control characters. But it is best to use the standard Maple forward slash, which is portable across operating systems.)

When using a GUI, it is not always immediately obvious what is the "current directory". On UNIX it will probably be the current directory when Maple was started. On Windows, it is specified in the shortcut used to run Maple, and the default for Maple 6 appears to be the root of the Maple directory tree. This default can be changed, but it is probably safest to specify an absolute file pathname that begins with a `/`, which refers to the root of the user's filespace on the current drive. (On Windows, the current drive is better defined than the current directory, but that too can be specified within a file pathname if desired.) I will use the directory `/tmp`, which is a directory that I like to ensure exists on computers that I use (partly for UNIX compatibility).

As an example, let us output a few random numbers. The function `rand` called with no argument generates random natural numbers with (up to) 12 digits, e.g.,

```
> rand();
```

$$427419669081$$

For output to a file, it is convenient to have the numbers left-justified rather than centred, which can be achieved by outputting them with `lprint` rather than `print`. The following execution group generates a convenient short data file containing 10 random numbers:

```
> writeto("/tmp/tmp.txt");           # open file
  to 10 do lprint(rand()) end do;  # write data
  writeto(terminal);               # close file
```

The function `appendto` appends output to the end of a file, whereas the function `writeto` overwrites any existing file.

11.3.2 Inputting from a file

Numerical data (only) separated by white space can be read most conveniently into a list or list of lists using the standard library procedure `readdata`. By default, `readdata` reads one datum per line into a list. An optional argument specifies the number of data per line, which are then read into a list of lists. Data are read as floats by default, accurate to the current value of `Digits`, but this can be changed by specifying `integer` as an optional second argument.

```
> readdata("/tmp/tmp.txt");
```

$$[.3211106933\,10^{12}, .3436330737\,10^{12}, .4742561436\,10^{12}, .5584587190\,10^{12}, .7467538305\,10^{12}, .3206222208\,10^{11}, .7229741218\,10^{12}, .6043056139\,10^{12}, .7455800374\,10^{12}, .2598119527\,10^{12}]$$

These floats can be converted to integers, but note the rounding error caused by using only 10 significant figures (the default value of `Digits`), compared with reading them directly as integers:

```
> convert(%, rational, exact);
```

[321110693300, 343633073700, 474256143600, 558458719000, 746753830500, 32062222080, 722974121800, 604305613900, 745580037400, 259811952700]

```
> readdata("/tmp/tmp.txt", integer);
```

[321110693270, 343633073697, 474256143563, 558458718976, 746753830538, 32062222085, 722974121768, 604305613921, 745580037409, 259811952655]

Accurate reading as floats requires a larger value of `Digits` (12 to be precise, since `rand` generates 12-digit random numbers):

```
> proc() Digits:=12; readdata("/tmp/tmp.txt") end proc();
```

$$[.321110693270\,10^{12}, .343633073697\,10^{12}, .474256143563\,10^{12}, .558458718976\,10^{12}, .746753830538\,10^{12}, .32062222085\,10^{11}, .722974121768\,10^{12}, .604305613921\,10^{12}, .745580037409\,10^{12}, .259811952655\,10^{12}]$$

11.3.3 Elementary statistics

As an introduction to data processing, let us compute a few elementary statistics of the above list of random numbers. The mean (μ) and standard deviation (σ) of a collection of n data x_i, $1 \le i \le n$, are defined to be

$$\mu = \frac{1}{n}\sum_{i=1}^{n} x_i, \quad \sigma = \sqrt{\frac{1}{n}\left(\sum_{i=1}^{n}(x_i - \mu)^2\right)}.$$

They can be computed using the following Maple code, which is very close to the above mathematics:

```
> x := readdata("/tmp/tmp.txt");
```

$$x := [.3211106933\,10^{12},\ .3436330737\,10^{12},\ .4742561436\,10^{12},\ .5584587190\,10^{12},\ .7467538305\,10^{12},\ .3206222208\,10^{11},\ .7229741218\,10^{12},\ .6043056139\,10^{12},\ .7455800374\,10^{12},\ .2598119527\,10^{12}]$$

```
> n := nops(x);  # number of data
```

$$n := 10$$

```
> mu := sum(x[i],i=1..n)/n;  # mean
```

$$\mu := .4808946408\,10^{12}$$

```
> sigma := sqrt( sum((x[i]-mu)^2,i=1..n)/n );
```

$$\sigma := .2270092972\,10^{12}$$

Alternatively, we could express the two summations more succinctly and more efficiently as follows:

```
> mu := `+`(op(x))/n;  # mean
```

$$\mu := .4808946408\,10^{12}$$

```
> sigma := sqrt( add((xi-mu)^2,xi=x)/n );
```

$$\sigma := .2270092972\,10^{12}$$

Finally, it is very easy in Maple to plot all the statistics. The data values must be converted into a list of $(x,\ y)$-coordinate pairs in order to plot them, where the ordinal number or index of a datum within the list of data can be used as its horizontal coordinate. There is no particular relationship among random numbers so there is no reason to join the data points together and they are best plotted as discrete points. In order to display the mean and standard deviation, note that numbers can be treated as constant functions or expressions, and their graphs are horizontal lines:

```
> plots[display]({
    # Data:
    plot([seq([i,x[i]], i=1..n)], index,
       style=POINT, symbol=CIRCLE, color=black),
    # Mean +/- standard deviation:
    plot({mu, mu+sigma, mu-sigma},
       index=1..n, color=black)},
    title="Data and mean +/- standard deviation");
```

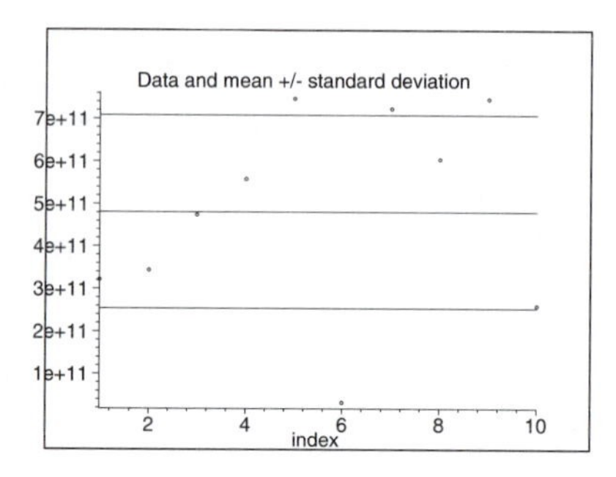

11.3.4 Other methods of input and output

Maple provides a number of other I/O (input/output) facilities. The functions `readline` and `writeline` are a pair of line-oriented I/O functions that read and write text strings, and the function `writedata` is the closest inverse of `readdata`. Maple also provides families of functions related to `scanf` and `printf` that are analogous to the functions with the same names in C (and C++), which provide very flexible input and output formatting. I will use `scanf` in Chapter 14 and I use `printf` occasionally throughout this book; we will return to some of the functions later. See the online help for complete details.

11.3.5 Aside: offline processing

If data is input from a file and output to a file then there is little point in running the program via an interactive "online" worksheet interface and in practice such programs would normally be run via a non-interactive "offline" command interface. This is sometimes also referred to as "batch" processing, which is an archaic term referring to a computer operator running non-interactive programs in "batches", or "background" execution, which refers to a program being run without interactive access to a terminal "in the background", usually at lower priority.

To run a Maple program non-interactively it must be in the form of a source code text file rather than a worksheet. The visible part of a whole worksheet can be converted to a text file via the File menu by selecting the *Export As* sub-menu and then selecting either the *Maple Text...* or *Plain Text...* item. The difference is that text is output as separate comment lines (each preceded by a # symbol) in *Maple Text* format but as a single line (paragraph) in *Plain Text* format. For small segments of code it may be simplest just to copy from the worksheet via the clipboard to a text editor and save the file from the editor. Either way, the resulting text file needs to be edited to remove the prompts and it is probably best to edit out any text (which is *necessary* if

Plain Text format is used). Output should be hidden before export or edited out. Alternatively, you can create the Maple source code directly in a text editor.

The source code filename is of no significance, although an extension of "`.txt`" or "`.mpl`" is probably a good idea, to indicate roughly what kind of file it is. Let us suppose it is called "`MapleProgram.txt`". Then it can be executed by Maple non-interactively by executing an operating system command of the form:

```
cmaple < MapleProgram.txt
```

on Windows, or:

```
maple < MapleProgram.txt
```

on UNIX. (On UNIX the command can optionally be followed by a `&` character to cause Maple to run as a background process). The `<` character tells the operating system to send the file "`MapleProgram.txt`" to Maple as its "standard input", just as if it had been typed in at the keyboard. When Maple reaches the end of the input file it automatically terminates, so there is no need for an explicit `quit` command.

These commands can be executed interactively at a suitable operating system command prompt (for example, in an "MS-DOS Prompt" or "xterm" window) or they can be executed via some scripting facility, such as "Windows Scripting Host" or a Perl program (e.g., to run Maple via a web server). However, there are a number of provisos to these instructions. Either the Maple installation binary directory must be in your execution path, which is certainly not the case by default on Windows, or the simple Maple executable filename must be replaced by its full pathname; e.g., on my Windows system it would be `"C:\Program Files\Maple 6\BIN.WNT\cmaple"` (*including* the double quotes, which are part of the pathname because it contains space characters). Similarly, it is necessary either to change directories so that the directory (folder) containing the input file is the current directory or to replace the input filename by its full pathname.

11.4 Interactive programs in Maple

An *interactive* program is one that interacts with its user *while it is running*, usually by reading input from the keyboard. Although Maple itself is interactive, it is unusual for Maple programs to be interactive. The usual way to control a Maple program is to implement it as a procedure and to input information as arguments when the procedure is called, but not to interact with the procedure while it is actually running. However, the latter is possible, and here are some trivial examples. Beware that it is very easy to forget that Maple is waiting for interactive input when experimenting with such interactive programs, and to get both yourself and Maple completely confused. If this happens then it may help to execute a `restart` command. This is one situation in which the Maple command-line interface may be easier to use than the worksheet interface.

The following program simulates a throw of a die; i.e., it generates a random integer between 1 and 6 inclusive, each time the *Return* key is pressed until one or more characters are input. The call `rand(1..6)` returns a *procedure*, which returns a random integer in the specified range when it is itself called. The call `readline(terminal)` reads one complete *line* of input (actually, a paragraph, terminated by *Return*, not *Shift-Return*) from the terminal (keyboard) and returns it as a Maple string. (The same remarks made above about the argument to `writeto` apply also to `readline`.) Hence, the following procedure continues to loop until non-empty input is read:

```
> die := proc()
     local QUIET, throw, r;
     QUIET := interface(quiet);  interface(quiet = true);
     throw := rand(1..6);  # returns a PROCEDURE
     printf("Throw a single die repeatedly ...\n");
     r := "";
     while r = "" do
        print(throw());
        printf("<Return> to throw again, \"anything\" <Return>
  to stop\n");
        r := readline(terminal)
     end do;
     interface(quiet = QUIET)
  end proc:
```

```
> die();
```

```
   Throw a single die repeatedly ...
                                    4

   <Return> to throw again, "anything" <Return> to stop

>

                                    4

   <Return> to throw again, "anything" <Return> to stop

>

                                    3

   <Return> to throw again, "anything" <Return> to stop
```

```
> q
```

The logic of such interactive loops is tricky, and I don't want to go into the details, only to illustrate that this is possible in Maple. The change to the interface parameter `quiet` is simply to stop the non-empty input that terminates the loop from being echoed, and is purely cosmetic, although it illustrates how to change an interface parameter locally (but see also the next section). The messages are output using `printf` to avoid the string quotes that would be output by `print` (or `lprint`) spoiling the output. This example illustrates the need for an explicit newline at the end of each message string and the need to escape double quote characters within a string. This way of using `printf` can output only fixed strings; normally the first argument to `printf` is a format string that contains format codes to output other (usually variable) data. The messages could alternatively be output as identifiers (in backquotes).

The following modification illustrates the use of the `readstat` function, which (optionally) changes the Maple prompt, reads one complete Maple *statement* (which must be properly terminated), executes it, and returns the result. The input this time is the number of dice to throw, whose values are then returned as an expression sequence. Note that the call of `throw` must be *unevaluated*, because `$` evaluates its operands before it constructs the sequence. Hence, without unevaluation, the values of `throw()` would all be the same. (Using `seq` would not require the unevaluation but is unwieldy just to generate several identical copies of some quantity, which here is an unevaluated function-call.) This procedure also illustrates the use of a *repeat* loop constructed using the `break` statement.

```
> dice := proc()
      local throw, n;
      throw := rand(1..6);
      printf("Throw a number \"n\" of dice repeatedly ...\n");
      do
         n := readstat("Number of dice: n; > ");
         if n = NULL or n <= 0 then break end if;
         print('throw()'$n)
      end do
  end proc:
```

```
> dice();
```

```
   Throw a number "n" of dice repeatedly ...

Number of dice: n; > 1;
```

$$3$$

```
Number of dice: n; > 3;
```

2, 1, 4

```
Number of dice: n; > 10;
```

4, 6, 1, 1, 1, 2, 4, 2, 1, 3

```
Number of dice: n; > ;
```

A problem with the above implementation is the need to terminate the input, which can seem rather unnatural. A solution is to read the input using `readline`, concatenate a statement terminator onto the end (using `cat`), and then parse the result, as follows. Since we are once again using `readline` we must again set the prompt explicitly:

```
> dice := proc()
     local PROMPT, throw, n;
     PROMPT := interface(prompt);
     interface(prompt = "Number of dice: n > ");
     throw := rand(1..6);
     printf("Throw a number \"n\" of dice repeatedly ...\n");
     do
        n := parse(cat(readline(terminal),";"));
        if n = NULL or n <= 0 then break end if;
        print('throw()'$n)
     end do;
     interface(prompt = PROMPT)
  end proc:
```

```
> dice();
```

```
Throw a number "n" of dice repeatedly ...
```

```
Number of dice: n > 2
```

6, 3

```
Number of dice: n > 3
```

6, 4, 3

```
Number of dice: n > 0
```

11.5 Error handling: `try`, `catch`

Well-written programs try to detect errors before they happen by validating input data. But there are situations in which it is either not possible or not convenient to do this. It may not be possible if an error could arise in code that is outside your control, such as a call to some library facility. It may not be convenient if detecting potential errors would involve a lot of complicated code. In such cases, the solution is to let the error happen and then try to catch it and recover.

The terms *throw* and *catch, control* are sometimes applied to non-local transfers of control within computer programs. By default, when an error occurs in Maple all the code currently executing is terminated and Maple returns to its "top level", usually an interactive prompt. This is an example of a non-local transfer of control, which is sometimes referred to as "throwing an error" or more generally "throwing an exception". An error is strictly a special case of a more general class of events called *exceptions*, which includes abnormal events that might not be errors. However, in Maple all exceptions are generated by executing the `error` statement; the point is that you can cause Maple to execute an `error` statement for whatever reason you wish, and you can provide information about that reason in the error message string. Programming jargon often refers to exceptions being *raised*, which comes from the analogy of signalling using flags that can be raised and lowered. I will describe only the main features of exception handling in Maple; see the online help for full and precise details.

It is possible to ensure that some code is executed regardless of whether an exception is thrown, and/or it is possible to catch some or all exceptions and continue. The mechanism for doing this in Maple is very similar to that in Java and involves a `try` statement, with the following syntax:

```
try
   main_statement_sequence
catch message_string_sequence_1 :
   error_handling_statement_sequence_1
   ...
catch message_string_sequence_n :
   error_handling_statement_sequence_n
finally
   final_statement_sequence
end try
```

This statement is somewhat analogous to a conditional statement, and the keywords `try`, `catch`, `finally`, and `end` act as statement delimiters, as do the keywords `then`, `elif`, `else`, and `end`. The `catch` and `finally` clauses are optional, although a `try` statement serves no useful purpose unless it has at least one of them.

The keyword `catch` can be followed by a sequence of zero or more strings, which will each be matched in turn against the message string generated by the `error` statement that raised the current exception. There can be more than one `catch` clause provided they all have different message strings. When *all* the characters in one of the catch strings match the characters at the *beginning* of the error message string, the exception is caught and the sequence of statements after the colon in that catch clause is executed to handle it. A `catch` clause with an empty *message_string_sequence* catches all exceptions (and should therefore be the last `catch` clause if there is more than one). For simple error handling, a single `catch` clause with no catch strings will often suffice. If an `error` statement with no arguments is executed within a `catch` clause then it simply re-raises the original exception, which provides a way to take some action and then allow the default effect of the exception to proceed.

The `finally` clause is guaranteed to be executed after the `try` clause, regardless of whether the `try` clause succeeds or fails. If any exception raised in the `try` clause is not caught then it is raised again after the `finally` clause has been executed. Hence, a `try ... finally` statement (with no `catch` clauses) can be used to ensure that when some change is made to the Maple environment it is reset even if an error occurs.

The final version of procedure `dice` in the previous section needs such protection. Invalid user input can cause any of several errors: Maple may not be able to parse invalid input syntax; if the parsed input `n` is not a number then the `if` statement will fail; if `n` is not an integer then the repetition `$n` will fail. Any such error would terminate execution of the procedure but leaves the prompt set to its new value, which would be very confusing if you tried to continue using Maple. (Try it!) We could check that `n` has a suitable numerical value but we have no control over errors arising in the call of `parse` (because the only way to find a syntax error is by parsing). Hence, we may as well treat all errors uniformly and protect against them by running the main body of the procedure in a `try` statement with the code to reset the prompt in the `finally` clause. This preserves the Maple environment even if an error occurs:

```
> dice := proc()
     local PROMPT, throw, n;
     PROMPT := interface(prompt);
     try
        interface(prompt = "Number of dice: n > ");
        throw := rand(1..6);
        printf("Throw a number \"n\" of dice repeatedly ...\n");
        do
           n := parse(cat(readline(terminal),";"));
           if n = NULL or n <= 0 then break end if;
           print('throw()'$n)
        end do
```

```
    finally
       interface(prompt = PROMPT)
    end try
  end proc:
```

```
> dice();
```

```
   Throw a number "n" of dice repeatedly ...
```

```
Number of dice: n > 2
```

6, 1

```
Number of dice: n > xyz
   Error, (in dice) cannot evaluate boolean: xyz <= 0
```

```
> # Note that the prompt has returned to normal here!
```

An alternative approach is to catch the error and handle it, which we can do using a single `catch` clause with no catch strings. Now, input errors do not terminate the loop, but simply lead to an error message and another input prompt. The best approach to error handling depends very much on the program being written!

```
> dice := proc()
     local PROMPT, throw, n;
     PROMPT := interface(prompt);
     interface(prompt = "Number of dice: n > ");
     throw := rand(1..6);
     printf("Throw a number \"n\" of dice repeatedly ...\n");
     do
        try  # Await valid input...
           n := parse(cat(readline(terminal),";"));
           if n = NULL or n <= 0 then break end if;
           print('throw()'$n)
        catch:
           printf("Invalid input - no input or 0 terminates -
  try again...\n")
        end try
     end do;
     interface(prompt = PROMPT)
  end proc:
```

```
> dice();
```

```
   Throw a number "n" of dice repeatedly ...
```

```
Number of dice: n > 2
```

6, 3

```
Number of dice: n > xyz
   Invalid input - no input or 0 terminates - try again...

Number of dice: n > 3
```

1, 4, 6

```
Number of dice: n > +
   Invalid input - no input or 0 terminates - try again...

Number of dice: n > 4
```

4, 3, 2, 5

```
Number of dice: n >


> # Note that the prompt has returned to normal here!
```

11.6 Debugging

Programmers don't make mistakes; however, bugs do creep inexorably into their code! This euphemistic reference to mistakes in programs as *bugs* dates back at least to the very earliest days of computers and there are various (possibly apocryphal) stories around concerning it. One is that a hardware problem in an electromechanical computer was traced to a moth caught between the contacts of a relay. Another is that the original bug was a moth that was attracted to the light source used to read punched paper tape, the random flight of the moth through the light beam causing inexplicable random errors in the behaviour of a program. However, Kernighan and Pike [14] quote a reference from the *Oxford English Dictionary* dating back to the end of the 19th century when the American inventor Thomas Edison (1847–1931) referred to a "bug in his phonograph". Nowadays, the term *bug* is used to refer to almost any kind of problem with almost anything, but in this book I will use it to refer to a mistake made at any point in the development of a computer program.

Debugging is the process of removing bugs from a program, and testing is the phase of program development in which any remaining bugs should be discovered. Programmers sometimes refer to *alpha* and *beta* versions of software: alpha usually means that the software is still under development and not intended for use; beta usually means that it is undergoing testing by a small group of experienced users who report problems back to the developers. Program testing and debugging is an important and subtle component of software engineering, which I will not pursue to any depth. Kernighan and Pike [14] devote a chapter to each of debugging and testing, and much of the rest of their book revolves around related issues, such as avoiding potential bugs in the first place. My advice is to take care when designing and implementing algorithms and think about the special cases that may arise. It is very tempting to brush special cases aside and produce code that "mostly works", but careless development may waste more time later tracking down obscure bugs than it ever saves. Careful and consistent layout of program source code makes it much easier to write, maintain, and check.

Many bugs arise from a failure to fully understand the operation of some library facility that you have not written yourself, sometimes because the documentation is not as clear as it might be. Others arise because you make an assumption that is not justified. For example, a mistake that is easily made in Maple is to assume that polynomials are in expanded form, but if a polynomial is constructed as a product this will not be the case by default and it must be explicitly expanded. Debugging requires some detective work and experimentation. Once you have detected incorrect behaviour, you need to try to deduce what might be causing it. Then, you need to perform an experiment to test whether that really is the cause, and if so you need to change the code to correct the problem. Sometimes, running a program with a carefully designed set of input data can shed light on the source of an error; for example, you might consider whether some error relating to powers occurs with input consisting of any power or only specific powers.

Develop and test a program in small pieces. This is very easy in Maple because almost any procedure can be executed interactively and does not require additional code to allow it to be run. It is much easier to track down mistakes in small segments of code, while you have the full details of the code in your mind, than in a large complete program, when you may have forgotten some of the details of subprograms that you wrote earlier. Meaningful choices of identifiers and informative comments make the code easier to understand.

There are many ways to get "debugging" information out of a program. One way that will work with almost any programming language is to insert some explicit output statement at what you think may be critical points in the program. In Maple, one could put some kind of print statement at appropriate places. Once the bug has been exterminated, and especially during development, it may be best not to remove such print statements but to "comment them out" by turning them into comments so that they can easily be re-activated if necessary. In Maple, a `#` character and everything following it on the same logical line (up to an explicit end-of-line character introduced by either *Return* or *Shift-Return*) is a comment that is not executed.

Most program development environments, including Maple, offer some facilities to assist with debugging. Their purpose is to make it easy to obtain the right amount of information about the operation of a specified part of a program. It is helpful to distinguish two levels of operation of a program: the higher more abstract level of the algorithm and the lower more concrete level of its implementation in a particular language. I will outline one way to get information at the algorithmic level and three ways to get information at the implementation level in Maple.

11.6.1 infolevel

The algorithms underlying computer algebra systems and their mathematical basis can be very sophisticated: consider the multitude of different techniques that must be implemented in a differential equation solver such as `dsolve`. Hence, it is usual for computer algebra systems to be able to provide information at an abstract algorithmic level, which I call "algorithmic tracing". This is normally turned off by default. In Maple, algorithm tracing can (in some cases) be turned on via the `infolevel` facility using an assignment of the form:

`infolevel[`*fn*`] := ` *level*

where *fn* is the name of a procedure or the identifier `all` and *level* is an integer in the range 1..5. This assignment turns on output of information from either the named procedure or all procedures, provided they support this facility. The significance of the levels is as follows:

level ≥ 1 gives information that the user must be told.

level ≥ 2 adds general information, including the technique or algorithm being used.

level ≥ 4 adds more detailed information about how the problem is being solved.

In fact, the global variable `infolevel` is initialized to a `table`, and assignments of the above form add entries to that table; see the online help for further details of how it works. Here is an example that outputs algorithmic tracing of the solution of an ordinary differential equation:

```
> infolevel[dsolve] := 5:
```

```
> x:='x':
  dsolve(x*diff(y(x),x$2) + x*diff(y(x),x) + y(x));

   Methods for second order ODEs:
```

```
Trying to isolate the derivative d^2y/dx^2...

Successful isolation of d^2y/dx^2

-> Trying classification methods

trying a quadrature

trying a symmetry of the form [xi=0, eta=F(x)]

   testing BRANCH 1 ->

   testing BRANCH 2 ->

   testing BRANCH 3 ->

   <- BRANCH 3 successful. Symmetry is: [0, 1/exp(Int((x-1)/x,x))]

linear_1 successful: a symmetry of the form [xi=0, eta=F(x)] was
found
```

$$\mathrm{y}(x) = \frac{(\mathbf{e}^x + \mathrm{Ei}(1, -x)\, x)\, \_C1}{\mathbf{e}^x} + \frac{\_C2\, x}{\mathbf{e}^x}$$

This kind of information may be very useful when trying to understand why `dsolve` produces a solution in a particular form, rather than some alternative but equivalent form. It may be genuinely useful for debugging if `dsolve` is called by some other procedure that is expecting it to return a result in a different form. It may also be very useful when just trying to understand how the Maple ODE solver works!

11.6.2 `printlevel`

A facility that I use a lot to debug Maple code is simply to increase the value of the variable `printlevel`. Its default value is 1, but increasing it causes Maple to display more information about the execution of programs. Essentially, it causes Maple to display the information that statements within control structures and procedures would display if they were executed at top level; the larger the value of `printlevel` the deeper the level from which information is displayed. The top level is level 0; every control structure adds 1 to the level of a statement and every procedure call adds 5; the execution of all statements at levels down to the value of `printlevel` is displayed. See the online help for precise details. For each appropriate procedure call, Maple displays the entry point with the actual parameter values and the exit point with the returned function value.

But beware that a large value of `printlevel` can produce a very great deal of information and it is easy to become overwhelmed. If that happens, the *Undo* button or the *Undo* item on the Edit menu, or the *Remove Output* item on the Edit menu, can be used to remove the excessive output. I often try increasing `printlevel` in steps of 10, and if that does not produce enough information I increase it to 100 and perhaps 1000. (It may be easier to experiment than to try to calculate an appropriate display depth.) Very often, a quick look through the internal variable values occurring within procedure executions shows that some variable is acquiring an unexpected value or the wrong data type, and a little thought then leads to the cause of the problem. But `printlevel` is a very blunt debugging instrument!

As an example, let us reconsider the following function definition, essentially from Chapter 8:

```
> Factorial := (n::nonnegint)->
    if n = 0 then 1 else n*procname(n-1) end if:
```

The smallest value of `printlevel` that produces any internal display at all is 5 and increasing it to 6 displays the value computed within the conditional statement in the first function-call:

```
> printlevel := 6:

> Factorial(3);

  {--> enter Factorial, args = 3
                                    6

  <-- exit Factorial (now at top level) = 6}
                                    6
```

Increasing `printlevel` to 12 displays the first two function-calls and the values computed within their conditional statements, etc.:

```
> printlevel := 12:

> Factorial(3);

  {--> enter Factorial, args = 3

  {--> enter Factorial, args = 2
                                    2

  <-- exit Factorial (now in Factorial) = 2}
                                    6

  <-- exit Factorial (now at top level) = 6}
                                    6
```

Remember to reset `printlevel` to its default level of 1 when you have finished debugging!

```
> printlevel := 1:
```

11.6.3 debug and trace

A slightly sharper debugging tool is provided by the function `debug`, which takes an arbitrary number of procedure names as arguments and turns on *execution tracing* for those procedures. The function `undebug` undoes this and turns execution tracing off again. (These functions have synonyms `trace` and `untrace`, but `trace` conflicts with the name of a function in the `linalg` package — unfortunately the word "trace" has two completely different meanings, both of which are relevant in Maple — so `debug` and `undebug` are safer names.)

```
> debug(Factorial);
```

Factorial

Execution tracing provides information similar to that provided by increasing the value of `printlevel` to a large value, but only for the specified procedures. The results of the statements that are executed are displayed only if the expression being evaluated, which contains the traced procedure call, is terminated with a semicolon, e.g.,

```
> Factorial(3);

   {--> enter Factorial, args = 3

   {--> enter Factorial, args = 2

   {--> enter Factorial, args = 1

   {--> enter Factorial, args = 0
                                  1

   <-- exit Factorial (now in Factorial) = 1}
                                  1

   <-- exit Factorial (now in Factorial) = 1}
                                  2

   <-- exit Factorial (now in Factorial) = 2}
                                  6

   <-- exit Factorial (now at top level) = 6}
                                  6
```

rather than a colon, e.g.,

```
> Factorial(3):

   {--> enter Factorial, args = 3

   {--> enter Factorial, args = 2
```

```
{--> enter Factorial, args = 1

{--> enter Factorial, args = 0

<-- exit Factorial (now in Factorial) = 1}

<-- exit Factorial (now in Factorial) = 1}

<-- exit Factorial (now in Factorial) = 2}

<-- exit Factorial (now at top level) = 6}
```

The debug function works by setting `option trace` within the procedure to be traced:

```
> print(Factorial):
```

proc(n::*nonnegint*)
option *operator*, *arrow*, *trace*;
if $n = 0$ **then** 1 **else** $n *$ procname$(n - 1)$ **end if**
end proc

Remember to turn off debugging of traced procedures when you have finished debugging!

```
> undebug(Factorial);
```

Factorial

11.6.4 Interactive debugging

The sharpest debugging tool is an interactive source code debugger, which allows the programmer to interact with a running program in terms of its actual source code. The best graphical source code debuggers provide multiple windows, each dedicated to particular information about the program, and a separate way to interact with the debugger. On Windows, the Maple debugger is currently not graphical and everything happens in the normal worksheet. If you are not familiar with this kind of debugger then you may find the learning curve too steep for occasional use.

The Maple interactive debugger refers to the source code of procedures by statement numbers, which can be displayed at any time as a side-effect of the `showstat` function, e.g.,

```
> showstat(Factorial):

Factorial := proc(n::nonnegint)

   1   if n = 0 then
```

```
  2    1
       else
  3      n*procname(n-1)
       end if
  end proc
```

or by executing the `showstat` debugger command once the debugger is running. (Note that Maple uses *statement* numbers rather than *line* numbers, which are ambiguous since Maple reformats the input.)

One of the facilities provided by an interactive debugger is the ability to set *breakpoints*, which are points in a program at which execution will be suspended and control returned to the debugger. This allows you to examine the state of the computation at that point and decide what to do next. In a source-code debugger, breakpoints are set in terms of points in the source code (rather than whatever low-level code is actually being executed). The Maple function `stopat` sets a breakpoint at the start of the procedure supplied as its argument and returns a list of all the procedures currently containing breakpoints. (It also takes two optional arguments that allow more control, see below.)

```
> stopat(Factorial);
```

$$[Factorial]$$

When execution stops as a breakpoint, Maple displays the context of the breakpoint (namely, the result of the last statement executed, the procedure it has stopped in, and the statement to be executed next) and changes the prompt to "`DBG> `". At the debugger prompt, Maple accepts both debugger commands and also normal Maple statements, the context for which is the environment in which execution has stopped, thus giving access to local variables within procedures. One of the main purposes of an interactive debugger is to allow code to be executed in steps, with some control over the size of those steps. The facilities provided by the Maple debugger are fairly standard. The main stepping commands are as follows:

step — executes the next statement at *any* nesting level and stops again; the smallest execution step.

into — like **step** but does not stop in a called procedure.

next — executes the next *complete* statement at the current nesting level and stops again; an intermediate step.

cont — executes up to the next breakpoint or normal termination; the largest execution step.

outfrom — like **cont** but stops at the end of the current statement sequence.

return — like **cont** but stops at the end of the current procedure.

quit — terminates execution of the current program and returns to the top level.

For example, if we run procedure `Factorial` with its default breakpoint set as above and continue once, that should cause the debugger to stop at the first recursive call, in which we expect that $n = 2$. Having checked that, we might then quit as follows:

```
> Factorial(3);
  Factorial:
     1*  if n = 0 then
           ...
         else
           ...
         end if
> cont
  Factorial:
     1*  if n = 0 then
           ...
         else
           ...
         end if
> n
  2
  Factorial:
     1*  if n = 0 then
           ...
         else
           ...
```

```
        end if
```

```
> quit
```

```
   Warning, computation interrupted
```

There are also functions and debugger commands to unset breakpoints (`unstopat`), set and unset watchpoints that stop execution when the value of a variable changes (`stopwhen` / `unstopwhen`), set and unset error watchpoints that stop execution when a particular exception is raised (`stoperror` / `unstoperror`), display the current stack of procedure calls (`where`), and display all stop points (`showstop`). For example, we could make procedure `Factorial` stop at its base case ($n = 0$), display its call stack, and then quit, as follows:

```
> stopat(Factorial, 1, n=0);
```

$$[Factorial]$$

```
> Factorial(3);

   Factorial:

      1?  if n = 0 then

             ...

          else

             ...

          end if

> where

   TopLevel: Factorial(3)

   [3]

   Factorial: n*procname(n-1)

   [2]

   Factorial: n*procname(n-1)

   [1]

   Factorial: n*procname(n-1)

   [0]
```

```
Factorial:

   1?  if n = 0 then

          ...

       else

          ...

       end if
```

```
> quit
   Warning, computation interrupted
```

11.7 Exercises

The array-based exercises below could use either table or rtable-based arrays, or in some cases both.

1. Write down a list of the six trigonometric functions (the online help will tell you exactly what they are), apply this list of functions (as if it were a single function) to the variable x (as its argument), and assign the result to the variable L. Use `map` to construct from L a list DL of the first derivative of each element of L and a list IL of the integral of each element of L. Use `select` together with the `type` predicate to select from each of DL and IL those elements that involve *only* trigonometric functions with no other algebraic structure (i.e., for which the surface type is a trigonometric function). Observe the results of mapping `whattype` over each of the lists DL and IL. (Make sure you understand the differences.)

2. Write and test a procedure called `selectops` that takes an arbitrary number of arguments, all but the last of which must be positive integers and the last of which must be an expression from which a subexpression can be selected (as for `select`). The procedure should return a copy of its last argument containing only the operands that were specified by index in the previous arguments, in the order specified and allowing repetitions (in cases such as a list where that makes sense). Any index out of range should be ignored. The procedure should work as follows:

```
> selectops(1,4,3,2,1, f(a,b));
```

$$\mathrm{f}(a, b, a)$$

```
> selectops(1,4,3,3,2, [a,b,c,d]);
```

$$[a,\ d,\ c,\ c,\ b]$$

```
> selectops(1,3, a+b+c);
```

$$a + c$$

```
> selectops(2,2, a*b*c);
```

$$b^2$$

3. Extend procedure `MAP` described in the text to handle rtable-based arrays.

4. Write and test a procedure called `MatAdd` that adds two matrices provided they have the same dimensions and reports an error otherwise. The procedure must determine the matrix dimensions from the array data structures (i.e., you should not use any packages).

5. The Hessian matrix of an expression f depending on n variables x_i, $1 \le i \le n$, is defined to be the $n \times n$ matrix of second partial derivatives with elements

$$H_{ij} = \frac{\partial^2}{\partial x_j\, \partial x_i} f, \quad 1 \le i,\ j \le n.$$

Write and test a procedure called `Hessian` that returns the Hessian matrix of an expression with respect to a list of variables. It should check that its arguments are of suitable types, and give the following results. (You should not use any packages.)

```
> Hessian(x^2+y^2, [x,y]);
```

$$\begin{bmatrix} 2 & 0 \\ 0 & 2 \end{bmatrix}$$

```
> Hessian(x*y, [x,y]);
```

$$\begin{bmatrix} 0 & 1 \\ 1 & 0 \end{bmatrix}$$

6. Write and test a procedure called `commutator` to compute the commutator of two square matrices of the same size efficiently using at most three loops. The commutator of two objects A and B is defined to be $A \cdot B - B \cdot A$.

7. Write and test a procedure called `DiagonalTree` that takes a single matrix argument A and returns the longest possible sequence of top-left square sub-matrices of A of dimensions 1, 2, 3, ... (i.e., square sub-matrices that contain only nearest-neighbour elements of A and have the same (1, 1) element as A). `DiagonalTree` must check that its argument is a matrix. (You should not use any packages.)

   Here is an example, assuming that the matrix A has been defined as follows:

$$A := \begin{bmatrix} a & b & c & d & e \\ f & g & h & i & j \\ k & l & m & n & o \end{bmatrix}$$

```
> DiagonalTree(A);
```

$$\begin{bmatrix} a \end{bmatrix}, \begin{bmatrix} a & b \\ f & g \end{bmatrix}, \begin{bmatrix} a & b & c \\ f & g & h \\ k & l & m \end{bmatrix}$$

8. Write and test a procedure called `minormatrix` that takes arguments A, i, j, where A must be a matrix and i and j must be positive integers, and returns the matrix A but with row i and column j deleted if they are within range. (You should not use any packages.)

   Here is an example, assuming that the matrix A has been defined as follows:

$$A := \begin{bmatrix} a & b & c & d & e \\ f & g & h & i & j \\ k & l & m & n & o \end{bmatrix}$$

```
> minormatrix(A, 2, 3);
```

$$\begin{bmatrix} a & b & d & e \\ k & l & n & o \end{bmatrix}$$

```
> minormatrix(A, 4, 4);
```

$$\begin{bmatrix} a & b & c & e \\ f & g & h & j \\ k & l & m & o \end{bmatrix}$$

9. Let M_{ij} denote the matrix A but with row i and column j deleted (as in the previous exercise). Then the *cofactor* A_{ij} of the element a_{ij} of a square $n \times n$ matrix A is defined to be $A_{ij} = (-1)^{(i+j)} \det(M_{ij})$. The determinant of A can then be defined recursively as $\det(A) = \sum_{j=1}^{n} a_{ij} A_{ij}$ for any $1 \leq i \leq n$, or similarly $\det(A) = \sum_{i=1}^{n} a_{ij} A_{ij}$ for any $1 \leq j \leq n$ (i.e., the summation may be over either the column or the row index). These formulae are called the *Laplace expansion* of the determinant of A by, respectively, the ith row or the jth column. Write and test a procedure called `determinant` to compute the determinant of a square matrix by recursive cofactor expansion. (You should not use any packages except to check results from your procedure.)

10. Extend the vector and matrix algebra facilities described in the text to rtable-based arrays.

11. Write and test a procedure called `HiLo` that runs the following simple interactive guessing game. The user inputs a natural number N and Maple generates a random natural number M in the range $0..N$. The user then inputs a guess at M and Maple responds with the information that the guess is too high or too low. When the user finally guesses M, Maple responds with the number of guesses required (and perhaps a congratulatory message). It could also output a graph of the guesses.

Chapter 12

Algebraic Programming

This chapter gives some insight into how a system such as Maple is able to perform some of its algebraic tasks while also illustrating some simple algebraic programming techniques. Much more sophisticated versions of the code that we will develop already exist in Maple. The first section is concerned with univariate, primarily polynomial, algorithms; the second section is concerned with the more general problem of multivariate polynomial algorithms. Although this chapter is based on some fairly sophisticated pure mathematics, the algorithms are not too difficult to explain and do not require a great deal of Maple code to implement, especially when the code is built as a hierarchy of well-defined tools, each calling on the services of those built previously.

12.1 Univariate polynomial algorithms

This section develops some of the tools that one needs in order to build a reliable numerical univariate polynomial equation solver. The first step is to reduce the problem to one of finding *simple* zeros, the second is to find *isolating intervals* for them, and the third is to *approximate* each real zero as accurately as required. The first two steps are algebraic and the last is numerical.

12.1.1 Squarefree factorization

A polynomial is called *irreducible* if it has no factors; i.e., it cannot be written as the product of two or more other polynomials (of degree at least 1). It is necessary to specify the coefficient domain for this to make sense. In Maple, it is assumed to be the field $\mathbb{Q}$ of rational numbers. Hence,

```
> p := x^2 + 2*x + 2;
```

$$p := x^2 + 2\,x + 2$$

is irreducible (over $\mathbb{Q}$), because it has only one factor:

```
> factor(p);
```

$$x^2 + 2\,x + 2$$

Note, however, that it is reducible over $\mathbb{Z}[i]$, the Gaussian integers:

```
> factor(p, I);
```

$$(x + 1 + I)\,(x + 1 - I)$$

It is worth remarking that factorization applies only to rings and makes no sense in fields. Irreducible means the same as prime, but prime is usually applied to integers whereas irreducible is usually applied to polynomials, both of which constitute rings. By default, the term *factorization* means irreducible factorization, i.e., expression as a product of irreducible (or prime) factors. However, there are other important kinds of factorization, and my purpose here is to implement *squarefree* factorization.

A polynomial is called *squarefree* if it has no *repeated* or *multiple* factors. Thus,

$$x^3 + 4\,x^2 + 6\,x + 4 = (x^2 + 2\,x + 2)\,(x + 2)$$

is squarefree, whereas

$$x^4 + 6\,x^3 + 14\,x^2 + 16\,x + 8 = (x^2 + 2\,x + 2)\,(x + 2)^2$$

is not squarefree. An irreducible polynomial is necessarily squarefree, but not vice versa.

An important property of a squarefree polynomial is that it has no repeated or multiple *zeros*. It is much easier and more reliable to find a simple zero than a multiple zero, hence it is much easier to find the zeros of a polynomial if it is squarefree than if it is not. Moreover, squarefree factorization is a relatively easy and inexpensive operation and each squarefree factor can then be solved on its own, thereby reducing the complexity of the problem. Hence, squarefree factorization is an important component when building a reliable univariate polynomial equation solver.

The squarefree factorization of a polynomial can be computed quite easily using only GCD (greatest common divisor) computation, division, and one differentiation. To see how it works, consider the following. If two polynomials were irreducibly factorized then their GCD would be the product of their common factors. For example, if

$$p = (x + 3)\,(x + 2)^2, \quad q = (x + 1)\,(x + 2)$$

then

$$\gcd(p,\, q) = x + 2.$$

If

$$p = (x - a)^n$$

then its derivative with respect to the variable x is

$$p' = n\,(x - a)^{(n-1)}.$$

Note that p can be expressed as

$$p = (x - a)\,(x - a)^{(n-1)}$$

so clearly

$$\gcd(p,\, p') = (x - a)^{(n-1)}.$$

This is the repeated factor of p but repeated one less time, which this computation would give even if there were a product of repeated factors. It is easy to prove in general, but here is an example (using Maple):

```
> p := (x+1)*(x+2)^2*(x+3)^3;
```

$$p := (x + 1)\,(x + 2)^2\,(x + 3)^3$$

```
> `p'` := factor(diff(p, x));
```

$$p' := 2\,(x + 2)\,(3\,x^2 + 11\,x + 9)\,(x + 3)^2$$

Hence, p with the power of each of its factors reduced by 1 is given by:

```
> p_1 := gcd(p, `p'`);
```

$$p\_1 := (x + 2)\,(x + 3)^2$$

This observation is the basis of at least two useful algorithms. Note that I have used the Maple factorizer above only to make the illustration clearer — explicit factorization is *not* really part of the algorithm!

We can build up the full squarefree factorization algorithm as follows. "*The squarefree factor*" of:

```
> p;
```

$$(x + 1)\,(x + 2)^2\,(x + 3)^3$$

is given by:

```
> q1 := p / p_1;
```

$$q1 := (x + 1)\,(x + 2)\,(x + 3)$$

The factors repeated at least twice can now be picked out as:

```
> q2 := gcd(p_1, q1);
```

$$q2 := (x+2)(x+3)$$

and these can be divided out of the squarefree factor to leave just the factors that occur raised to the power 1 as:

```
> p1 := q1 / q2;
```

$$p1 := x+1$$

Now we can repeat this analysis with p replaced by:

```
> p_1;
```

$$(x+2)(x+3)^2$$

Dividing out the repeated factors gives *p_1* with the power of each of its factors reduced by 1, which is p with the power of each of its remaining factors reduced by 2:

```
> p_2 := p_1 / q2;
```

$$p\_2 := x+3$$

The factors repeated at least three times can now be picked out as

```
> q3 := gcd(p_2, q1);
```

$$q3 := x+3$$

The following procedure performs this algorithm as far as possible to pick out the squarefree (but not necessarily irreducible) factors of a polynomial that are raised to different powers. This squarefree factorization is the first step of a complete irreducible factorization. As you see, it is not very complicated! It returns "the squarefree factor" via an optional argument `SFF`. (Note that it does not call the Maple function `factor`.) The procedure returns an unexpanded polynomial, which is possible (easily) only because Maple does not automatically expand products.

```
> SqFree := proc(Poly::polynom, Var::symbol, SFF::name)
     # Return the squarefree factorization of a polynomial
     # Poly (with respect to the variable Var).
     local p_i, q1, qi, qi1, P, i;
     P := 1;
     p_i := gcd(Poly, diff(Poly, Var));
     q1 := normal( Poly / p_i );  qi := q1;
     if nargs >= 3 then SFF := q1 end if;  # optional arg
     for i while qi <> 1 do
        qi1 := gcd(p_i, q1);  qi := normal( qi / qi1 );
        if qi <> 1 then  # Return only non-trivial factors
           P := P * qi^i
```

```
        end if;
        p_i := normal( p_i / qi1 );  qi := qi1
    end do;
    p_i * P
  end proc:
```

Here is a test example:

```
> p := 123*(x^2-1)*(x^2-4)^2*(x^2-9)^3;
```

$$p := 123\,(x^2 - 1)\,(x^2 - 4)^2\,(x^2 - 9)^3$$

```
> p := expand(p);
```

$$p := 123\,x^{12} - 4428\,x^{10} + 62730\,x^8 - 440340\,x^6 + 1577475\,x^4 - 2630232\,x^2 + 1434672$$

```
> SqFree(p, x, 'squarefreefactor');
    # NOT an irreducible factorization!
```

$$123\,(x^2 - 1)\,(x^2 - 4)^2\,(x^2 - 9)^3$$

```
> squarefreefactor;
```

$$x^6 - 14\,x^4 + 49\,x^2 - 36$$

```
> factor(%);  # check it really is squarefree
```

$$(x - 1)\,(x - 2)\,(x - 3)\,(x + 3)\,(x + 2)\,(x + 1)$$

12.1.2 Sturm sequences

Sturm sequences provide a way of locating the real zeros of (real) polynomials. Suppose p is a *squarefree* univariate polynomial and p' is its derivative. Then the Sturm sequence of p is the sequence of polynomials p_i, $0 \le i \le k$, defined by:

$$p_0 = p, \quad p_1 = p',$$
$$p_i = -\mathrm{rem}(p_{i-2}, p_{i-1}), \quad 2 \le i \le k,$$
$$p_k = \textit{non-zero constant}.$$

This is a "polynomial remainder sequence" that, apart from the negative sign in front of the remainder, is the same as the sequence constructed by the Euclidean GCD algorithm.

The last element is essentially $\gcd(p, p')$, which must be a non-zero constant because p is assumed to be squarefree. The following procedure computes the Sturm sequence (as a Maple list for convenience of further manipulation) for any univariate polynomial:

```
> Sturm := proc(p::polynom)
    local x, p0, p1, p2, S;
    x := indets(p); # set of indeterminates (or variables)
    if nops(x) <> 1 then
       error "poly must be univariate"
    end if;
    x := op(x);  # single indeterminate
    p1 := p / gcd(p, diff(p, x));  # squarefree factor
    p2 := diff(p1, x);  S := p1, p2;  # initial sequence
    while has(p2, x) do  # remainder still contains x
       p0 := p1;  p1 := p2; p2 := -rem(p0, p1, x);
       S := S, p2
    end do;
    [S]
  end proc:
```

Here is an example polynomial with zeros at $x = 1, 3, 5$:

```
> p := (x-1)*(x-3)*(x-5);
```

$$p := (x - 1)\,(x - 3)\,(x - 5)$$

```
> expand(p);  S := Sturm(%);
```

$$x^3 - 9\,x^2 + 23\,x - 15$$

$$S := [x^3 - 9\,x^2 + 23\,x - 15,\ 3\,x^2 - 18\,x + 23,\ -8 + \frac{8}{3}\,x,\ 4]$$

Suppose a is a real number that is not a zero of a polynomial p. Then the *variation* of p at a is defined to be the number of variations of sign (ignoring any zeros) in the elements of the Sturm sequence evaluated at a. For example, the above Sturm sequence at $x = 0$ is:

```
> subs(x = 0, S);
```

$$[-15,\ 23,\ -8,\ 4]$$

The sequence of signs is clearly shown by:

```
> map(signum, %);
```

$$[-1,\ 1,\ -1,\ 1]$$

so the variation of p at $x = 0$ is 3. The following procedure computes the variation at a of any Sturm sequence S:

```
> Var := proc(a::numeric, S::list(polynom))
    local x, Sa, v, el, sgn1, sgn2;
    x := indets(S);
    if nops(x) <> 1 then
       error "polys must be univariate"
    end if;
    Sa := subs(op(x) = a, S);  # evaluate at a
    if Sa[1] = 0 then
       error "poly is zero at %1", a
    end if;
    v := -1;
    for el in Sa do
       sgn2 := signum(el);
       if el <> 0 and sgn2 <> sgn1 then
          v := v + 1;  sgn1 := sgn2
       end if
    end do;
    v
  end proc:
```

```
> Var(0, S);
```

$$3$$

```
> Var(1, S);
```

```
Error, (in Var) poly is zero at 1
```

Now recall that

```
> 'p' = p;
```

$$p = (x-1)\,(x-3)\,(x-5)$$

and note that

```
> map(Var, [0, 2, 4, 6], S);
```

$$[3, 2, 1, 0]$$

i.e., the variation decreases each time a zero of p is passed. This motivates:

Sturm's Theorem
If a and b are two real numbers that are not zeros of a polynomial p, such that $a < b$, then the number of real zeros of p in the interval $(a, b]$ is $\mathrm{Var}(a, p) - \mathrm{Var}(b, p)$.

This is true because each time a zero of p is passed the sign of p changes, and it can be proved that this alone causes a change in the variation of the Sturm sequence. For a neat proof, see Knuth's *Seminumerical Algorithms* [19]. The following procedure counts the number of real zeros of p in any allowed range:

```
> NZeros := proc(S::list(polynom), R::range(numeric))
     if lhs(R) >= rhs(R) then error "invalid range" end if;
     Var(lhs(R), S) - Var(rhs(R), S)
  end proc:
```

Let us apply it to our example polynomial p with simple known zeros:

```
> 'p' = p;
```

$$p = (x - 1)(x - 3)(x - 5)$$

```
> NZeros(S, 0 .. 2);
```

$$1$$

```
> NZeros(S, 0 .. 4);
```

$$2$$

```
> NZeros(S, 2 .. 6);
```

$$2$$

```
> NZeros(S, 0 .. 6);
```

$$3$$

```
> NZeros(S, 0 .. 1);
   Error, (in Var) poly is zero at 1
```

Now we can use Sturm's Theorem to divide up any given interval into isolating intervals, each of which contains precisely one zero of p, by recursively halving intervals until they contain one zero and discarding intervals that contain no zeros, as implemented in the following procedure:

```
> Isolate := proc(S::list(polynom), R::range(numeric))
     local NZ, a, b, m;
     NZ := NZeros(S, R);
     if NZ = 1 then R
     elif NZ > 1 then
        a := lhs(R);  b := rhs(R);  m := (a+b)/2;
        Isolate(S, a..m), Isolate(S, m..b)
     end if
  end proc:
```

```
> Isolate(S, -8..8);
```

$$0..2,\ 2..4,\ 4..8$$

In practice, care needs to be taken about hitting zeros, and the algorithm could be implemented a lot more efficiently. Each isolating interval can be made as small as desired by some purely numerical method, such as interval halving, which we will develop in the next section. Given an interval that is guaranteed to contain all the real zeros, the technique outlined here will find them all reliably to any required accuracy. There are fairly simple formulae for upper bounds on the magnitudes of real zeros, which give suitable initial intervals; see the exercises.

Rather than starting from the squarefree factor of a polynomial, a full squarefree factorization would usually be performed, and then the zeros of each squarefree factor would be computed separately and each zero returned together with its multiplicity, which is the power to which the squarefree factor appears in the polynomial. This sketch gives some idea of how the `fsolve` function might work when applied to a single univariate polynomial. Non-polynomial equations are much more difficult to solve automatically, although the numerical routine presented in the next section will still reliably find simple (or odd-multiplicity) zeros within specified isolating intervals.

12.1.3 Finding zeros by interval halving

The zeros of a univariate expression (polynomial or transcendental) can be found to any desired accuracy by various purely numerical algorithms (given suitable information about their approximate locations), among which a very simple yet reliable algorithm is *interval halving* or *bisection*. Provided the initial interval contains a single simple (i.e., non-multiple) zero then that zero can be found by halving the interval and then repeating the algorithm on the half-interval containing the zero. In general, if the signs of the values of a function at opposite ends of an interval are different then the function has an odd number of zeros within the interval, which is the basis for choosing the right sub-interval. In the case of a polynomial, the algorithm should be applied only to squarefree factors, which ensures that the zeros are simple, and we can determine an isolating interval for each zero from a Sturm sequence.

The following procedure illustrates a Maple implementation of interval halving. Let us take a "top-down" approach to developing this procedure and start with a flexible user interface. The user interface procedure `solveh` (solve by halving) accepts either an expression or a mapping as its first argument; if it is an expression then `solveh` checks that it is univariate and converts it to a mapping for internal use. Procedure `solveh` accepts an optional fourth argument that specifies the maximum absolute error; if it is omitted then this is set to a default value (10^{-6}) for internal use. This implementation *assumes* that the hardware floating-point evaluation function `evalhf` can be used to evaluate the input expression or mapping. It would be better to test this using a `try` statement and if it fails to use the much more flexible but slower software floating-point evaluation function `evalf` instead, but for simplicity I will not illustrate that here. (See the printed *Maple 6 Programming Guide* [22] for details.) Procedure `solveh` numerically evaluates the interval end-points and checks that neither of them is a zero of the function. Then it checks that there really is a zero within the interval by checking that the function values have different signs at the two end-points and if so it calls a subsidiary procedure `solveh1` with the end-points guaranteed to be in increasing numerical order.

```
> solveh := proc(f_arg::{algebraic, procedure},
        a_arg::constant, b_arg::constant, abserr_arg::positive)
    # Solve f(x) = 0, a <= x <= b (or b <= x <= a),
    # by interval halving.
    # This is the user-interface procedure.
    local f, x, a, b, fa, fb, abserr;
    if not type(f_arg, procedure) then  # unapply it
       x := indets(f_arg, 'name');
       if nops(x) <> 1 then
          error "expression must be univariate"
       end if;
       f := unapply(f_arg, op(x))
    else f := f_arg end if;
    # Provide a default absolute error:
    if nargs = 4 then abserr := abserr_arg
    else abserr := 1e-6 end if;
    a := evalhf(a_arg);  b := evalhf(b_arg);
    # evalhf cannot perform general function application...
    fa := f(a);  fa := evalhf(fa);
    fb := f(b);  fb := evalhf(fb);
    if fa = 0 then a elif fb = 0 then b
    elif evalb(fa > 0) = evalb(fb > 0) then NULL  # No soln
    elif a < b then solveh1(f, a, fa, b, fb, abserr)
    elif b < a then solveh1(f, b, fb, a, fa, abserr)
    # otherwise a = b so no solution
    end if
 end proc:
```

The expression

```
evalb(fa > 0) = evalb(fb > 0)
```

is true if either both relations are true or both are false, and so is a succinct and efficient way of implementing the condition:

```
((fa > 0) and (fb > 0)) or ((fa <= 0) and (fb <= 0))
```

Since we have already checked that $fa \neq 0$ and $fb \neq 0$, this expression will be true only if the function values have the same sign at both ends of the interval and hence there is no zero within the interval.

Here is a naive and explicitly recursive implementation of the subsidiary procedure:

```
> solveh1 := proc(f, a, fa, b, fb, abserr)
     # Solve f(x) = 0, a <= x <= b, fa <> 0, fb <> 0,
     # by recursive interval halving.
     local x, fx;
     x := 0.5*(a+b);  # mid-point
     if abs(b-a) < abserr then return x end if;
     fx := f(x);  fx := evalhf(fx);
     if fx = 0 then x
     elif evalb(fa > 0) <> evalb(fx > 0) then
        solveh1(f, a, fa, x, fx, abserr)
     else
        solveh1(f, x, fx, b, fb, abserr)
     end if
  end proc:
```

It works, but only to fairly low precision:

```
> solveh(sin, 2, 4);
```

$$3.141592501$$

```
> solveh(sin(x), 2, 4);
```

$$3.141592501$$

```
> solveh(sin, 2, 4, 1e-9);
```

```
Error, (in sin) too many levels of recursion
```

```
> solveh(a*sin(x), 2, 4);
```

```
Error, (in solveh) expression must be univariate
```

The algorithm can also be implemented iteratively (i.e., without explicit recursion). The fastest numerical evaluation in Maple is obtained by calling the hardware floating-point evaluation function `evalhf` in such a way that each call performs as much numerical computation as possible (to minimize the overhead of converting between the floating-point formats used by Maple and by the hardware). The following version of `solveh1` creates a purely numerical and non-recursive interval-halving local procedure called `solveh2` and then calls `evalhf` *once only* to evaluate it. User-defined functions can be evaluated by `evalhf`, but it does not support lexical scoping. Hence, the trick using `subs` is necessary before `evalhf` is called. (In earlier versions of Maple this trick was used routinely to achieve the effect of lexical scoping before it was provided automatically in most situations.)

```
> solveh1 := proc(f, a, fa, b, fb, abserr)
     # Solve f(x) = 0, a <= x <= b, fa <> 0, fb <> 0,
     # by iterative interval halving making maximum
     # use of fast hardware floating-point computation.
     local solveh2;  global _f;
     solveh2 := proc(aa, faa, bb, fbb, abserr)
        # a purely numerical procedure
        local a, b, fa, fb, x, fx;
        a := aa;  b := bb;  fa := faa;  fb := fbb;
        while abs(b - a) > abserr do
           x := 0.5*(a+b);  fx := _f(x);  # mid-point
           if fx = 0 then return x
           elif (fa > 0) <> (fx > 0) then
              # Boolean evaluation of each
              # relation (>) forced by evalhf
              b := x;  fb := fx
           else a := x;  fa := fx end if
        end do;
        0.5*(a+b)  # mid-point
     end;
     # Complete the definition of solveh2:
     solveh2 := subs(_f = f, eval(solveh2));
     # and execute it using only hardware floating point:
     evalhf(solveh2(a, fa, b, fb, abserr))
  end proc:
```

This version runs significantly faster and does not suffer from stack overflow (too many levels of recursion):

```
> solveh(sin, 2, 4);
```

$$3.14159250259399414$$

```
> solveh(sin, 2, 4, 1e-9);
```

$$3.14159265393391252$$

```
> solveh(sin, 2, 4, 1e-14);
```

$$3.14159265358979312$$

```
> evalhf(Pi);
```

$$3.14159265358979312$$

However, there are significantly more efficient algorithms than simple interval halving.

12.2 Multivariate polynomial algorithms

This section is concerned with the more general problem of simplifying or solving systems of multivariate polynomials. A very powerful technique is to construct a Gröbner basis for the polynomial system, which requires the ability to perform generalized polynomial division of one multivariate dividend polynomial by a sequence of multivariate divisor polynomials. This is a generalization of the Euclidean division that is performed for univariate polynomials by the standard Maple functions `quo` and `rem`. The theory and algorithms used in this section are based fairly closely on *Ideals, Varieties, and Algorithms* by Cox *et al.* [4] with some modifications by my colleague Leonard Soicher for use in a Queen Mary course on Advanced Algorithmic Mathematics.

12.2.1 Generalized polynomial division

The theory

Consider the problem of dividing multivariate polynomials that have coefficients in some field, which is (not surprisingly) less straightforward than dividing univariate polynomials. It arises in the theory of ideals of rings of multivariate polynomials. An *ideal* of a polynomial ring R is said to be *generated* by a set of polynomials in R if any element of the ideal can be expressed as an R-linear combination of the generators, which means that the coefficients in the linear combination are arbitrary elements of R. Thus, if an ideal I is generated by the polynomials $f_1, \ldots, f_s$ then any polynomial f in I can be expressed as $f = q_1 f_1 + \ldots + q_s f_s$ where the q_i are arbitrary polynomials in R. An ideal in a ring of univariate polynomials can always be generated by a single polynomial, which is the GCD of any larger set of generators. Hence, it is necessary only to divide by a single generator to determine whether a polynomial is a member of a univariate polynomial ideal. But in the multivariate case this is not true, and in general it is necessary to divide a multivariate polynomial f by a sequence $f_1, \ldots, f_s$ of generators; if there is no remainder then f is a member of the ideal generated by $f_1, \ldots, f_s$. The converse holds only if $f_1, \ldots, f_s$ is a Gröbner basis for the ideal. Maple provides good support for Gröbner bases, for which it uses the alternative spelling Groebner, but I have intentionally avoided using any facilities from the Maple `Groebner` package, which is documented in the online help. (But you are invited later to try it as an exercise.)

Let R be the ring of polynomials in the variables $x_1, \ldots, x_n$ over (i.e., with coefficients in) some field $\mathbb{K}$. A monomial in this ring is any product of powers of the variables x_i and a term is a product of a coefficient in $\mathbb{K}$ and a monomial. The multidegree (mdeg) of a monomial or term is the sequence of powers of the variables x_i that it contains. Various orderings can be imposed on monomials and a suitable monomial ordering must be chosen and then kept fixed. In the chosen ordering, the first term in a polynomial is called the leading term (lt), the corresponding monomial is called the leading monomial (lm) and the multidegree of a polynomial is the multidegree of its leading monomial. Given this background, we can define precisely what we mean by generalized division of multivariate polynomials. Given a fixed ordering on the monomials in R and an ordered s-tuple $f_1, \ldots, f_s$ of polynomials in R, every polynomial f in R can be written as:

$$f = q_1 f_1 + \ldots + q_s f_s + r,$$

where the q_i and r are polynomials in R. The polynomial r is the *remainder* and satisfies the condition that either $r = 0$ or r is a $\mathbb{K}$-linear combination of monomials, none of which is divisible by the leading monomial of any f_i. The polynomials q_i are the quotients and satisfy the condition that if $q_i f_i \neq 0$ then $\text{mdeg}(q_i f_i) \leq \text{mdeg}(f)$.

A first implementation

Here is a version of the algorithm essentially as given by Cox *et al.* to compute the quotients and remainder, implemented as a fairly direct translation to a Maple procedure. It is somewhat analogous to a combination of the standard Maple functions quo and rem, except that it requires a *list* of divisors and a *list* of variables, and it returns both the quotients (as a list) and the remainder together as an expression sequence, rather than returning one directly and one by assignment to an optional argument.

```
> GeneralizedPolynomialDivision :=
    proc(dividend::polynom, divisors::list(polynom),
       vars::list(name), ord::monord)
    # INPUT: 1) dividend (F), a polynomial in R, where
    #    R is a ring of multivariate polynomials over a field.
    # 2) divisors ([f[1],...,f[s]]), an ordered list of
    #    NON-ZERO polynomials in R.
    # 3) vars ([x[1],...,x[n]]), an ordered list of variables.
    # 4) OPTIONAL ordering: 'plex' (Pure LEXicographic, the
    #    default) or 'tdeg' (Total DEGree then lexicographic).
    # OUTPUT: quotient and remainder
    #    ([q[1],...,q[s]],r), polynomials in R such that
    # a) F = q[1]f[1] + ... + q[s]f[s] + r,
    #    if q[i] <> 0 then mdeg(F) >= mdeg(q[i]f[i]) for all i,
```

```
   # b) r = 0 or LM(f[i]) does not divide any monomial of r
   #    for all i.
   local ORD, lt, f, s, q, r, p, i, Q, division_occurred;
   if member(0, divisors) then
      error "divisors must all be non-zero"
   end if;

   ORD := `if`(nargs >= 4, ord, 'plex');
   lt := p->
      # Leading term of multivariate polynomial
      # with respect to ordering ORD
      if p::`+` then op(1,sort(p, vars, ORD))
      else p end if;

   f := expand(divisors);  # ALL polys MUST be expanded!
   s := nops(f);  # no of divisors
   q := vector(s, 0);  # q[i] = 0, i = 1..s
      # q is an array for efficient updating.
   r := 0;  p := expand(dividend);
   while p <> 0 do
      # Loop invariant: F = q[1]f[1] + ... + q[s]f[s] + p + r
      ASSERT(dividend = add(expand(q[i]*f[i]),i=1..s) + p + r,
         "loop invariant");
      i := 1;  division_occurred := false;
      while i <= s and not division_occurred do
         if divide(lt(p), lt(f[i]), 'Q') then
            # Division step takes place: lt(p) | lt(f[i])
            q[i] := q[i] + Q;
            p := p - expand(Q*f[i]);
            division_occurred := true
         else
            i := i + 1
         end if
      end do;
      if not division_occurred then
         # Remainder step takes place
         r := r + lt(p);
         p := p - lt(p)
      end if
   end do;
   convert(q, list), r
end proc:
GPD := GeneralizedPolynomialDivision: # short synonym
```

I used a "top-down" approach to implement this algorithm. By that, I mean that initially I assumed the existence of the following: a function called `lt` to compute the leading term of a polynomial in some specified monomial ordering, a way to test divisibility of two multivariate polynomial terms, and a way to compute their quotient if they were divisible. (The rest of the code is straightforward.) I then searched the Maple online help for anything related to division, quotients, divisibility, etc., and by so doing I found the standard function `divide`, which checks divisibility and computes the quotient as required, so I modified the code to use `divide`, as shown above.

Maple provides the standard function `lcoeff`, which returns the leading coefficient of a multivariate polynomial and can also return its leading term via an optional argument. However, the online help does not state what ordering it uses and there appears to be no way to control the ordering, so it is not appropriate here. I therefore searched the online help again for anything related to leading, ordering, sorting, etc. and concluded that the only relevant standard facility was the function `sort`, which will sort a polynomial (in place) according to either one of two standard orderings or a user-defined ordering. I therefore implemented the function `lt` by calling `sort` appropriately. If `lt` is applied to a polynomial term (not a sum) then it returns the term unchanged; otherwise, it sorts the polynomial and returns the first term. The default sort order is lexicographic, but a sort order can be specified as an optional fourth argument to the division procedure in terms of the two identifiers used by Maple, namely either `'plex'` (pure lexicographic) or `'tdeg'` (total degree then lexicographic). The following type definition is used (as a shorthand) to check that an acceptable sort order has been used:

```
> `type/monord` := '{identical(plex), identical(tdeg)}':
```

I have called the procedure `GeneralizedPolynomialDivision`, which is a reasonably descriptive identifier but is also rather long, so I have also assigned the abbreviation `GPD`. Because I have assigned the original long name of the procedure rather than the procedure itself to the variable `GPD` (i.e., I have not applied `eval` to the original procedure name), error messages refer to the long descriptive name (as illustrated below). Attempting to divide by zero would cause an error, so the procedure first checks for any zero divisors. Polynomials must be in expanded form, so I explicitly apply `expand` to the input polynomials and to any products that arise during the algorithm.

Loop invariants and ASSERT

A loop invariant for an algorithm is an assertion that depends on quantities that change within a loop but which nevertheless remains true for every iteration of that loop. Loop invariants provide a way to prove the correctness of an algorithm and consequently they also provide a way to check the correctness of its implementation. For the generalized polynomial division algorithm, a loop invariant is $f = (\sum_{i=1}^{s} q_i f_i) + p + r$. Maple provides a function called `ASSERT` that takes as its required first argument an expression that can be evaluated to a Boolean value. If this expression evaluates to false then execution can be terminated by generating an error (that cannot be caught) and the optional second string argument, if provided, is output as the error message. Assertion checking is off by default (`assertlevel = 0`), and when it is off the overhead of `ASSERT` calls in Maple code is minimal. `ASSERT` can be used for various purposes including checking loop invariants as in the above procedure.

Assertion checking must be explicitly turned on as follows to activate it when testing code:

```
> kernelopts(assertlevel = 1);
```

$$0$$

Since assertion checking does not show up any errors in the procedure, the effect of turning on assertion checking is not shown below. While I was developing the code, I changed the assertion in a trivial way so that it should be false to ensure that the assertion mechanism was working correctly. Having assured myself that the assertion code was correct, I reinstated the correct assertion before testing the rest of the procedure as follows:

```
> GPD(x^2+1, [], [x]);
```

$$[\,],\ x^2+1$$

```
> GPD(x^2+1, [0], [x]);
```

```
Error, (in GeneralizedPolynomialDivision) divisors must all be
non-zero
```

```
> GPD(x^2+1, [x], [x]);
```

$$[x],\ 1$$

```
> GPD(x^2+1, [x^2], [x]);
```

$$[1],\ 1$$

```
> GPD(x^2+1, [x^3], [x]);
```

$$[0],\ x^2+1$$

```
> GPD(x^2+1, [x+1], [x]);
```

$$[x-1],\ 2$$

```
> GPD(x^2+1, [x^2+1], [x]);
```

$$[1], 0$$

In order to be able to check the results of the following more complicated bivariate polynomial tests I show the input as well as the output in pretty-printed form:

```
> x^2*y+x*y^2+y^2, [x*y-1, y^2-1];  GPD(%, [x,y]);
```

$$x^2\,y + x\,y^2 + y^2,\ [x\,y - 1,\ y^2 - 1]$$
$$[x + y,\ 1],\ x + y + 1$$

```
> x^2*y+x*y^2+y^2, [y^2-1, x*y-1];  GPD(%, [x,y]);
```

$$x^2\,y + x\,y^2 + y^2,\ [y^2 - 1,\ x\,y - 1]$$
$$[x + 1,\ x],\ 2\,x + 1$$

```
> x*y^2-x, [x*y+1, y^2-1];  GPD(%, [x,y]);
```

$$x\,y^2 - x,\ [x\,y + 1,\ y^2 - 1]$$
$$[y,\ 0],\ -x - y$$

```
> x*y^2-x, [y^2-1, x*y+1];  GPD(%, [x,y]);
```

$$x\,y^2 - x,\ [y^2 - 1,\ x\,y + 1]$$
$$[x,\ 0],\ 0$$

An improved implementation

The most important property of a computer program is that it should work correctly and reliably, and writing clearly and simply and staying close to the theory is a good way to achieve that. Then, if necessary, attention can be paid to improving such things as the speed of execution and memory usage.

There are several aspects of the previous implementation that could be improved. The leading term of p is recomputed in every iteration of the inner loop, even though p does not change. It should be computed once only before entry into the loop and saved in a local variable. What is worse, the leading term of each f_i is recomputed in every iteration of the inner loop, even though none of the f_i change at all. These values should also be computed once only before entry into the outer loop and saved in a local variable. These improvements are important; they are easy to implement and should improve the efficiency considerably for non-trivial problems (although I have not tested this).

These improvements introduce two more local variables. However, the local variable `division_occurred` is not really necessary. The inner loop could be terminated by a `break` and the "`not division_occurred`" test replaced by "`i > s`", which will be true only if no division occurred. In fact, I have taken this modification a step further so that the inner loop only tests divisibility and has an empty body. It terminates either when all divisions have failed or when one has succeeded. The value of `i` after loop termination indicates whether division occurred. The code resulting from this change is slightly more streamlined from a programming point of view. The following implementation also accepts either a list or a set of divisor polynomials; the latter will prove convenient later when the divisors will be the elements of a Gröbner basis.

```
> GeneralizedPolynomialDivision :=
    proc(dividend::polynom, divisors::{list,set}(polynom),
       vars::list(name), ord::monord)
    # INPUT and OUTPUT as for previous version.
    local ORD, lt, f, s, q, r, p, i, Q, ltf, ltp;
    if member(0, divisors) then
       error "divisors must all be non-zero"
    end if;

    ORD := `if`(nargs >= 4, ord, 'plex');
    lt := p->
       # Leading term of multivariate polynomial
       # with respect to ordering ORD
       if p::`+` then op(1,sort(p, vars, ORD))
       else p end if;

    # Ensure f is a list to preserve order
    # of elements between f and ltf:
    f := `if`(divisors::set, [op(divisors)], divisors);
    f := expand(f);  # ALL polys MUST be expanded!
    ltf := map(lt, f);
    s := nops(f);  # no of divisors
    q := vector(s, 0);  # q[i] = 0, i = 1..s
       # q is an array for efficient updating.
    r := 0;  p := expand(dividend);
    while p <> 0 do
       # Loop invariant: F = q[1]f[1] + ... + q[s]f[s] + p + r
       ASSERT(dividend = add(expand(q[i]*f[i]),i=1..s) + p + r,
          "loop invariant");
       ltp := lt(p);
```

```
      for i to s while not divide(ltp, ltf[i], 'Q') do end do;
      if i <= s then
         # Division step takes place: lt(p) | lt(f[i])
         q[i] := q[i] + Q;
         p := p - expand(Q*f[i])
      else
         # Remainder step takes place
         r := r + ltp;
         p := p - ltp
      end if
   end do;
   convert(q, list), r
end proc:
```

This implementation produces exactly the same test results as the previous version. Note that the previous abbreviation GPD now refers to the new version, because GPD evaluates to the *name* of the procedure rather than to the procedure itself.

12.2.2 Gröbner bases

An application of generalized polynomial division is the computation of a Gröbner basis for a multivariate polynomial ideal generated by a given finite set of polynomials. Gröbner bases have various uses: determining equivalence of two ideals, determining whether a polynomial is a member of an ideal, or computing the common zeros of the members of an ideal (i.e., solving a system of coupled multivariate polynomial equations). For the latter application it is necessary to use the right kind of monomial ordering, called an *elimination ordering*; lexicographic is an elimination ordering, which is why I chose it as the default.

In the previous section I implemented generalized polynomial division as a single procedure, which contains a local procedure to extract the leading term of a polynomial. In this section I will take a more open tool-based approach. This raises the question of how to pass information among the tools. Passing all information via arguments can be unwieldy, but relying on global variables is potentially dangerous because there is no control over them. In particular, I do not want every tool to have to handle the choice of monomial ordering, so I will make this a global variable. But there is no way to ensure that it is always set to an appropriate value, so care is required. A serious implementation would use a module; then the monomial ordering could be local to the module but effectively global to all the procedures within the module (via lexical scoping) and its value could be reliably controlled. I will illustrate such use of a module in Chapter 14 and in more detail in Chapter 15. But to avoid confusing the issue here I will simply use a global variable.

A Gröbner basis (GB) can be computed from an arbitrary set of multivariate polynomials (regarded as generators of an ideal) by an algorithm first given by Bruno Buchberger in his Ph.D. thesis in 1965. (Buchberger coined the name "Gröbner basis" in honour of his Ph.D. supervisor W. Gröbner, 1899–1980.) Buchberger's algorithm requires two main ingredients: one is generalized polynomial division, which we have already developed, and the other is S-polynomials.

S-polynomials

The *S-polynomial* of two polynomials f, g is defined by:

$$S(f, g) = \begin{cases} 0 & \text{if } f = 0 \text{ or } g = 0 \\ \left(\frac{\operatorname{lcm}(\operatorname{lm}(f), \operatorname{lm}(g))}{\operatorname{lt}(f)}\right) f - \left(\frac{\operatorname{lcm}(\operatorname{lm}(f), \operatorname{lm}(g))}{\operatorname{lt}(g)}\right) g & \text{otherwise} \end{cases}$$

where lt denotes leading term, lm denotes leading monomial, and lcm denotes least common multiple. The purpose of $S(f, g)$ is to eliminate the leading terms of f and g while using only polynomial arithmetic. The denominators are chosen to ensure cancellation of the leading terms and the numerator is chosen to be the lowest degree monomial that is divisible by both $\operatorname{lt}(f)$ and $\operatorname{lt}(g)$, thus ensuring that the two factors in parentheses are polynomial. The numerical factors are largely irrelevant since we assume that all polynomial coefficients are field elements.

Here is the procedure that we used earlier to extract the leading term of a multivariate polynomial, except that now the variable `ORD` representing the monomial ordering is global:

```
> lt := proc(poly, vars)
     # Leading term of multivariate polynomial
     # with respect to ordering ORD.
     global ORD;
     if poly::`+` then op(1,sort(poly, vars, ORD))
     else poly end if
  end proc:
```

Since we need both the leading terms and the leading monomials of f and g, an efficient way to compute the leading monomials is to divide the leading terms by their coefficients. Hence, we need a procedure to extract the coefficient from a single term. This requires some care because either the coefficient or the monomial in a term may be implicit. Here is one approach; in general all the work is done by the call of `remove`, but the second test is required in case the term is a single variable.

```
> termcoeff := (term, vars)->
     # Coeff of a term.
     if term::`*` then remove(has,term,vars)
```

```
    elif has(term,vars) then 1
    else term end if:
```

The following tests illustrate all three branches of the conditional statement:

```
> termcoeff(3*x*y, [x,y]);
```

$$3$$

```
> termcoeff(x*y, [x,y]);
```

$$1$$

```
> termcoeff(x, [x,y]);
```

$$1$$

```
> termcoeff(3, [x,y]);
```

$$3$$

Now it is easy to implement a procedure to compute S-polynomials reasonably efficiently:

```
> Spoly := proc(f, g, vars)
     # S-polynomial of polynomials f and g.
     local ltf, ltg, lcmfg;
     if f = 0 or g = 0 then return 0 end if;
     ltf := lt(f,vars);  ltg := lt(g,vars);
     lcmfg := lcm(ltf/termcoeff(ltf,vars),
                  ltg/termcoeff(ltg,vars));
     ltf := lcmfg/ltf;  ltg := lcmfg/ltg;
     expand(ltf*f - ltg*g)
  end proc:
```

Before trying to test it we must ensure that the global variable `ORD` is set appropriately:

```
> ORD := 'plex';
```

$$ORD := plex$$

```
> x^2*y-1, x*y^2-1;  Spoly(%, [x,y]);
```

$$x^2\,y - 1,\, x\,y^2 - 1$$
$$-y + x$$

Buchberger's algorithm

The following simple version of Buchberger's algorithm includes additional polynomials into the original set until the result is a Gröbner basis. It does this by considering each distinct pair of polynomials, computing its S-polynomial, and then reducing it with respect to the rest of the polynomials (i.e., computing its remainder in a generalized polynomial division). If the result is non-zero then it is included in the set. This process is repeated until the set no longer changes. Probably the trickiest aspect of implementing this algorithm is to keep track of which polynomials have been considered, and one way to do this is explicitly to maintain a set of all pairs of polynomials, which I will call P. However, I will represent the Gröbner basis under construction as an expression sequence and only convert it to a set immediately before returning it. This should be slightly more efficient that representing it as a set internally and the elements included in it must always be distinct as a consequence of the reduction step of the algorithm.

We need to compute the set of all (unordered) pairs of elements of a set. This is a special case of computations such as the power set computation that we considered in Chapter 9, but the general approach used there is overkill and a naive and inefficient double sequence computation will suffice, since we need to perform this construction only once. Here is the code and some tests:

```
> Pairs := (S::set)->
    # Return the set of all pairs of elements of S.
    {seq(seq(`if`(t=s,NULL,{s,t}), t=S), s=S)}:
```

```
> Pairs({});
```

$$\{\}$$

```
> Pairs({1,2,3});
```

$$\{\{1, 2\}, \{1, 3\}, \{2, 3\}\}$$

It might seem that maintaining a set of pairs of basis elements is inefficient, but expressions in Maple are represented uniquely, so in fact we are only maintaining a set of pairs of *pointers* to polynomials that we need anyway. Hence, there is no reason to believe that this implementation will be grossly inefficient, despite the fact that it is remarkably straightforward. It reduces each S-polynomial with respect to the basis computed so far by calling `GeneralizedPolynomialDivision` and extracting the remainder as the second element of the sequence returned.

```
> GroebnerBasis := proc(F::set(polynom), vars::list(name),
        ord::monord)
```

```
    # Return a Groebner Basis for the ideal generated by F.
    local P, G, r, g;
    global ORD;
    ORD := `if`(nargs >= 3, ord, 'plex');

    # Set of all pairs of basis elements:
    P := Pairs(F);
    # The initial basis sequence:
    G := op(F);
    while P <> {} do
       r := P[1];        # an arbitrary pair
       P := P[2..-1];  # P minus {r}
       r := Spoly(op(r), vars);
       r := GeneralizedPolynomialDivision
               (r, [G], vars, ORD)[2];
       if r <> 0 then
          for g in G do
             P := P union {{g,r}}
          end do;
          G := G,r
       end if
    end do;
    {G}
 end proc:
```

Here is a test example considered by Cox *et al.*:

```
> F := {x^3-2*x*y, x^2*y-2*y^2+x};
```

$$F := \{x^2\,y - 2\,y^2 + x,\; x^3 - 2\,x\,y\}$$

The GB using lexicographic ordering (our default) with $x > y$ is

```
> GroebnerBasis(F, [x,y]);
```

$$\{4\,y^3,\; x - 2\,y^2,\; x^2\,y + x - 2\,y^2,\; x^3 - 2\,x\,y,\; -4\,y^5,\; x^2\}$$

and with $y > x$ is

```
> GroebnerBasis(F, [y,x]);
```

$$\{-\frac{1}{2}\,x^2,\; -2\,y^2 + y\,x^2 + x,\; -2\,y\,x + x^3\}$$

We see that these two GBs are different. The GBs using graded lexicographic (or total degree) ordering as used by Cox *et al.* are

```
> GroebnerBasis(F, [x,y], 'tdeg');
```

$$\{-2\,x\,y,\, -2\,y^2 + x,\, x^2\,y - 2\,y^2 + x,\, x^3 - 2\,x\,y,\, x^2\}$$

```
> GroebnerBasis(F, [y,x], 'tdeg');
```

$$\{-2\,y\,x,\, -2\,y^2 + x,\, y\,x^2 - 2\,y^2 + x,\, x^3 - 2\,y\,x,\, x^2\}$$

which are actually the same:

```
> evalb(% = %%);
```

true

but slightly different from that given by Cox *et al.* In general we should expect different GBs if we use different orderings. We have computed a GB different from that computed by Cox *et al.* using the *same* ordering because in general GBs are not unique and depend on precisely how they are computed, but they can be made unique as follows.

Reduced Gröbner bases

A *reduced GB* is one in which every polynomial is *reduced* with respect to the rest, which means that it is replaced by its remainder in a generalized polynomial division, and if that remainder is zero then the polynomial is discarded. Moreover, every polynomial in the basis is made monic, meaning that it is divided by its leading coefficient. (This is the appropriate definition for polynomials over a field; for polynomials over a more general coefficient domain a different definition would be necessary and indeed the Maple `Groebner` package does not return a monic GB.) A reduced GB is of minimal size and moreover is *unique*. Hence, two ideals are identical if their reduced GBs are identical.

Here is a procedure to reduce a GB. It constructs the output GB initially as a sequence for ease of appending elements. It reduces each element of the input GB with respect to all the other elements by calling `GeneralizedPolynomialDivision` and extracting the remainder as the second element of the sequence returned, as does `GroebnerBasis`. If the result is non-zero then it divides it by its leading coefficient and appends it to the output GB sequence. This is a simple implementation. There are some obvious potential improvements that could be made, such as using the new GB as it is computed rather than continuing to use the old GB, but I will not pursue such improvements here.

```
> GBreduce := proc(G::set(polynom), vars::list(name),
        ord::monord)
     # Reduce the Groebner Basis G.
     local newG, i, r;
     global ORD;
     ORD := `if`(nargs >= 3, ord, 'plex');
```

```
    newG := NULL;  # new GB as sequence
    for i to nops(G) do
       r := GeneralizedPolynomialDivision
          (G[i], subsop(i=NULL,G), vars, ORD)[2];
       if r <> 0 then
          newG := newG, r/lc(r,vars)
       end if;
    end do;
    {newG}
 end proc:
```

Computation of the leading term of a multivariate polynomial with respect to the chosen ordering is delegated to the following procedure that simply calls the procedures `lt` and `termcoeff` that we developed earlier:

```
> lc := (poly, vars)->
    # Leading coeff of multivariate polynomial
    # with respect to ordering ORD.
    termcoeff(lt(poly, vars), vars):
```

Here is a simple test:

```
> GroebnerBasis({x^2+1,x,y}, [x,y]);
```

$$\{-1, x, y, x^2+1\}$$

```
> GBreduce(%, [x,y]);
```

$$\{1\}$$

Usually, when we compute a GB we want a reduced GB. The following procedure computes a reduced GB naively by first computing a general GB and then reducing it (although a more sophisticated approach would combine these two operations):

```
> GB := proc(F::set(polynom), vars::list(name),
       ord::monord)
    # Return a REDUCED Groebner Basis
    # for the ideal generated by F.
    global ORD;
    ORD := `if`(nargs >= 3, ord, 'plex');
    GBreduce(GroebnerBasis(F, vars, ORD), vars, ORD)
 end proc:
```

Now let us compute *reduced* GBs for the example considered by Cox *et al.* For comparison, I will also recompute the GBs that we obtained before directly from Buchberger's algorithm:

```
> GroebnerBasis(F, [x,y]);  # lex, x > y
```

$$\{-2\,x\,y,\, x-2\,y^2,\, x^2\,y+x-2\,y^2,\, x^3-2\,x\,y,\, x^2,\, 2\,y^3\}$$

```
> GB(F, [x,y]);  #reduced
```

$$\{x-2\,y^2,\, y^3\}$$

```
> GroebnerBasis(F, [y,x]);  # lex, y > x
```

$$\{-\frac{1}{2}\,x^2,\, -2\,y^2+y\,x^2+x,\, -2\,y\,x+x^3\}$$

```
> GB(F, [y,x]);  #reduced
```

$$\{y^2-\frac{1}{2}\,x,\, y\,x,\, x^2\}$$

```
> GroebnerBasis(F, [x,y], 'tdeg');  # grlex, x > y
```

$$\{-2\,x\,y,\, -2\,y^2+x,\, x^2\,y-2\,y^2+x,\, x^3-2\,x\,y,\, x^2\}$$

```
> GB(F, [x,y], 'tdeg');  #reduced
```

$$\{y^2-\frac{1}{2}\,x,\, x\,y,\, x^2\}$$

This is identical to the reduced GB computed by Cox *et al.*, as it should be. For this particular example, the variable ordering makes no difference, whether or not the GB is reduced:

```
> GroebnerBasis(F, [y,x], 'tdeg');  # grlex, y > x
```

$$\{-2\,y\,x,\, 2\,y^2-x,\, y\,x^2-2\,y^2+x,\, x^3-2\,y\,x,\, x^2\}$$

```
> GB(F, [y,x], 'tdeg');  #reduced
```

$$\{y^2-\frac{1}{2}\,x,\, y\,x,\, x^2\}$$

Application to ideal membership

Once again, here are a couple of examples from Cox *et al.*, who use total degree (also called graded lexicographic) ordering, so we will do the same. Let I be an ideal of the ring $R = \mathbb{C}[x, y, z]$ generated by the two polynomials:

```
> f1 := x*z-y^2;  f2 := x^3-z^2;
```

$$f1 := x\,z - y^2$$
$$f2 := x^3 - z^2$$

First, we want to know whether the polynomial:

```
> f := -4*x^2*y^2*z^2+y^6+3*z^5;
```

$$f := -4\,x^2\,y^2\,z^2 + y^6 + 3\,z^5$$

is in I; i.e., do there exist polynomials $q_1, q_2 \in R$ such that $f = q_1 f_1 + q_2 f_2$? If generalized polynomial division of f by f_1, f_2 gives remainder zero then we have an affirmative answer and the quotients give the coefficients in the R-linear combination, but if the remainder is non-zero then in general we have no answer at all. Let us try a couple of trial divisions:

```
> GPD(f, [f1,f2], [x,y,z], 'tdeg');
```

$$[-4\,y^2\,z\,x - 4\,y^4,\ 0],\ -3\,y^6 + 3\,z^5$$

```
> GPD(f, [f2,f1], [x,y,z], 'tdeg');
```

$$[0,\ -4\,y^2\,z\,x - 4\,y^4],\ -3\,y^6 + 3\,z^5$$

The non-zero remainders give us no information. So we need to compute a GB and divide by that instead of by the basis used to specify the ideal. The GB is

```
> G := GB({f1,f2}, [x,y,z], 'tdeg');
```

$$G := \{x^3 - z^2,\ x\,z - y^2,\ x\,y^4 - z^4,\ y^6 - z^5,\ x^2\,y^2 - z^3\}$$

which is (of course) exactly as given by Cox *et al.*, and dividing now gives remainder zero:

```
> GPD(f, G, [x,y,z], 'tdeg');
```

$$[0,\ -4\,y^2\,z\,x - 4\,y^4,\ 0,\ -3,\ 0],\ 0$$

Hence, f is in the ideal I, but how do we find the polynomials q_1, q_2 in R such that $f = q_1 f_1 + q_2 f_2$? The above division shows how to express f as an R-linear combination of three elements of the GB, none of which is either f_1 or f_2. We could in principle construct polynomials q_1, q_2 with undetermined coefficients, then equate coefficients of monomials and solve for the undetermined coefficients. However, further experimentation (which Maple makes very easy!) and some guesswork inspired by the knowledge that a solution exists leads quickly to a solution via division with a different variable ordering, such as the following (although various variable orderings work equally well):

```
> GPD(f, [f1,f2], [z,y,x]);
```

$$[3\,z^2\,x^2 - y^2\,z\,x - y^4,\ -3\,z^3],\ 0$$

Second, we want to know whether the polynomial:

```
> f := x*y-5*z^2+x;
```

$$f := y\,x - 5\,z^2 + x$$

is in the ideal I. Dividing by the GB gives:

```
> GPD(f, G, [x,y,z], 'tdeg');
```

$$[0,\ 0,\ 0,\ 0,\ 0],\ x\,y - 5\,z^2 + x$$

and so the answer is a definitive *no*, this polynomial f is *not* in the ideal I.

12.3 Exercises

The exercises in this and the remaining chapters are more open ended than previously and I will leave more of the detail for the reader to decide.

1. Suppose $p = \sum_{r=0}^{n} a_r\,x^r$ is a polynomial in x of degree n with coefficients a_r and α is a zero of p; i.e., $p = 0$ at $x = \alpha$. Several polynomial root bounds (i.e., bounds on the magnitude of α) are known; two are attributed to Cauchy (1829) and the following is attributed to Knuth [19]:

$$|\alpha| \leq 2 \max \left\{ \left| \frac{a_r}{a_n} \right|^{1/(n-r)},\ 0 \leq r < n \right\}.$$

   Implement this bound and test it for some random polynomials.

2. Combine Knuth's polynomial root bound and the tools developed in the text to implement a complete solver that will return to any specified accuracy all the real roots of a univariate polynomial equation together with their multiplicities. Check its performance against the standard Maple solvers.

3. Compare the reduced GBs computed by procedure `GB` developed in the text with those computed by the `gbasis` function in the Maple `Groebner` package for the simple commutative examples given in the online help for `gbasis`. Note that the latter does not return a monic basis, so you will need to perform some minor conversion in order to make reliable comparisons. Apart from that, the bases should be the same.

4. Write and test a procedure called `IdealEqual` that takes as arguments two multivariate polynomial ideals, represented by sets of generators, and returns `true` if they are identical and `false` otherwise. It should compute reduced GBs for the two ideals and compare them. This should work with any term ordering provided it is the same throughout the computation.

5. Write and test a procedure called `IdealMember` that takes as arguments a multivariate polynomial and a multivariate polynomial ideal, represented by a set of generators, and returns `true` if the polynomial is a member of the ideal and `false` otherwise. It should compute a reduced GB for the ideal and then divide the polynomial by the GB, as illustrated in the text. It should return `true` if the remainder is zero and `false` otherwise. This should work with any term ordering provided it is the same throughout the computation.

6. Write and test a procedure called `IdealSolve` that takes as arguments a set or list of multivariate polynomial expressions or equations and a set or list of variables and attempts to return the solution set of the corresponding system of equations. It should do this by computing a reduced GB for the polynomial ideal generated by the first argument using *lexicographic ordering* (with respect to the implied variable ordering if the variables were given as a list). Then it should find the zeros of the polynomials in the GB in increasing lexicographic order, which should be possible because lexicographic ordering is an *elimination ordering.* Compare your results with those obtained using the standard Maple solvers.

Chapter 13

Spreadsheets

Spreadsheets are conventionally used for several purposes. At their simplest they are used to tabulate data that could be equally well represented as a table in a word or text processor. This leads into simple database applications, in which data in more than one table can be related and data can be sorted and searched. Spreadsheets also provide an alternative computational model intended to match the way some financial calculations (such as income tax) are done on paper, which was, I believe, the original motivation. This computational model spans the range from "interactive calculator" to "edit-debug-execute" and parallels the way that Maple also spans this range. The introduction of spreadsheets into Maple adds algebraic capabilities to the list of spreadsheet applications.

The first section introduces in more detail the operation of spreadsheets in general and Maple spreadsheets in particular. The programming aspect of spreadsheets works by allowing a formula in one spreadsheet cell to reference the values stored in other cells. One of the things that make spreadsheets convenient to work with is their facility for automatically updating formulae when they are copied or moved among cells, which is explained in Section 13.2. A further generalization of this facility allows data to be entered automatically into rows or columns provided the data are simply related as a "series". These standard spreadsheet facilities can be put to good use to investigate mathematical sequences, as described in Section 13.3, and to tabulate non-standard function values, as described in Section 13.4. Spreadsheets are normally manipulated interactively, but it is usually also possible to manipulate them via an external programming interface, and the external programming interface to Maple spreadsheets is outlined in Section 13.5. The last section consists of a case study based on an investigation of the motion of a large-amplitude simple pendulum, which is used as a vehicle to illustrate various applications of spreadsheets to simple numerical computation.

13.1 Introduction

A spreadsheet is an intelligent table. You can ignore the intelligence and just use a spreadsheet as a convenient way to display tabular information, or you can use the intelligence to provide an alternative model for computer programming, which can be very convenient when the input and/or output data are naturally tabular. Maple provides support for spreadsheets and allows the entries to be symbolic, which is potentially very powerful. However, the facilities for manipulating and formatting spreadsheets are not as sophisticated as those provided by a dedicated spreadsheet package such as Microsoft Excel. Fortunately, it is possible to have the best of both facilities (on Windows only), because Maple includes an interface that allows Microsoft Excel 2000 to call Maple as an add-in to perform computations, effectively giving an Excel interface to the Maple computational engine. See the online help entry `updates[Maple6,Excel]` for further details. I will consider only the use of spreadsheets within Maple itself; Maple spreadsheets are available on all platforms but only within Maple worksheets and not via the command-line interface.

A spreadsheet must first be inserted into a worksheet before any spreadsheet facilities can be used. A spreadsheet can be inserted at the cursor by selecting the *Spreadsheet* item from the Insert menu; it is inserted in its own execution group at the left of the worksheet. If desired, the group delimiter can be deleted (by selecting it and then pressing the *Delete* or *Backspace* key), and the position of the spreadsheet can be changed by regarding it as a paragraph consisting of a single (large) character; in particular, its horizontal position can be changed by putting the text cursor on the paragraph marker immediately following the spreadsheet (at the bottom right) and using either the justification buttons on the context bar or the justification items on the Format menu. (Plots inserted using the Insert menu behave very similarly, except that they are centred by default.) I normally delete the group delimiter and centre spreadsheets, which is how they will mostly appear in this book.

Clicking on a spreadsheet makes it *active* and the Spreadsheet menu on the menu bar then becomes active; it is also available as a context menu by *right*-clicking on a spreadsheet (*Option*-clicking on a Macintosh). A Maple worksheet can contain multiple spreadsheets (as can an Excel workbook). The spreadsheets must have different names and by default they are named SpreadSheet001, SpreadSheet002, etc. The name can be accessed (and changed) via the *Properties...* item on the Spreadsheet menu. There is no way to control the size of the spreadsheet itself and you just use as much or as little of it as necessary. However, each spreadsheet appears within its own independently scrollable window, and you can control the size and shape of the window by dragging the square box at the bottom right corner where the two scroll bars meet. (Alternatively, but much less easily, you can control the size and shape of the window in almost the same way that you control the size and shape of a plot, by clicking *just outside* the spreadsheet and dragging one of the square black "handles".) By default, the spreadsheet shows a border, within which the rows of the spreadsheet are labelled 1, 2, 3, ..., 100 and the columns are labelled A, B, C, ..., Z, AA, AB, AC, ..., AZ. One normally starts using cells from the top-left corner, moving right and/or down to use more cells. The border display can be turned off via the Spreadsheet menu, and the spreadsheet window can be made to contain an integral number of cells via the *Resize to Grid* item on the Spreadsheet menu.

The cells of a spreadsheet can play the roles of variables in a programming language. Instead of giving them names, they are referenced by their "coordinates". The horizontal coordinate is given first and is the column label; the vertical coordinate is given second and is the row label. (Unfortunately, this is the opposite of the convention for accessing array elements.) Thus, the top-left cell of a spreadsheet is referenced as A1. However, a spreadsheet cell reference in Maple must be preceded by a tilde to distinguish it from a normal Maple variable name, so the top-left cell in a Maple spreadsheet is actually referenced as `~A1`.

A spreadsheet cell can contain either constant data, which can constitute the input to a computation, or formulae, which constitute the program text. However, it is the *value* of the formula that is displayed in a formula cell, which constitutes the output of the computation. When a cell is activated by clicking on it (or otherwise selecting it using the usual keyboard controls such as the cursor keys) the spreadsheet context bar shows the formula or expression in the cell. In the case of constant input data, the expression and its value are the same. If you want to use a spreadsheet trivially to display fixed information in a tabular layout then simply select each cell as appropriate and enter its data. But note that it will be processed as if it had been typed after a normal Maple input prompt. Numbers and strings are constants and evaluate to themselves, identifiers will be evaluated and may evaluate to themselves, and comments will be ignored. To put text into a Maple spreadsheet cell it is probably best to make it into a symbol by enclosing it in backquotes (if necessary) and possibly also to prevent it being evaluated by enclosing the symbol in forward quotes, because the double quotes around strings are displayed (as usual) and can be ugly and distracting.

For example, Figure 13.1 shows the top-left corner of a spreadsheet and indicates the coordinate references of the four top-left cells; the content of each of these cells is an unassigned identifier (without the leading tilde, because that would cause a "circular reference" error). Spreadsheets will be shown essentially as they appear by default in a Maple worksheet, except that I normally change the window size, and in all but this first spreadsheet I avoid showing irrelevant empty cells.

	A	B	C	D
1	*A1*	*B1*		
2	*A2*	*B2*		
3				
4				

Figure 13.1: SpreadSheet001.

13.2 Copying and moving cells

Spreadsheets are often used to build up tables of results that are all computed in essentially the same way. It would be very tedious to have to type similar values or formulae into a large number of cells, so spreadsheets provide ways to copy (or move) cells. A single cell can be selected by clicking on it and a rectangular range of cells can be selected by dragging diagonally from one corner to the opposite corner of the range. The selection can then be copied or moved using normal GUI techniques; to move or copy cells by dragging it is necessary to drag some point very near the edge of the range. A rectangular range of cells is normally denoted by its top-left and bottom-right cell coordinates separated by a colon; thus, the four top-left cells A1, A2, B1, B2 (illustrated in Figure 13.1) would be denoted as the range A1:B2.

If the formula in a cell remained exactly the same when it was copied to another location then the new value would be exactly the same as the old value, whereas what we often want is a different but related formula and value. Usually, the references to other cells made within a formula should either stay exactly the same or they should preserve their *relative* locations. The latter is most often required, so normal cell references are relative, meaning that they automatically change so as to preserve the relation to the cell containing them when the formula is copied or moved to another cell. Either the column or the row component of a cell reference (or both) can be made *absolute* by preceding it with a `$` symbol, which means that it will not change at all. It is very important to understand and use correctly relative and absolute cell references in order to use spreadsheets efficiently for non-trivial computations.

As a first simple example, let us construct a truth table for the Boolean `and` operator. The first column (A) and the first row (1) will contain the two possible operand values, `true` and `false`, for the two operands and the four inner cells (B2:C3) will contain the four possible values of the `and` operator. Provided we get the cell references correct in one of the four inner cells we can then just copy it to the remaining three cells. The first cell reference (left operand) is always to the first column, so the column reference should be the absolute reference `$A`, but the row can change and should be the same as the current cell (i.e., it should be relative). The second cell reference (right operand) is always to the first row, so the row reference should be the absolute reference `$1`, but the column can change and should be the same as the current cell (i.e., it should be relative). Hence, the formula in cell B2 should be `~$A2 and ~B$1`. Having defined this formula with the correct absolute and relative cell references, we can then copy it to cells B3, C2, and C3 and expect it to change as appropriate. The copied formulae are shown in the spreadsheet in Figure 13.2, where I have enclosed each formula in backquotes in order to show the formula rather than its value.

	A	B	C
1			
2		~$A2 and ~B$1	~$A2 and ~C$1
3		~$A3 and ~B$1	~$A3 and ~C$1

Figure 13.2: SpreadSheet002.

We can easily see that the copying has produced the correct cell references if initially we put their own coordinates in the cells in the first column and row, instead of `true` and `false`, as in the spreadsheet in Figure 13.3.

	A	B	C
1		*B1*	*C1*
2	*A2*	*A2* **and** *B1*	*A2* **and** *C1*
3	*A3*	*A3* **and** *B1*	*A3* **and** *C1*

Figure 13.3: SpreadSheet003.

This is correct: the first operand corresponds to the first column and the current row; the second operand corresponds to the first row and the current column. Finally, we can change the constant input data in the first row and column, re-evaluate the spreadsheet, put the name of the Boolean operator being investigated (`and`) in the top-left cell (A1) and hide the border to give the truth table shown in Figure 13.4.

Remember that, in order to use the binary infix operator `and` as an identifier, it is necessary to "escape" its operator syntax by enclosing it in backquotes (even though it contains no non-identifier characters). Essentially no formatting is (currently) available for Maple spreadsheets; for that it is necessary to use Excel.

and	*true*	*false*
true	*true*	*false*
false	*false*	*false*

Figure 13.4: SpreadSheet004.

Copying cell contents manually to more than a few other cells is tedious, so spreadsheets provide various facilities for copying to all the cells in a range, which is called *filling*. Maple spreadsheets provide two ways to fill a *selected range* of cells. Simple filling automatically copies the contents of all the cells along one edge of the range to all other cells in the range and is initiated by choosing the required direction of filling. This is most easily done from the *Fill* sub-menu of the Spreadsheet menu (which is active only when a cell range is selected). The copying performed is the same as would have been performed by copying one cell at a time and relative cell references are updated during the copying. Alternatively, series filling can be used to generate an arithmetic series automatically by adding some fixed increment to each copy in sequence. This is most easily done by clicking the *Fill* button at the left of the spreadsheet context bar; it can also be done by selecting the *Detailed...* item on the *Fill* sub-menu of the Spreadsheet menu.

As a simple illustration, let us use filling to produce a simple integer multiplication table, analogous to the truth table above. Let us start by inserting a new spreadsheet into a worksheet and then inserting in cell A1 an asterisk (*) as an identifier (in backquotes). Next, let us enter the constant 1 into cell A2. Then, we can select cells A2:A4 by dragging the mouse over them, click the *Fill* button at the left of the spreadsheet context bar, enter 1 in the *Step Size* box and click *OK*. (We could alternatively use the Spreadsheet menu on the main menu bar or the spreadsheet context menu.) Similarly, we can enter the integers 1, 2, 3 into the row range B1:D1.

Next, we enter the product formula `~$A2*~B$1` into cell B2, by analogy with the formula that we developed above for the truth table. We simply want to copy this formula to every cell in the range B2:D4. This has to be done in two steps, first copying it within its own column or row and then copying the whole column or row over the rectangular range. For example, we could first select the sub-column B2:B4 and select *Down* from the *Fill* sub-menu of the Spreadsheet menu, then select the whole range and select *Right* from the *Fill* sub-menu. If we then resize and centre the spreadsheet, the result should be the (small) integer multiplication table shown in Figure 13.5.

	A	B	C	D
1	*	1	2	3
2	1	1	2	3
3	2	2	4	6
4	3	3	6	9

Figure 13.5: SpreadSheet005.

13.3 Working with sequences

The only sequence that a Maple spreadsheet can generate automatically is an arithmetic sequence; for any other kind of sequence we must do some of the work. As a first trivial example, consider a sequence of factorials, e.g.,

```
> seq(n!, n=0..5);
```

$$1, 1, 2, 6, 24, 120$$

The naive way to generate such a sequence using a spreadsheet would be to construct the sequence 0..5 as a column (say, or equivalently as a row) of values and then from those values compute the factorials in the next column. We could proceed as follows. Enter the column headings n and $n!$ into cells A1 and B1. Enter the initial value 0 into cell A2, select cells A2 down to A10 or so, click the *Fill* button (which shows direction down by default), enter a step size of 1 and a stop value of 5, and execute the fill (which automatically stops at cell A7 with the value 5). Enter the formula `~A2!` into cell B2, select cells B2 down to B7, and select *Down* from the Spreadsheet *Fill* sub-menu. This gives the spreadsheet shown in Figure 13.6.

However, if you really want a table of all the factorials up to some limit, it is much more efficient to compute each new factorial from the previous factorial by using the recursive definition:

$$n! = n(n-1)! \text{ for } n > 0, \quad 0! = 1.$$

In a spreadsheet, we can implement the base case $0! = 1$ first by entering the values 0 and 1 into cells A2 and B2, respectively. Then we can enter the recursive formulae `~A2+1` and `~A3*~B2` into cells A3 and B3, respectively, to generate n as $(n-1)+1$ and $n!$ as above, and copy the pair of cells A3:B3 down the range A3:B7. The resulting spreadsheet appears exactly as shown before in Figure 13.6.

A more interesting example is to construct a table of the Fibonacci sequence, which can be defined recursively as:

$$f_1 = 1, \quad f_2 = 1, \quad f_n = f_{n-1} + f_{n-2} \text{ for } n > 2.$$

	A	B
1	n	$n!$
2	0	1
3	1	1
4	2	2
5	3	6
6	4	24
7	5	120

Figure 13.6: SpreadSheet006.

Let the first column contain the values of n, which we can generate either as a fill sequence or recursively, and the second column the values of f_n. We must enter the two base case values of 1 into cells B2 and B3, and then we can enter the recursive formula `~B2+~B3` into cell B4, which we can then copy down the column as far as we wish by using simple fill. This generates the first two columns of the spreadsheet shown in Figure 13.7.

It is now easy to investigate further properties of the Fibonacci sequence. For example, we can leave the base case values symbolic as in the third column. To do this, the names f_1 and f_2 are entered into cells C2 and C3, respectively, but otherwise the input for column C is the same as column B. The ratio of successive pairs of elements of the Fibonacci sequence is known to converge to the so-called "Golden Ratio", i.e.,

$$\lim_{n\to\infty} \frac{f_n}{f_{n-1}} = \frac{1}{2}(\sqrt{5}+1) = 1.6180.$$

These successive ratios are computed in the fourth and fifth columns. Cell D3 contains the formula `~B3/~B2` and cell E3 contains the formula `~D3` (i.e., just a copy of the value of cell D3). These cells can then be copied down the spreadsheet. The difference between columns D and E is that I selected the whole of column E (by clicking on the column label), selected *Properties...* from the Spreadsheet menu, and then selected *Floating-point Evaluation.* The spreadsheet in Figure 13.7 shows that the ratio does indeed appear to be converging to the predicted value, and moreover we see that the convergence is oscillatory; i.e., successive values are alternately above and below the limit.

13.4 Tabulating data

It is quite common that the solution of some mathematical problem cannot be expressed conveniently as a simple formula; perhaps the problem cannot be

	A	B	C	D	E
1	n	f_n	f_n	$\frac{f_n}{f_{n-1}}$	$\frac{f_n}{f_{n-1}}$
2	1	1	f_1		
3	2	1	f_2	1	1.0000
4	3	2	$f_1 + f_2$	2	2.0000
5	4	3	$2 f_2 + f_1$	$\frac{3}{2}$	1.5000
6	5	5	$2 f_1 + 3 f_2$	$\frac{5}{3}$	1.6666
7	6	8	$5 f_2 + 3 f_1$	$\frac{8}{5}$	1.6000
8	7	13	$5 f_1 + 8 f_2$	$\frac{13}{8}$	1.6250
9	8	21	$13 f_2 + 8 f_1$	$\frac{21}{13}$	1.6153
10	9	34	$13 f_1 + 21 f_2$	$\frac{34}{21}$	1.6190
11	10	55	$34 f_2 + 21 f_1$	$\frac{55}{34}$	1.6176

Figure 13.7: SpreadSheet007.

solved in terms of standard functions or perhaps such a solution exists but is very complicated. As explained in Chapter 5, in such cases it is often better to forget about exact solutions and compute numerical approximations instead. But how do we present approximate numerical solutions? As either a graph or a table, or both. And there are two obvious ways to construct a data table in Maple, as either a matrix (a technique that we have already used several times and which I will use again in the next section) or a spreadsheet.

As a simple example, consider the following function defined in terms of a definite integral:

```
> f := x -> int(exp(x*y^3),y=0..1);
```

$$f := x \to \int_0^1 \mathrm{e}^{(x\, y^3)}\, dy$$

For all real values of x the integrand is real and positive, so the integral must also be real and positive, and there are no singularities for finite values of x. In other words, this is a fairly simple and well-behaved function. So, let us see what it actually looks like:

```
> plot(f(x), x=-5..5);
```

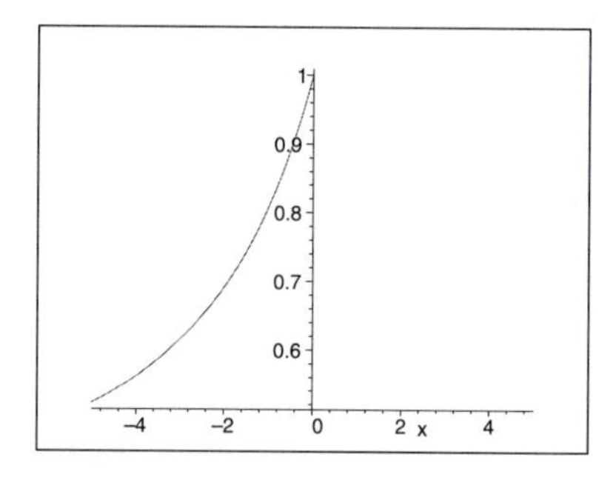

This is not what we would expect, since there is no *a priori* reason to expect the graph to be missing for positive x. So why has this happened? The expression that we are actually plotting is this:

```
> f(x);
```

$$\frac{1}{3}\,\frac{\dfrac{2}{3}\,\dfrac{\pi\sqrt{3}}{\Gamma(\frac{2}{3})} - \Gamma(\frac{1}{3}, -x)}{(-x)^{(1/3)}}$$

Maple has "evaluated" the integral in terms of the incomplete Gamma function and a cube root function, both of which have branch-point singularities (which cancel) at the origin. Hence, this exact symbolic representation of the integral is in many ways much more complicated than was the original integral, and Maple seems to be unable to evaluate the exact representation for $x > 0$. (This is analogous to the way that a naive algorithm to compute real cube roots numerically might fail when attempting to compute the logarithm of a negative number.) There are therefore advantages to evaluating this integral numerically, rather than first evaluating it symbolically. Hence, let us redefine the function as an inert integral:

```
> f := x -> Int(exp(x*y^3),y=0..1);
```

$$f := x \to \int_0^1 \mathbf{e}^{(x\,y^3)}\,dy$$

This inert implementation actually plots much faster than the active version and evaluates correctly (numerically) for all values of x:

```
> plot(f(x), x=-5..5);
```

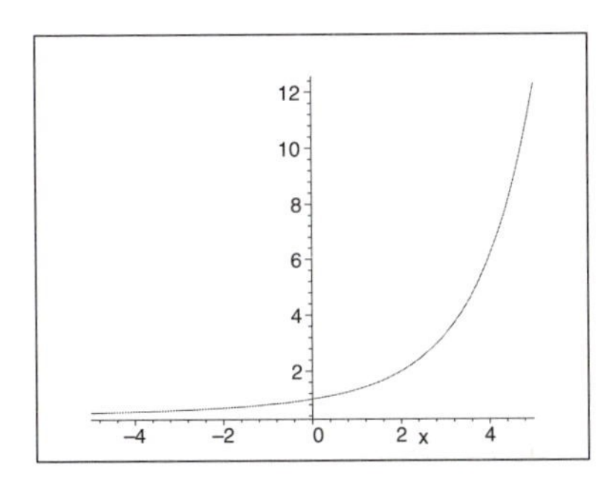

This is a typical example of cases where an exact symbolic solution is not the best way to proceed even when it can be found! (Incidentally, attempting to plot the mapping f rather than the expression $f(x)$ in the active case appears not to terminate.)

In passing, it appears from the graph above that this function has the curious property that $f(2) = 2$, but in fact

```
> 'f(2)' = evalf(f(2));
```

$$f(2) = 1.993144340$$

Although the function f is best evaluated numerically, it has a sufficiently simple definition that the standard Maple numerical equation solver `fsolve` can handle it, and the value of x at which $f(x) = 2$ is

```
> fsolve(f(x)=2, x);
```

$$2.007602821$$

Hence, this inert-integral definition of f does not produce a "strictly numerical" function (see Chapter 3 and below) that `fsolve` cannot handle.

Now let us return to tabulating data. This is very easy to do using a spreadsheet, as shown in Figure 13.8. Here is the recipe. Insert a spreadsheet into the worksheet, insert `x` into cell A1 and `f(~A1)` into cell B1. For purposes of illustration I will use very widely spaced values of x: insert the number -4 into cell A2, select A2 and several cells in the column below it, click the context-bar button to fill a range of cells, and select fill down in steps of 2 stopping at 4. Next, select cell B2, select spreadsheet *Properties...* (either from the Spreadsheet menu on the menu bar or by *right*-clicking to pop up the context menu) and select Evaluation type *Floating-point.* (You can also select the floating-point precisions if you want, but I left them at their default settings.) Select the range B1:B6 and then select fill down. Note that the integrals all evaluate to floating-point numbers, without needing any explicit calls to `evalf`. This completes our small table of data. You could perform some of these operations in a different order if you wanted; for example, you could fill the cell range B2:B6 with formulae first and then select floating-point evaluation. The disadvantage of doing it this way around is that you then need to narrow the rows back to a sensible size for simple numbers, which does not happen automatically.

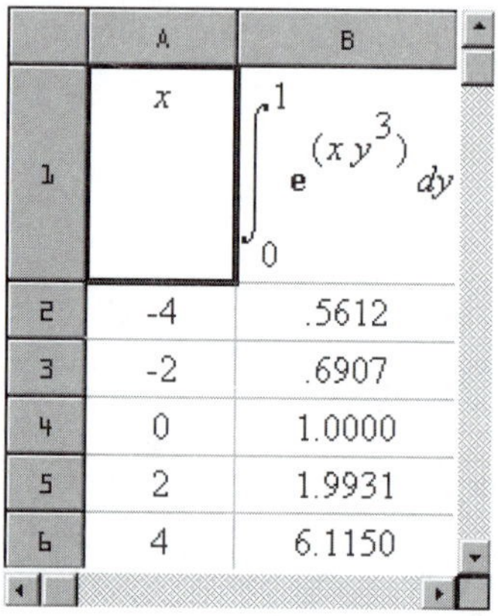

	A	B
1	x	$\int_0^1 e^{(x\,y^3)}\,dy$
2	-4	.5612
3	-2	.6907
4	0	1.0000
5	2	1.9931
6	4	6.1150

Figure 13.8: SpreadSheet008.

13.5 Programming spreadsheets

Spreadsheets are primarily intended for interactive manipulation rather than programming in a conventional sense. By contrast, the traditional way to "program" the above table of data in Maple would be to construct a matrix as follows (which I will also save for subsequent use):

```
> array([[x,f(x)],
        seq([x,evalf[5](f(x))],x=[-4,-2,0,2,4])]);  M:=%:
```

$$\begin{bmatrix} x & \int_0^1 \mathbf{e}^{(x\,y^3)}\,dy \\ -4 & .56120 \\ -2 & .69073 \\ 0 & 1. \\ 2 & 1.9931 \\ 4 & 6.1151 \end{bmatrix}$$

However, the `Spread` package provides conventional programmability for spreadsheets; for details see the online help and for further examples see the help topic `examples,spread`. The package contains the following functions:

```
> with(Spread);
```

> [*CopySelection*, *CreateSpreadsheet*, *EvaluateCurrentSelection*, *EvaluateSpreadsheet*, *GetCellFormula*, *GetCellValue*, *GetFormulaeMatrix*, *GetMaxCols*, *GetMaxRows*, *GetSelection*, *GetValuesMatrix*, *InsertMatrixIntoSelection*, *IsStale*, *SetCellFormula*, *SetMatrix*, *SetSelection*]

As usual, the first step is to insert a spreadsheet into a worksheet. The function `CreateSpreadsheet` does that, and returns its identifier, which is needed to perform further operations on the spreadsheet. One easy way to get data into the spreadsheet is simply to import a *tabular* data structure, such as the matrix `M` that we created above, by calling the function `SetMatrix`, and then to *evaluate* the spreadsheet by calling the function `EvaluateSpreadsheet`. The spreadsheet contained in the execution group in Figure 13.9 shows the result. The first statement in the execution group creates an empty spreadsheet, which Maple inserts into the worksheet immediately before where the normal return value from the function-call appears (or would appear). Note that the spreadsheet is created and inserted into the worksheet as a *side-effect* of the function-call (like printing). The second statement imports data but leaves the spreadsheet unevaluated and the changed cells are cross-hatched to indicate this. The final statement evaluates the spreadsheet and the cross-hatching disappears.

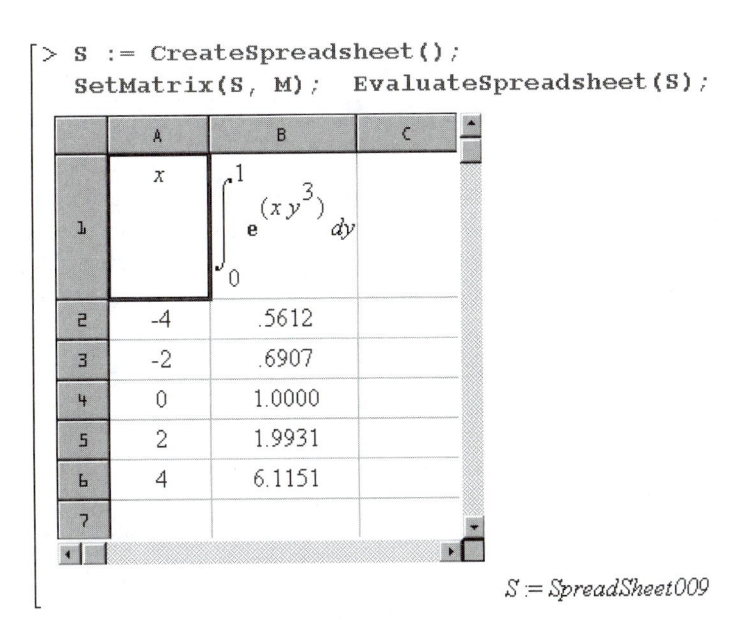

Figure 13.9: SpreadSheet009.

Alternatively, we can enter data directly into a spreadsheet as follows, where I have used a module only to localize a few variables. The resulting spreadsheet appears exactly as in Figure 13.9 so I will not show it again.

```
> module()
     local S, x, i;
     S := CreateSpreadsheet();
     SetCellFormula(S, 1, 1, x);
     SetCellFormula(S, 1, 2, f(x));
     i := 2;
     for x in -4,-2,0,2,4 do
        SetCellFormula(S, i, 1, x);
        SetCellFormula(S, i, 2, 'evalf[5]'('f'(~A||i)));
        i := i + 1
     end do;
     EvaluateSpreadsheet(S)
  end module:
```

In the first example of setting up a spreadsheet programmatically I simply inserted pre-computed values rather than formulae to be evaluated by the spreadsheet, which is perhaps cheating. In the second example I have inserted spreadsheet formulae, but it is not possible programmatically to change spreadsheet or cell *properties*, so the only programmatic way to obtain numerical evaluation is explicitly to apply `evalf`. In the second example, it is necessary to uneval the `evalf` so that it is applied within the spreadsheet context. I have also unevaled the function f for the purely cosmetic reason of stopping the unevaluated integral being pretty-printed and so increasing the row spacing. Currently, the programmatic spreadsheet interface does not provide complete control.

Note that outputting to a spreadsheet is the only situation in Maple in which output goes to some part of a worksheet other than where the code is executed and in principle provides a way to direct output to a particular location; all operations that relate to a spreadsheet cause changes within that spreadsheet wherever it happens to be located (which has potential applications not strictly related to spreadsheets).

For most purposes, I recommend using spreadsheets for interactive manipulation and matrices for programmatic manipulation of tables of data.

13.6 Case study: the simple pendulum

A spreadsheet can be a convenient environment for performing one-off interactive numerical investigations. A *large-amplitude* pendulum provides an example of a very simple mechanical problem that does not have a simple analytical solution. (It can be solved in terms of elliptic functions — see, for example, the last chapter of the classic text on *Methods of Mathematical*

Physics by Jeffreys and Jeffreys [13] — but these are not elementary functions and moreover Maple does not explicitly introduce them into the solution, as we will see, even though Maple supports various elliptic functions.)

A *simple pendulum* is idealized as a bob having a point mass m swinging at the end of a massless straight rigid rod of length l that hangs from a frictionless pivot. Suppose that, at some instant of its swing, the pendulum makes an angle θ with the vertical, as illustrated in Figure 13.10. If the acceleration due

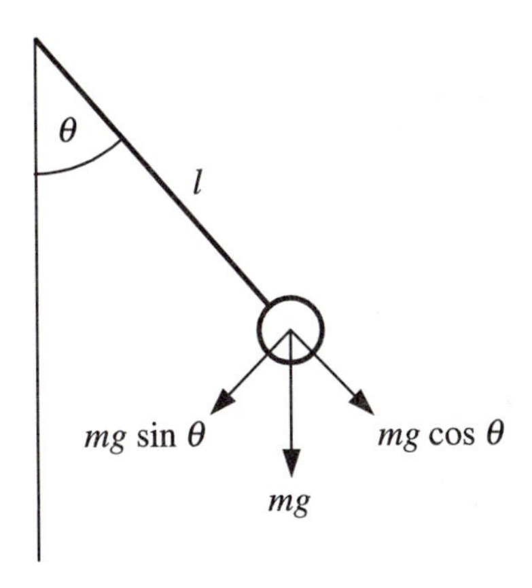

Figure 13.10: A simple pendulum.

to gravity is denoted by g then the weight of the bob can be resolved parallel and perpendicular to the rod, as in Figure 13.10. The component parallel to the rod is balanced by the force in the rod. The component perpendicular to the rod provides a restoring force that pulls the pendulum back toward its vertical rest position. The restoring *torque* on the pendulum is therefore $-mg\sin(\theta)l$ in the direction of increasing θ, which must be equal to its angular acceleration $\ddot{\theta}$ times its moment of inertia, leading to the equation of motion:

$$\ddot{\theta}\,ml^2 = -mg\sin(\theta)l.$$

The mass cancels out to give the following mathematical model.

13.6.1 Exact solutions for small and large amplitude

A simple pendulum of length l swinging under the influence of gravity (acceleration g) is described by the following ordinary differential equation, where θ is the angle the pendulum makes with the downward vertical at time t:

```
> restart;
```

```
> DE := diff(theta(t),t$2) = -g*sin(theta(t))/l;
```

$$DE := \frac{\partial^2}{\partial t^2}\,\theta(t) = -\frac{g\sin(\theta(t))}{l}$$

If the pendulum starts at rest at angle A, which is the *amplitude* of the swing, then this implies the initial conditions:

```
> IC := theta(0) = A, D(theta)(0) = 0;
```

$$IC := \theta(0) = A,\ \mathrm{D}(\theta)(0) = 0$$

For small θ, $\sin(\theta)$ is close to θ and the differential equation can be approximated as:

```
> ApproxDE := eval(DE, sin = (x -> x));
```

$$ApproxDE := \frac{\partial^2}{\partial t^2}\,\theta(t) = -\frac{g\,\theta(t)}{l}$$

which, with the specified initial conditions, has the solution:

```
> dsolve({ApproxDE, IC});
```

$$\theta(t) = A\cos(\frac{\sqrt{g\,l}\,t}{l})$$

This represents *simple harmonic motion* with angular frequency ω and period T given by:

```
> omega := sqrt(g*l)/l;  T := 2*Pi/omega;
```

$$\omega := \frac{\sqrt{g\,l}}{l}$$

$$T := 2\,\frac{\pi\,l}{\sqrt{g\,l}}$$

where we can put the value of T into a more natural form using the following trick:

```
> T := sqrt(T^2);
```

$$T := 2\,\pi\,\sqrt{\frac{l}{g}}$$

Thus, the period is independent of the amplitude A, provided it is small. But suppose it is not. Then how does the period depend on the amplitude A? Maple gives the *general* solution of the large-amplitude pendulum equation as:

```
> dsolve(DE);
```

$$\int^{\theta(t)} \frac{l}{\sqrt{l\,(2\,g\cos(\_a) + \_C1\,l)}}\,d\_a - t - \_C2 = 0,$$
$$\int^{\theta(t)} -\frac{l}{\sqrt{l\,(2\,g\cos(\_a) + \_C1\,l)}}\,d\_a - t - \_C2 = 0$$

The integrals are of a type known as *elliptic integrals*, although `dsolve` does not appear to recognize this, and (not too surprisingly) seems to be unable to impose the initial condition; i.e., I gave up waiting while the following function-call used progressively more memory without terminating:

```
> dsolve({DE, IC});

  Warning, computation interrupted
```

13.6.2 Approximate numerical solution

The problem is to determine the period of a simple pendulum with a given amplitude A that is not small. In other words, if we start the pendulum from rest at an angle A, how long will it take to return to this position? This determines its period T. By the obvious symmetry of the problem (in the absence of any damping) the pendulum will pass through the vertical after a quarter of its total period, $T/4$, which is easier to compute. If we had an exact solution for the displacement $\theta(t, A)$ as a function of time t and amplitude A and we could invert or solve it for $t(\theta, A)$, then we would immediately have that $T/4 = t(0, A)$. However, Maple is not able to give us such an analytic solution $\theta(t, A)$ directly, and even if it were it might well not be able to invert it. Can we solve this problem numerically? We could use numerical evaluation of the exact elliptic integral solution given above, but we might as well use the more direct approach of numerical integration of the original ODE, as introduced in Chapter 5.

Without loss of generality, we can replace the constant g/l by a single constant K, to give:

```
> DE := subs(g = K*l, DE);
```

$$DE := \frac{\partial^2}{\partial t^2}\,\theta(t) = -K\sin(\theta(t))$$

Giving K a numerical value does not make it any easier to solve the problem symbolically although it is essential in order to solve the problem numerically. For this example, let us arbitrarily set

```
> K := 1:
```

For small amplitude, the period is

```
> T := subs(g = K*l, T);
```

$$T := 2\,\pi$$

Now we want to integrate the ODE, starting with the initial condition $\theta = A$ used above, which is what introduces the amplitude A into the problem. For small amplitude, the quarter period is

```
> T/4:   % = evalf(%);
```

$$\frac{1}{2}\,\pi = 1.570796327$$

Now let us try to compute the period for a larger amplitude such as:

```
> A := 1:
```

The `dsolve` function with the following optional arguments returns a list of equations, the right side of each of which is a numerical procedure to evaluate the left side at the time specified as its argument:

```
> dsolve({DE, theta(0) = A, D(theta)(0) = 0}, theta(t),
     numeric, output = listprocedure);
```

$$[t = (\mathbf{proc}(t)\ \ldots\ \mathbf{end\ proc}),\ \theta(t) = (\mathbf{proc}(t)\ \ldots\ \mathbf{end\ proc}),$$
$$\tfrac{\partial}{\partial t}\,\theta(t) = (\mathbf{proc}(t)\ \ldots\ \mathbf{end\ proc})]$$

We are interested only in the displacement θ and the following `subs` provides a reliable way to extract the procedure that computes it:

```
> Theta := subs(%, theta(t));
```

$$\Theta := \mathbf{proc}(t)\ \ldots\ \mathbf{end\ proc}$$

Note that `Theta` is a "strictly numerical" function of the type discussed in Chapter 3, e.g.,

```
> Theta(t);

  Error, (in dsolve/numeric_solnall_rkf45) cannot evaluate boolean:
  abs(t)-max(.1e-7*abs(t),.1e-7) < 0
```

and so must be used (e.g., plotted) with care!

Physically, the maximum simple pendulum amplitude is $\pi = 3.1416$ (beyond which the pendulum rotates rather than oscillates), in relation to which 1 is not huge. We therefore expect the period for amplitude $A = 1$ to be fairly close to the small-amplitude period, so let us investigate the displacement θ at times around the small-amplitude quarter-period $\frac{\pi}{2}$ using a plot:

```
> plot(Theta, 0..Pi, labels=["t","theta"]);
```

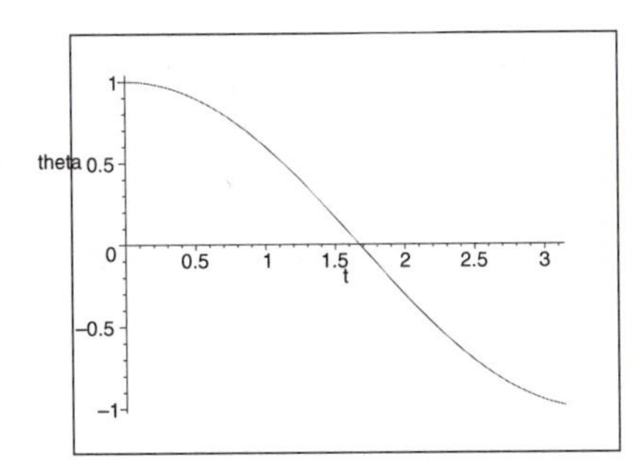

Noting the time at which θ passes through zero suggests that the quarter-period is about 1.7, which is indeed not far from the small-amplitude quarter-period $\frac{\pi}{2} = 1.57$. We might expect to be able to compute the quarter-period accurately using `fsolve`, but

```
> fsolve(Theta, 0..Pi);
```

$$\mathrm{fsolve}(\Theta(r),\, r,\, 0..\pi)$$

This means that `fsolve` is unable to find the zeros of this numerical function Θ.

Aside: why does `fsolve` not work?

If you want to find our why `fsolve` does not work in this situation then you need to use some of the debugging techniques introduced in Chapter 11. For example, you might hope that setting

```
> infolevel[fsolve] := 5:
```

would give you some useful information, but in fact it gives no information at all. The next thing to try is this:

```
> printlevel := 10:
```

```
> fsolve(Theta, 0..Pi);
```

```
{--> enter fsolve, args = Theta, 0 .. Pi
```

$$sol := \{\text{“cannot evaluate boolean: \%1”}\}$$

$$sol := \{\mathrm{fsolve}(\Theta(r),\, r,\, 0..\pi)\}$$

```
<-- exit fsolve (now at top level) = fsolve(Theta(r),r,0 .. Pi)}
```

$$\text{fsolve}(\Theta(r),\ r,\ 0..\pi)$$

By judiciously increasing the value of `printlevel` and reading the code for `fsolve`, `Theta` (see Chapter 8) and their sub-procedures, you should be able to ascertain why `fsolve` is trying and failing to evaluate a Boolean expression. This often happens when a procedure intended to be used purely numerically is used in a symbolic context. It happens here because, as we saw earlier, `Theta` is a "strictly numerical" function. Tracking down and repairing this deficiency could be regarded as an advanced Maple programming exercise; it really is an excellent way to become a Maple expert if you are so inclined! But there are quicker ways to solve our current problem.

```
> printlevel := 1:
```

13.6.3 Numerical computation of the period

Spreadsheets provide a convenient medium for "trial-and-error" solutions. To solve this problem, we need two columns, the first listing values of time t and the second the corresponding displacement $\Theta(t)$, as in the spreadsheet shown in Figure 13.11. We know that the solution is between 1.6 and 1.8, so let us start with those two times, at which $\Theta(t)$ has opposite signs and hence its graph crosses the time axis somewhere in between. Then we use the mid-point, t = 1.7, for which the value of $\Theta(t)$ shows that the solution lies between 1.7 and 1.6. Taking their mid-point shows that the solution lies between 1.65 and 1.7, and taking their mid-point shows that the solution is very close to 1.675, because here $\Theta(t)$ is zero to more than 4 decimal places. The values of $\Theta(t)$ are all computed by starting with the formula `Theta(~A2)` in cell B2 and then copying that formula down the $\Theta(t)$ column as necessary.

	A	B
1	t	$\Theta(t)$
2	1.6000	.0718
3	1.8000	-.1195
4	1.7000	-.0239
5	1.6500	.0239
6	1.6750	$-.5840\ 10^{-5}$

Figure 13.11: SpreadSheet010.

We can very easily automate most of this calculation, which is called a *bisection* algorithm. We must manually choose the first two time values to be on either side of the solution and compute the corresponding values of $\Theta(t)$ in rows 2 and 3. Then we can compute the next time value by bisection using the formula `(~A3+~A2)/2` in cell A4. The subsequent time values require a simple decision based on the signs of the preceding values of $\Theta(t)$, which we can automate by using the formula `` `if`(~B4*~B3 < 0, ~A4+~A3, ~A4+~A2)/2 `` in cell A5. With the formula `Theta(~A5)` in cell B5, we can then copy the cell range A5:B5 down the spreadsheet until either the difference between successive time values or the value of $\Theta(t)$ is small enough that we have attained sufficient accuracy, when we say that the solution algorithm has *converged.* The resulting spreadsheet is identical to the previous one, so I show the corresponding *input formulae instead of the output results* in the spreadsheet in Figure 13.12.

	A	B
1	*t*	*'Theta(t)'*
2	1.6000	*Theta(~A2)*
3	1.8000	*Theta(~A3)*
4	*(~A3+~A2)/2*	*Theta(~A4)*
5	*`if`(~B4*~B3 < 0,~A4+~A3,~A4+~A2)/2*	*Theta(~A5)*
6	*`if`(~B5*~B4 < 0,~A5+~A4,~A5+~A3)/2*	*Theta(~A6)*

Figure 13.12: SpreadSheet011.

The bisection algorithm should converge a little faster if we take account of the actual values of $\Theta(t)$, rather than only their signs. In this way we should be able to estimate where the graph cuts the time axis, assuming it is locally straight. First, let us do this "by hand" to build up the spreadsheet in Figure 13.13. If, approximately as shown, $\Theta(1.6) = .07$ and $\Theta(1.8) = -.12$ then the solution is near to 1/3 of the distance between these two points, so let us now try $t = 1.67$. This is good, but the solution is a little closer to 1.8, so let us try $t = 1.68$. These two estimates suggest that a better solution would be half way between them, so we try $t = 1.675$, which leads us to exactly the same solution as before and in the same number of steps.

To be more systematic and precise, we can use what is called the *chord* algorithm, because it approximates the graph of the function whose zero we seek by its chord. Suppose we know that $f(x)$ has a zero between $x = a$ and $x = b$. Then, provided $f(x)$ is locally a sufficiently smooth function, it is reasonable to approximate its graph by the straight line between the two points $(a, f(a))$ and $(b, f(b))$, as shown in Figure 13.14. The two triangles are similar, meaning that they have the same angles and hence the lengths of their sides are in the same ratios, which we can use to find x as follows:

```
> op(solve((x-a)/(-f(a)) = (b-x)/f(b), {x}));
```

	A	B
1	t	$\Theta(t)$
2	1.6000	.0718
3	1.8000	-.1195
4	1.6700	.0047
5	1.6800	-.0048
6	1.6750	-.5826 10^{-5}

Figure 13.13: SpreadSheet012.

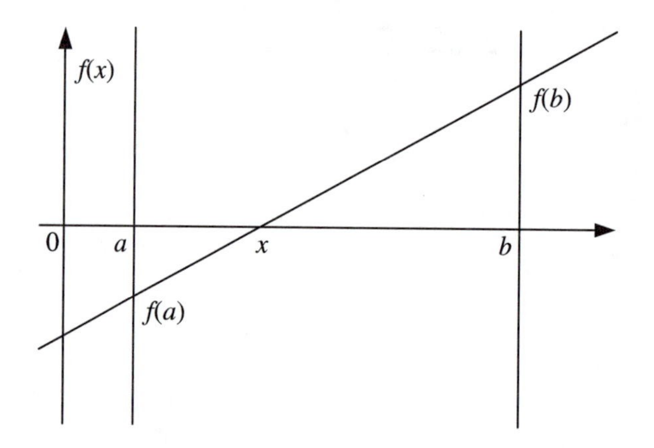

Figure 13.14: The chord algorithm.

$$x = \frac{-\mathrm{f}(b)\,a + \mathrm{f}(a)\,b}{-\mathrm{f}(b) + \mathrm{f}(a)}$$

Hence, instead of simple bisection, we can use the formula:

```
(~A3*~B2-~A2*~B3)/(~B2-~B3)
```

in cell A4. This immediately gives $\Theta(t) = 0$ to about 4 decimal places, as shown in the spreadsheet in Figure 13.15, so let us increase the display precision to 8 decimal places and use the analogous formula:

```
`if`(~B4*~B3 < 0, (~A4*~B3-~A3*~B4)/(~B3-~B4),
        (~A4*~B2-~A2*~B4)/(~B2-~B4))
```

in cell A5. This immediately gives $\Theta(t) = 0$ to 6 decimal places, and one further iteration gives $\Theta(t) = 0$ to 8 decimal places. Hence, at the price of a very small increase in complexity, the chord algorithm converges significantly faster, at least for this example.

	A	B
1	t	$\Theta(t)$
2	1.60000000	.07184062
3	1.80000000	-.11955052
4	1.67507204	-.00007492
5	1.67499383	$.70147241\ 10^{-7}$
6	1.67499390	$.15111187\ 10^{-9}$

Figure 13.15: SpreadSheet013.

We conclude from the spreadsheet in Figure 13.15 that the quarter-period for amplitude $A = 1$ is 1.6749939.

13.6.4 Numerical computation of the amplitude

Now suppose we want to solve the inverse problem, namely to find the amplitude that gives a specified period, again assuming the amplitude is not small. This is more difficult, because the amplitude is buried inside the numerical solution of the ODE, so each time we change the amplitude we need to re-solve the ODE. It is convenient to incorporate the necessary steps into a procedure:

```
> Theta0 := proc(A::numeric, T::numeric)
     dsolve({DE, theta(0) = A, D(theta)(0) = 0}, theta(t),
        numeric, output = listprocedure);
     subs(%, theta(t))(T)
  end proc;
```

$$\Theta 0 := \mathbf{proc}(A{::}numeric,\ T{::}numeric)$$
$$\quad \mathrm{dsolve}(\{\theta(0) = A,\ DE,\ \mathrm{D}(\theta)(0) = 0\},\ \theta(t),\ numeric,$$
$$\quad\quad output = listprocedure);$$
$$\quad \mathrm{subs}(\%,\ \theta(t))(T)$$
$$\mathbf{end\ proc}$$

Since we have already determined that an amplitude $A = 1$ gives a quarter-period $T/4 = 1.6749939$, we can use the fact that the following function value should be (very close to) zero as a check:

```
> Theta0(1, 1.6749939);
```

$$.871900001503389532\,10^{-8}$$

Now we want to fix the quarter-period $T4$ and vary A until $\theta = 0$. Let us try to find the amplitude A that gives a quarter-period:

```
> T4 := 2:  A := 'A':
```

We expect a larger amplitude to give a longer period. (An amplitude of π corresponds to the pendulum pointing vertically up, which is an unstable equilibrium and so corresponds to an infinite period. Jugglers rely on this fact!) Let us guess a range and plot the pendulum displacement θ at the intended quarter-period for a range of amplitudes greater than 1:

```
> plot('Theta0(A, T4)', A = 1..2, theta);
```

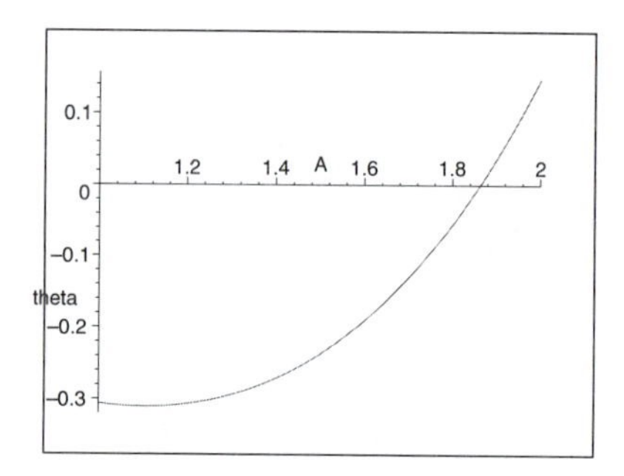

This plot is slow, but it shows us that our theory is correct and that the required amplitude is between 1.8 and 1.9. This approximation can be refined using any of the spreadsheet techniques described in the previous section. No new concepts are involved beyond the fact that we are seeking the zero of a function of two variables, the second of which is held fixed. Hence, the details are left as an exercise.

13.7 Exercises

1. Construct as spreadsheets the following tables:

   (a) truth tables for the Boolean operators `not` and `or`;

   (b) the multiplication table for the finite field $\mathbb{Z}_7$, namely the integers modulo the prime 7.

2. Euler's numerical integration formula for a first-order ordinary differential equation $\frac{dy(x)}{dx} = f(x, y(x))$ is

$$y(x_{n+1}) = y(x_n) + h\, f(x_n, y(x_n))$$

   where $h = x_{n+1} - x_n$. Use this formula in a spreadsheet to find an approximate numerical solution in the case that $f(x, y(x)) = y(x)$ with the initial condition $y(0) = 1$ and compare it with the exact solution. Note that this is the simplest possible numerical integration formula and consequently in general the least accurate. Consider different fixed numerical values for the step length h and also investigate the case that h remains symbolic, which leads to an approximate algebraic solution.

3. Compute numerically the amplitude required to give the simple pendulum described in the text a quarter-period of 2. Use one or more of the spreadsheet techniques described, namely bisection by hand, bisection using spreadsheet formulae, educated guessing, and the chord algorithm using spreadsheet formulae.

4. Animate the swinging of a large-amplitude pendulum. Choose an amplitude A, where $0 < A < \pi$, and compute the corresponding quarter-period (or vice versa) using any convenient technique. Split the quarter-period into a suitable number of time steps so that there is a frame of the animation showing both the maximum and minimum displacements (i.e., the amplitude and zero, respectively) and at equal time steps in between. Use symmetry to extend the animation so that if it is allowed to run continuously it looks like a smoothly swinging pendulum. (But do not actually compute more frames than necessary!)

5. (Beware that this question involves applied mathematics beyond the scope of this book.) Solve the pendulum problem exactly in terms of elliptic integrals. Hence, by using Maple to evaluate the functions involved, verify the numerical results presented in the text and those that you obtained as your solution to exercise 3, above.

Chapter 14

Text Processing

Text processing is data processing where the data consist of text. The text may be anything: a novel, a book on computer programming, the text of a computer program, or just phrases, words, or arbitrary characters. Maple is not the most obvious language to choose for writing text processing programs, but it is entirely possible to process text using Maple and text processing is an important aspect of computer programming. Maple is no worse than (say) C for text processing as far as the actual language is concerned, although Maple lacks some of the useful text-processing library functions that are included in the standard C support environment and a Maple program will run much slower than an equivalent C program. Both Maple and C lack the sophisticated pattern matching facilities of a language such as Perl. (It would be a useful programming exercise to implement some of these facilities in Maple.) *Software Tools* by Kernighan and Plauger [15] provides a good introduction to text processing (and programming in general).

The first section sets the scene by describing the simplest way to represent a text document as a computer file and discusses the two main problems that afflict even this simplest computer representation of text. Then, as an introduction to text processing itself, the second section develops Maple code to count the number of lines, words, and characters in a text document. The third section introduces the rudiments of text reformatting by explaining how to fill lines of text to a specified length. The last section provides both some light entertainment and an opportunity to exploit some core Maple facilities in the unusual setting of a stochastic text rewriter. This chapter is based on algorithms and code (in languages other than Maple) described by Brian Kernighan and his co-authors.

14.1 Text files

The term "text file" implies a *simple* computer representation of written human language (as opposed to the sophisticated representation possible in a word-processor file). Any such representation must include the letters, digits, and punctuation that appear in normal prose together with blank or *space* characters. However, formatting is minimal: it must include information about how text is broken into lines but not much else. This is sufficient information to store a simple natural language document as a computer file. It should be possible to print it directly on a simple printer or display it on a computer screen with essentially no processing, and the output should be readable and intelligible.

The representation also includes sufficient information to store the text of a programming language (such as Maple). By introducing appropriate annotation conventions called *mark-up*, which is itself normal text although it usually includes a few uncommon characters, a text file can represent much more information, such as sophisticated formatting, multiple fonts, and mathematics. Examples include the mark-up languages LaTeX, RTF (Rich Text Format), and HTML (HyperText Markup Language), and Maple can export worksheets in all these formats. Moreover, the format used to store Maple worksheets is itself an annotated text file. (You can open a `.mws` file in a simple text editor, such as Notepad on Windows. After some introductory annotation you will find the text of the worksheet, which is human-readable although the mark-up is somewhat distracting. Other information, such as inline plots, is encoded in a way that is not human-readable. You can easily guess the meaning of some of the Maple worksheet mark-up.)

Letters, digits, and punctuation are called *graphic* (i.e., visible) characters and the minimal formatting represented directly in a text file uses what are called *control* characters, which are not directly visible. The space character is regarded as a graphic character, even though it is manifestly visible only when surrounded by other graphic characters. A text file consists of a sequence of *bytes*, each of which consists of 8 *bits* (binary digits). The problem is precisely how to represent graphic and control characters as bytes. An encoding that is widely used at present is ASCII (American Standard Code for Information Interchange), which was introduced in 1963 and encodes 128 characters. Of these, the upper 96 (encoded by the integers 32–127) are graphic characters and the lower 32 (encoded by the integers 0–31) are control characters, most of which are not used in text files. (ASCII was originally designed as an encoding for text transmission, so it includes control characters that are not necessary for text storage.) The ASCII encoding uses 7 bits, so one byte per character suffices and in fact leaves one bit unused. ASCII was accepted in 1973 with minor national variability as the international standard encoding ISO-7. In continental Europe, the 128 unused upper 8-bit encodings are often used to represent local additions; for example, the Latin-1 encoding adds accented letters and diphthongs plus other special symbols, but use of this encoding is not completely standard.

The main problem with graphic characters is that there is a huge number of them used by all the natural languages of the world. The ASCII character set suffices for English text, but not for text in most other languages and not for mathematical or other special symbols. A current hot topic in computing is *internationalization* (abbreviated as *i18n*, because there are 18 letters between the initial i and the final n). Various *ad hoc* schemes have been used to encode other character sets but a new standard called Unicode is emerging that uses two bytes (16 bits) per character. For example, Unicode is specified to be the character encoding used in Java. Unicode includes Latin-1, and hence ASCII, as a subset in its first 256 character positions, and the single-byte encodings of the Unicode (strictly UCS) Transformation Format UTF-8 (in which the high order bit is cleared to indicate a single-byte encoding) are *identical* to ASCII. For a little more background on character encodings, see *The Practice of Programming* by Kernighan and Pike [14], whom I will refer to henceforth as K&P. (This book is an excellent guide to serious computer programming, focusing on C, C++, and Java.) See also Appendix C.2 of *The LaTeX Web Companion* [10] and the URL http://www.cl.cam.ac.uk/~mgk25/unicode.html. Maple appears to support whatever single-byte character encoding is provided by the computer on which it is running (and ignores any character encoding > 255). As far as I am aware, the user has no control over this.

The main problem with control characters is how the ends of lines of text are represented, and to a lesser extent how the end of the text itself (i.e., end of file) is represented. The non-standard encoding of text files can be a problem in a heterogeneous computing environment; for example, if you write a text file using some non-ASCII characters on a Macintosh or UNIX system and try to read it on Windows then the non-ASCII characters may appear differently from what was intended and the line breaks may be lost.

The line break problem arises because the standard representation for the end of a line on Windows is the pair of ASCII control characters carriage return (CR) followed by line feed (LF), which is often abbreviated to CRLF, whereas on Macintosh it is just CR and on UNIX it is just LF. This problem can be handled by making a distinction between *text* and *binary* files. When reading a text file, the actual sequence of control characters used to indicate the end of a line of text is converted into a canonical newline character that the program reading the file sees; when writing a text file the inverse conversion is performed. In practice this usually means that UNIX format is used internally and is mapped to the actual format used on the current platform. This is what Maple does. Hence, in Maple and most other modern programming languages, a program processing a text file can use the single newline character `"\n"`, which is the ASCII linefeed character LF, to denote the end of a line. Of course, this translation would corrupt a file such as an executable program file that was not a text file, so it is important to access files in the correct mode, either text or binary, where binary means that no translation is performed. Increasingly, computer software handles issues of character encoding and line ending convention automatically; for example, the GNU Emacs editor since version 20 has done so with a high degree of success.

Maple allows files to be opened in either text or binary mode by specifying one of the identifiers `TEXT` or `BINARY`. It also allows files to be opened for reading, writing, or both by specifying one of the identifiers `READ` or `WRITE`. However, in most cases the default mode is correct. This is `TEXT` and whichever of `READ` or `WRITE` is implied by the operation being performed, so normally one can just access files by name without explicitly opening them. We will, however, occasionally need to close files explicitly. For further details of file access see the online help.

14.2 Counting lines, words, and characters

One of the simplest classic text processing tasks is to implement the standard UNIX program called `wc`. The name stands for *word count*, but it is part of UNIX culture that more than three characters is too long for a program name and two character names are preferred! The program actually counts the number of lines, words, and characters in a text file (or its standard input). My presentation is based on the description and algorithm in *The C Programming Language* by Kernighan and Ritchie [17].

14.2.1 Counting words and characters within a line

The easiest way to avoid the line break ambiguity is to let Maple handle it and never explicitly examine the ends of lines, although it would be possible to implement in Maple the Kernighan and Ritchie `wc` program more closely than I will do here. It is good program design to split a programming task into subtasks, so let us begin by counting the words and characters in a text *string* that is guaranteed not to contain any line breaks; i.e., the string corresponds to a single line of text. The Maple function `length` returns the number of characters in a string (and other "lengths"; see the online help), so we may as well use it, although we could explicitly count the characters exactly as Kernighan and Ritchie do. The only task that we must implement explicitly is the word count. For this, we must first decide what we mean by a *word*. As a first attempt at a definition we might decide that words are sequences of letters separated from each other by "white space". Within a line, white space appears as a sequence of one or more space characters, although some of those space characters may be represented by ASCII "horizontal tabulator" (usually abbreviated to just *tab*) characters. So we can regard white space within one line as consisting of a sequence of one or more spaces and tabs. In Maple (and most other modern programming languages) the tab character can be represented as `"\t"`. (It is generally better to use this "escape sequence" than to press the *Tab* key on your keyboard because that will probably be interpreted as a command to the application you are running to do something; for example, in the Maple worksheet GUI [on all platforms] *Tab* [*Shift+Tab*] moves the text cursor to, respectively, the next [previous] Maple input.) But what about punctuation, which is neither part of a word nor white space. Perhaps we should define non-word characters to be spaces, tabs, and punctuation and define words to be sequences of characters separated by sequences of non-word characters. But what about digits? It really depends on exactly what you want to count and why, but digits are usually regarded as parts of words.

Following Kernighan and Ritchie, let us first regard a word as any sequence of characters that does not include a space, tab, or end of line and write a procedure to count the words and characters within a single-line string:

```
> wc_line := proc(line::string)
     # Count words and characters
     # in a "single-line" input string.
     local c, nw, out;
     nw := 0;  out := true;
     for c in line do
        if c = " " or c = "\t" then
           out := true
        elif out then
           out := false;
           nw := nw + 1
        end if
     end do;
     nw, length(line)
  end proc:
```

This procedure uses the local variables `c` to represent the current character, `nw` to count the number of words, and `out` to store state information. The `for`-loop assigns each character of the input string in turn to the variable `c` and can be imagined also to move a pointer or *cursor* along the input line. The variable `out` is true if the cursor was *out* of a word and so is initialized to true. The count `nw` is initialized to zero. The word-count algorithm is the following. If the next character is a non-word character then the cursor is now out of a word and so `out` is set to `true`. Otherwise, the cursor is in a word; if it was previously out of a word then `out` is set to `false` and the word count `nw` is incremented. The procedure returns a *sequence* of the numbers of words and characters. An expression sequence provides a very flexible way in Maple to return multiple data from a procedure.

It is tricky when designing such an algorithm to ensure that it works in all cases, including unexpected ones. Algorithms normally fail at their limits, which in this case are the beginning and end of the string. Let us test the extreme case first, in which the string is empty so the beginning and end coalesce:

```
> wc_line("");
```

$$0, 0$$

This is clearly correct: there are no words and no characters in an empty string. The next tests should include small numbers of words that both do and do not begin and end at the ends of the string, e.g.,

```
> wc_line(" Hello World! ");
```

$$2, 14$$

Let us now consider a few generalizations of this approach. Instead of considering only *space* and *tab* as word separators, we might allow *space* and all control characters. If we can assume the ASCII character encoding then this is easy to implement, because *space* is the first graphic character and all the earlier characters are control characters, so the test for a non-word character can be simply `c <= " "`.In addition to being more flexible than the previous test it should be slightly more efficient, since it involves only one test.

But how many words are there in the following nonsense text?

```
> wc_line("Nonsense -- with dashes, it’s abbreviated isn’t
  it , hyphenated words and non-standard spacing ! .");
```

$$16, 98$$

One way to have more control is to specify a *set* of non-word characters, which could include punctuation characters. If we regard apostrophes as word characters then an abbreviation containing an apostrophe would count as a single word; if we regard hyphens as non-word characters then a hyphenated word would count as two words. To treat a double-hyphen, intended to represent a single en-dash, differently from a single hyphen would require more complex code, which I will not pursue here. The set of non-word characters is best assigned to a global variable rather than "hard-wired" into the procedure definition. The best way to construct the set of non-word characters depends on whether you want to assume a particular character encoding and, if so, on its properties. In the worst case, you can simply enumerate all non-word characters explicitly.

Let us assume the ASCII character encoding, and consider the characters that count as parts of words to be the apostrophe, digits, and upper and lower case letters:

```
> word_characters := {"’", $"0".."9", $"A".."Z", $"a".."z"}:
```

Then all other characters are non-word. In Maple, not all control characters can be input using backslash escape sequences, and the ASCII NUL character encoded as zero cannot be represented at all. (This is presumably because a string in the C language cannot contain a NUL because it is used to denote the end of a string, and the Maple kernel is essentially written in C.) However, all characters other than NUL can be generated by converting from their integer encodings, so we can now construct the set of representable non-word ASCII characters like this:

```
> non_word_characters :=
    {$convert([1],bytes)..convert([127],bytes)}
      minus word_characters;
```

non_word_characters := {“ ”, “=”, “>”, “)”, “<”, “ \t”, “!”, “-”,
“,”, “.”, “□”, “□’, “□”, “□”, “□”, “□”, “□”, “\a”, “\b”, “ \n”,
“\v”, “ \f”, “\r”, “□”, “□”, “□”, “□”, “□”, “□”, “□”, “□”, “□”,
“□”, “□”, “□”, “□”, “\e”, “□”, “□”, “□”, “□”, “\“”, “#”, “$”,
“%”, “&”, “(”, “*”, “+”, “/”, “:”, “;”, “?”, “@”, “[”, “\\”, “]”,
“^”, “_”, “`”, “{”, “|”, “}”, “ ~”}

Before revising the word counting procedure, let us pause to consider the overall implementation. We will have a procedure that uses the value of a global variable `non_word_characters` that is computed from the value of another variable `word_characters`. However, `non_word_characters` need be global only to `wc_line` and it need not be "more global" than that, although it makes sense for it to be globally accessible. The variable `word_characters` is used to initialize `non_word_characters` and not for any other purpose. It would be confusing if it were globally accessible, since changing it would have no effect. It is appropriate to wrap the procedure and variables in a module and to export only `non_word_characters` and `wc_line`. This keeps `word_characters` completely local to the module and makes it inaccessible outside. It is used solely internally to initialize the module. The procedure `wc_line` and variable `non_word_characters` are exported from the module, which gives them the correct level of globality. Here is the revised module-based implementation:

```
> text_processing := module()
    local word_characters;
    export non_word_characters, wc_line;
    option package;
    description "Text processing";

    word_characters :=
      {"'", $"0".."9", $"A".."Z", $"a".."z"};
    non_word_characters :=
```

```
      {$convert([1], bytes)..convert([127], bytes)}
         minus word_characters;

   wc_line := proc(line::string)
      # Count words and characters
      # in "single-line" input string.
      local c, nw, out;
      nw := 0;  out := true;
      for c in line do
         if member(c, non_word_characters) then
            out := true
         elif out then
            out := false;
            nw := nw + 1
         end if
      end do;
      nw, length(line)
   end proc;

end module;
```

$$
\begin{array}{l}
\textit{text\_processing} := \mathbf{module}()\\
\mathbf{local}\ \textit{word\_characters};\\
\mathbf{export}\ \textit{non\_word\_characters},\ \textit{wc\_line};\\
\mathbf{option}\ \textit{package};\\
\mathbf{description}\ \text{“Text processing”};\\
\mathbf{end\ module}
\end{array}
$$

We have effectively written a small Maple package, so I have included the module option `package` and also included a `description` string, neither of which is necessary. Now we can access the new procedure using either the old "table oriented" package notation:

```
> text_processing[wc_line]("double-trouble");
```

$$2, 14$$

or the new "object oriented" module notation:

```
> text_processing:-wc_line("double-trouble");
```

$$2, 14$$

Note that the old procedure is still available and interprets a hyphen as a word character:

```
> wc_line("double-trouble");
```

$$1, 14$$

This is why we get a warning if we enable our new package to use short names:

```
> with(text_processing);
```

```
Warning, the assigned name wc_line now has a global binding
```

$$[non\_word\_characters,\ wc\_line]$$

which re-defines (or over-writes) the old procedure definition:

```
> wc_line("double-trouble");
```

$$2, 14$$

The new version handles the nonsense input as we want:

```
> wc_line("Nonsense -- with dashes, it's abbreviated isn't
  it , hyphenated words and non-standard spacing ! .");
```

$$13, 98$$

14.2.2 Counting lines, words, and characters in a file

Let us now return to our main task of counting the lines, words, and characters in a file. But first we need to generate a very short text file to use as test data. You can, of course, use any text editor to do this. But now we hit another ambiguity, which is that of file location. When a file is specified by just its name this implies that it is in the current directory or folder, but this concept is a somewhat ambiguous when using a GUI.

When running Maple on Windows, the current directory depends on several factors. If Maple was started explicitly via a shortcut (e.g., in the Start menu or on the desktop), then the current directory is what was specified in the "Start in:" field of the shortcut. This can be accessed by *right*-clicking the shortcut and selecting "Properties". By default, it is the root directory of the Maple installation, which is not a good place to put your own files, but (at least, on your own machine) you can change the current directory to be a directory of your choice. If Maple is associated with the file type `.mws` and you start Maple by opening a worksheet (e.g., from Windows Explorer), then the current directory is the directory containing the worksheet file. You can either remember these rules (which are not specific to Maple) or always use fully qualified pathnames to specify files.

The only way to avoid ambiguity about the location of files is always to use a fully qualified pathname, but that is not fully portable. (Maple converts the canonical directory or folder separator character `/` into whatever the underlying operating system uses, but it does not attempt any other directory mapping.) I will compromise by using the directory `/tmp`. This is a standard directory on UNIX that is intended for any user's temporary files.

Windows computers usually have a directory called `temp` in the boot partition (usually `C:`), which is used by many applications for temporary files and which the user can also use. However, on my Windows computers, I always create a directory called `tmp` in the root of the disk partition that I use for most of my own files, and often in all partitions. Regardless of the platform you are using, I will assume that either you will ensure that the directory `/tmp` exists (on the current drive) or that you will change references to `/tmp` to a directory that does exist.

Since this is a book about using Maple, let us use Maple to generate a text file to process. The function `writeline` takes a sequence of strings as its arguments, the first of which must be a valid file name. It writes the remaining sequence of strings to the file, one string per line, and returns a count of the number of characters output, e.g.,

```
> writeline("/tmp/words.txt",
     "This is a file",
     "containing a few words",
     "to test my Maple",
     "word count code.");
```

$$72$$

This count treats the end of each line as a single newline character, regardless of the platform, and is therefore portable. The function `writeline` creates and opens the specified file if it does not exist and opens the file and deletes its contents if it does exist. It then writes the specified output but does not close the file. Subsequent use of `writeline` to output to the same file appends the output to the end until the file is explicitly closed like this:

```
> close("/tmp/words.txt");
```

Note that output is normally buffered in memory and not actually written to the file until necessary. Closing the file forces the output to be written to it and so makes it accessible to other applications, and updates any other information about the file, such as its size. On Windows, once the file is closed, the Windows Explorer Properties dialogue shows that the file contains 76 bytes because each newline is actually stored as two characters instead of one. On UNIX and Macintosh platforms, the actual byte count should agree with the character count reported by Maple. In the spirit of portability, let us regard line breaks as single characters for the character count; the physical byte count is implied by the line count and the representation of the ends of lines on the current platform.

Here is a procedure to write lines to a text file and *then close it*, which will be useful later:

```
> writefile := proc()
     local count;
```

```
    count := writeline(args);
    close(args[1]);
    count
 end proc:
```

The Maple function that is the inverse of `writeline` is `readline`. It attempts to read the "next" line of text from the file specified as its argument and, if there is a next line, returns it as a string without any newline characters. When it gets to the end of the file it returns the integer 0 instead of a string. The first time it is called on a file, or after it has reached the end of the file, it reads the first line of the file. Hence, all lines of a text file can be read by using `readline` in a loop that is terminated when `readline` returns 0. This leads to the main structure of the following procedure:

```
> wc := proc(file::string)
    # Count lines, words and "characters" in a text file.
    local counts, nl, line;
    counts := [0,0];  nl := 0;
    do
       line := readline(file);
       if line = 0 then break end if;
       printf("%s\n", line);  # for testing
       counts := counts + [wc_line(line)];
       nl := nl + 1
    end do;
    counts := counts + [0,nl];  # add newlines
    nl, op(counts)
 end proc:
```

```
> printf("\nLines: %d, words: %d, characters: %d",
    wc("/tmp/words.txt"));

  This is a file
  containing a few words
  to test my Maple
  word count code.

  Lines: 4, words: 15, characters: 72
```

This procedure maintains two counters in the variables `counts` and `nl`. The variable `nl` counts the number of lines read in a straightforward manner. The variable `counts` holds a *list* of two values representing the number of words and characters, which is updated by list addition (which just adds corresponding elements in lists that must be of the same length). The pair of values added to `counts` is the pair of values returned by applying `wc_line` to the value returned by `readline`. Finally, the number of lines is added to the number of characters to include newlines, and the procedure returns a sequence of the number of lines, words, and characters. This makes it very convenient to call the procedure `wc` as part of the argument sequence of another procedure such as `printf`.

For purposes of testing, the procedure outputs each line as it reads it. There are various ways to do this, but I have used the function `printf`, which is analogous to the standard C library function of the same name. The first argument to `printf` is a format string, which consists of text to be output with formatting specifications embedded in it. A formatting specification consists of a `%` character followed by various codes, the full set of which is described in the online help for `printf`. The formatting specifications indicate exactly how to output the values of the subsequent arguments. The specification `%s` means output a string exactly as it is stored internally and `%d` means output an integer in normal decimal notation. A `\n` escape within the format string indicates a newline. The test output using the particular input file as shown above appears to be correct.

14.3 Text formatting

Another simple classic text processing task is to implement the standard UNIX program called `fmt`. The name is an abbreviation of *format.* (Remember that more than three characters is too long for a UNIX program name!) The purpose of `fmt` is primarily to fill lines of text with complete words up to some maximum line length, by joining lines and re-breaking them as necessary. The task can be seen as splitting the input into single white-space-delimited words and then rebuilding output lines that consist of as many of those whole words as fit within the specified maximum line length (although it need not necessarily be implemented exactly like that). A program such as `fmt` provides a very simple introduction to the vastly more sophisticated text processing performed by programs such as TeX, `troff`, and their relatives and by interactive word processors such as Microsoft Word and, to a lesser extent, the Maple worksheet interface. The UNIX `fmt` program was particularly useful before GUIs were commonplace and when text editors were line rather than screen oriented. On UNIX systems in the 1980s it was common to write e-mail using an interface that provided only very rudimentary input editing that did not include line filling. Hence, it was very convenient to be able to "pipe" an outgoing e-mail through `fmt` to tidy it up before actually sending it. Similarly, it was convenient for writing simple text documents that did not merit the use of TeX or troff (which were big and slow programs by the standards of the 1980s).

K&P present a version of `fmt` written in Awk. (The name "Awk" comes from the surnames of its designers: Aho, Weinberger, and Kernighan.) That version treats blank lines as paragraph breaks, which it preserves. This actually makes the task much more difficult. The following simple Maple version treats all white space the same, and so effectively ignores paragraph structure. It is sufficient for formatting a single paragraph, although it turns all text into one single paragraph, and it suffices for reformatting the output from the Markov mangler program to be described in the next section.

The Maple function `fscanf` with format specification `%s` provides a convenient way to read a file one white-space-delimited word at a time. It is based on the standard C library function of the same name. The Maple version returns the integer 0 when it reaches the end of the file; otherwise, it returns a list of the values that it is required to read, in this case a list of one string. It is necessary to explicitly close the file once it has been read to its end, so that it can be read again if necessary. The formatting algorithm outputs each word immediately after it has been input. However, it keeps a record of the length of the line that it has output so far. If outputting a space followed by the current word on the current line would exceed the allowed maximum line length then the current line is terminated by outputting a newline character `\n` immediately before outputting the current word. Otherwise, the current word is output preceded by a space unless it is the very first word of the text.

```
> fmt := proc(file::string)
     # Format text file to maximum line length MaxLen.
     local MaxLen, Len, word, L;
     MaxLen := 60;
     Len := 0;
     do
        word := fscanf(file, "%s");  # next "word"
        if word = 0 then close(file); return end if;
        word := op(word);  # fscanf returns list
        L := length(word);
        if Len + 1 + L > MaxLen then
           printf("\n%s", word);  Len := L
        else
           if Len > 0 then printf(" ") end if;
           printf("%s", word);  Len := Len + 1 + L
        end if
     end do
  end proc:
```

Here are a couple of tests; the first input text consists of a single long line. (The reason for this particular choice of nonsense input will be explained in the next section.)

```
> writefile("/tmp/Text.txt",
     "Show your face and conceal your eyes and I will be
  mystified.  Show your eyes and your face will be obvious.");
```

It is formatted as follows:

```
> fmt("/tmp/Text.txt");

   Show your face and conceal your eyes and I will be
   mystified. Show your eyes and your face will be obvious.
```

The second input text is identical to the first but broken up haphazardly:

```
> writefile("/tmp/Text1.txt",
      "  Show your face and conceal",
      "",
      " your eyes and I will be mystified.  ",
      "",
      "  Show your eyes and your ",
      "",
      "",
      "",
      " face will be obvious.",
      "");

                              123
```

It is formatted exactly as before:

```
> fmt("/tmp/Text1.txt");

   Show your face and conceal your eyes and I will be
   mystified. Show your eyes and your face will be obvious.
```

Here is a Maple translation of K&P's Awk implementation of `fmt`, which uses two local procedures called `addword` and `printline` that are very similar to the Awk versions: the first adds a word to the output line and the second prints the output line. The rest of the body of procedure `fmt` is a Maple translation of the remaining *three lines* of the Awk program. The fact that this takes considerably more than three lines of Maple is an indication of the power of Awk for simple text processing (which is essentially what it was designed for). It is also a consequence of the fact that the Maple function `sscanf` does not process its input string sequentially in the same way that `fscanf` does, and so cannot be used to extract the words from each input line. I have therefore used an explicit loop based on the code used earlier in procedure `wc`. This implementation makes essential use of Maple's lexical variable scoping: procedures `addword` and `printline` access the variable `lineout` as a global variable, although this variable is local to procedure `fmt`, as are the two procedures `addword` and `printline` themselves.

```
> fmt := proc(file::string)
     # Format text file to maximum line length MaxLen
     # preserving blank lines.
     local MaxLen, addword, printline, linein,
```

```
      lineout, inword, word, c;
   MaxLen := 60;

   addword := proc(w)
      if length(lineout) + 1 + length(w) > MaxLen then
         printline()
      end if;
      if length(lineout) = 0 then
         lineout := w
      else
         lineout := cat(lineout," ",w)
      end if
   end proc;

   printline := proc()
      if length(lineout) > 0 then
         printf("%s\n", lineout);  lineout := ""
      end if
   end proc;

   lineout := "";
   do
      linein := readline(file);
      if linein = 0 then  # end of file
         close(file);  printline();  return
      elif linein = "" then  # blank line
         printline();  printf("\n")
      else
         inword := false;  word := "";
         for c in linein do
            if c = " " or c = "\t" then
               if inword then
                  addword(word);
                  word := "";
                  inword := false
               end if
            else
               word := cat(word, c);
               inword := true
            end if
         end do;
         addword(word)
      end if
   end do
end proc:
```

Here is a test; the first input paragraph consists of a single long line and the second paragraph is broken up haphazardly:

```
> writefile("/tmp/Text2.txt",
    "Show your face and conceal your eyes and I will be
  mystified.  Show your eyes and your face will be obvious.",
    "",
    "  Show your face and ",
    "conceal your eyes and",
    "   I will be mystified.",
    "Show your eyes and your",
    "   face will be obvious.");
```

227

Both paragraphs are formatted identically and kept distinct:

```
> fmt("/tmp/Text2.txt");

   Show your face and conceal your eyes and I will be
   mystified. Show your eyes and your face will be obvious.

   Show your face and  conceal your eyes and I will be
   mystified. Show your eyes and your face will be obvious.
```

14.4 A Markov chain algorithm

This section is based on Chapter 3 of K&P, entitled "Design and Implementation", in which the authors remark that the choice of programming language is relatively unimportant to the overall design of a program. To illustrate this, they design a text processing program and then write it in C, Java, C++, Awk, and Perl. Here I will write it in Maple. My Maple implementation is based on K&P's Perl implementation except for the code to read words of input text, which is based on their C implementation. Maple provides almost all of the high-level Perl facilities that are relevant to this program.

The program turns a passage of text into nonsense that reads well by using a Markov chain randomization algorithm to preserve its essential statistical structure, which I will call *Markov mangling*. Any particular phrase (a sequence of words, called a *prefix*) in the input text may appear more than once, so if a particular phrase appears n times then there are up to n different (*suffix*) words that follow it. The Markov mangling algorithm chooses the suffix word following the prefix phrase in the output at random from all those that have followed it in the input. Two-word prefixes work well. The Markov mangling algorithm, following K&P, is this:

> Set the prefix w_1, w_2 to the first two words in the text;
> print w_1 w_2;

repeat
  randomly choose the suffix w_3, one of the successors
    of prefix w_1, w_2 in the text;
  print the suffix w_3;
  replace the prefix w_1, w_2 by w_2, w_3;
until w_3 is the last word in the text.

The algorithm must have access to the whole input text before it can start to produce output, which implies an input phase that stores all the input text in some appropriate data structure followed by an output phase as described by the above algorithm. The data structure must allow new data to be added easily as it arrives essentially randomly. K&P recommend a hash table in which the access keys are the prefixes and the elements are the corresponding collections of suffixes. During the input phase it must be possible to add new elements and to add new suffixes to existing elements; during the output phase it must be possible to access the elements quickly and to extract one of the suffixes randomly. An appropriate implementation in Maple is a table indexed by prefixes, the elements of which are sequences of suffixes. Maple can handle this data structure well since Maple tables are represented as efficient hash tables and Maple expression sequences are easily extended and efficiently accessed by indexing. Moreover, in Maple distinct expressions, including text strings, are stored once only. In standard Markov terminology, a prefix and all its possible suffixes is called a state, so the table that stores all the states is called a *state table* (`statetab`). Because both Perl and Maple provide suitable data structures as standard features, the implementations in both language can be quite short and fairly straightforward; a C implementation is much longer and more complicated, mainly to support appropriate data structures.

The question of what a word is arises once again. It is desirable to leave punctuation attached to words so as not to lose it, although this can lead to incorrect output punctuation. But generally it works well to regard words as sequences of characters delimited by white space. Hence, as before, the Maple function `fscanf` with format specification `%s` provides a convenient way to read a file one white-space-delimited word at a time. It returns the integer 0 when it reaches the end of the file; otherwise, it returns a list of the values that it is required to read, in this case a list of one string.

The implementation is easiest if we use a special marker string to start and stop the algorithm. This must be a string that cannot be read as an input word, so a string consisting entirely of white space will suffice. For purposes of documentation, and following K&P, let us assign the string consisting of a single newline character to the variable `NONWORD`. Then the initial prefix is set explicitly to `NONWORD, NONWORD` and the final suffix is set explicitly to `NONWORD`. This is an example of K&P's general rule that it is better to handle special cases in data than in code if possible.

Here is an implementation of the Markov mangler as a Maple procedure:

```
> markov := proc(file::string)
    # Apply a Markov chain mangler
    # to the specified text file.
    local MAXGEN, NONWORD, statetab, prefix, suffix, n;
    MAXGEN := 1000;  # maximum words generated
    NONWORD := "\n";
    statetab := table();

    # Build state table:
    prefix := NONWORD $ 2;  # initialize
    do
       suffix := fscanf(file, "%s");  # next "word"
       if suffix = 0 then close(file); break end if;
       suffix := suffix[1];  # extract single word from list
       if assigned(statetab[prefix]) then
          statetab[prefix] := statetab[prefix], suffix
       else
          statetab[prefix] := suffix
       end if;
       prefix := prefix[2], suffix  # next prefix
    end do:
    statetab[prefix] := NONWORD;  # terminate

    # Generate output:
    prefix := NONWORD $ 2;
    to MAXGEN do
       suffix := statetab[prefix];  # suffices
       if suffix = NONWORD then return end if;
       n := nops([suffix]);  # no. of suffices
       if n > 1 then
          n := rand(1..n)();
          suffix := suffix[n]  # random suffix
       end if;
       printf("%s ", suffix);
       prefix := prefix[2], suffix  # advance chain
    end do
 end proc:
```

The local variable `MAXGEN` sets the maximum number of output words that can be generated. This is to avoid a possible infinite loop; normally the output stops after the last word of the input has been output as a suffix. The local variable `statetab` holds the Markov state table. It need not be initialized to an empty table since the first assignment to one of its elements automatically initializes it, but the explicit initialization provides useful documentation. The function `fscanf` will automatically open the specified file for reading as a text file but will not close it, so the procedure must explicitly close it after having read to the end of the file. If a particular prefix already exists then that element of `statetab` has an assigned value, in which case the new suffix is appended to the current sequence of suffices; otherwise, the new suffix alone forms a new table element (equivalent to a sequence of unit length). The Maple library function-call `rand(1..n)` returns a procedure that generates a random integer in the specified range, which corresponds to the range of indices for the current suffix sequence. This procedure is executed immediately to generate a random index in the correct range. The test whether there is only one suffix is included only for efficiency; it seems pointless to include the overhead of generating a random integer in the range 1..1! The first loop is terminated by a `break` to allow the procedure to continue and execute the second loop, which is (normally) terminated by a `return` since there is no code to execute after that loop.

There are various ways to output the sequence of random suffixes. K&P output them one per line (i.e., separate them with newlines). I prefer to separate them with spaces, which is more immediately readable. K&P recommend re-processing the output with a simple formatter, such as that described in the previous section. For the purpose of illustrating the operation of the Markov mangler I have reformatted the output in the same way as the rest of this book (namely, using LaTeX).

My Maple implementation contains 29 lines of code (excluding comments and blank lines). This is about half as long again as K&P's Awk and Perl implementations but less than half their C++ implementation and much shorter than their C and Java implementations. I could have saved a few lines of Maple code but it would not have changed the position of Maple within this "league" table.

To illustrate the algorithm, K&P use two sentences paraphrased from *The Mythical Man Month* by Frederick Brooks [3]. I will use a similar pair of sentences, partly so that I can compare my output with that obtained by K&P and partly because these two sentences have enough repetition within very few words to constitute a useful but very succinct illustration of Markov mangling. The precise input text is the paragraph used above as an illustration of text formatting. Here are a few typical output examples:

```
> markov("/tmp/Text.txt");
```

```
Show your eyes and your face will be mystified. Show your face
will be mystified. Show your eyes and your face will be obvious.
```

```
> markov("/tmp/Text.txt");
```

```
Show your face and conceal your eyes and I will be obvious.
```

```
> markov("/tmp/Text.txt");
```

```
Show your eyes and I will be mystified. Show your face will be
obvious.
```

```
> markov("/tmp/Text.txt");

   Show your eyes and I will be obvious.


> markov("/tmp/Text.txt");

   Show your face will be obvious.
```

14.5 Exercises

1. Incorporate procedure `wc` into the `text_processing` module with its name included in the `export` declaration. Control printing of input lines by another exported variable, the value of which is set to `false` by default. Test the whole module.

2. Write a Maple procedure called `reverse` that takes a list or string as its single argument and returns a reversed copy of it.

3. There is a standard UNIX program called `tr`, and a similar facility built into the Perl language, that translates or deletes characters. Write and test an analogous Maple procedure called `tr` that takes at least one argument that must be a string and returns a possibly modified version of this string. The optional second argument must be a string or list of characters to be replaced, the third a string or list of characters to replace them with. Each character in the second argument is to be replaced by the corresponding character in the third argument if there is one, or deleted otherwise. Generalize the procedure to allow its second and third arguments to be strings or lists of strings (of length greater than 1).

4. Revise procedure `markov` so that the number of words in the prefix can be set to any number via an optional second argument but defaults to 2, and test it. (This generalization is almost trivial in view of the way the version in the text is written.)

Chapter 15

Object Orientation and Modules

This final chapter is about modules, which are one of the major new facilities introduced in Maple 6. They can be used in several ways and we have already used them in a fairly trivial way to provide a local context or block structure in which to execute some code. Maple has the concept of a statement sequence but it does not have the concept of a *compound statement* or *block* that most other modern programming languages (e.g., C, C++, Java, Pascal, Modula) have inherited from Algol 60 or LISP, in which local variables can be declared. Modules provide something closer to blocks than existed in earlier versions of Maple although they are still not the same as blocks in other languages. A Maple module is very similar to a procedure that is defined and executed in the same statement, except that the latter returns a value. This is even closer to the behaviour of blocks in other languages except that the procedural syntax is rather convoluted.

Two other applications of modules, which are the subject of this chapter, are as "objects" and "packages". The former application provides a way of emulating some of the facilities provided by modern object oriented programming languages. The latter provides a better way of implementing packages than the table-based approach used in earlier versions of Maple and is similar to the way that modules are used in the language Modula. (In fact, Maple syntax is quite similar to Modula syntax, Maple 6 more so than earlier versions.) These two applications rely quite heavily on lexical scoping.

The first 12 sections are concerned primarily with object oriented programming. Using the implementation of facilities for manipulating block or partitioned matrices as a theme, successive sections develop the concepts from a simple record-based implementation to a fairly complete object-oriented implementation. In passing, I illustrate some matrix manipulations that are not covered elsewhere in this book. I use the table-based matrix representation and the `linalg` package. There is no compelling reason to prefer this over the new rtable-based implementation although it makes some of the more symbolic code (such as the illustration in Section 15.2) easier. I leave it as an exercise to re-implement the code in this chapter using rtables. During this development process the code becomes large enough to make it advantageous to export it to a text file, maintain it with a text editor outside of Maple, and read it back in to use it.

Section 15.13 makes the transition to packages via operator overloading. A large body of code is most useful if it can be easily accessed in the same way as the facilities in the standard Maple library. Section 15.14 describes how to create and use a new library "repository" and the following section describes how to implement a module-based package and save it to a repository. Any program needs user documentation to be of much use and the final section explains how to create Maple online help pages.

15.1 Introduction to object orientation

One of the more recent developments in programming languages is *object orientation*, and commonly used modern programming languages such as C++ and Java are *object oriented.* Object orientation is an alternative programming paradigm, in the same way that spreadsheets are. Object oriented programming (OOP) covers several ideas, not all of which are supported by all object oriented (OO) languages. The main idea is to encapsulate as an *object* each distinct set of data *together with the procedures that operate on that data.* Hence, an object can be thought of as a set, the *members* of which are either variables or procedures. This is a generalization of the concept of user-defined types that has been around for a long time; they are called *records* in Algol 68, Pascal, and Modula; *structures* in C and C++. In OO terminology, variables in an object are often called the *properties* or *fields* of the object and procedures are usually called *methods.* Members of an object may be accessible outside the object, in which case they are called *public*, or they may be accessible only inside the object, in which case they are called *private.* The interface to an object consists of its public members. All the implementation details of an object can be hidden as private information, which means that they can be changed without upsetting the interface to other parts of a program.

Maple *modules* can be used to represent objects. Most OO languages use the syntax *`object.member`* to refer to a member of an object, but in Maple this syntax is used for non-commutative multiplication, so Maple uses the syntax *`module:-member`* to refer to a member of a module. However, the fact that Maple supports some aspects of OOP does not make Maple an OOP language. In an OOP language, an object is an instance of a *class*, which specifies the structure of the object, whereas in Maple the structure of a module is specified directly when the module is written and a module is not an instance of any explicit class. It is not possible to ignore object orientation completely in an OOP language, whereas in Maple it is entirely possible to ignore object orientation, and any code written for versions of Maple prior to Maple 6 almost certainly does! (There is one exception, namely the `Domains` package, which has been part of Maple for some time and provides support for parameterized domains, which is an aspect of OOP.) An important feature of classes in an OOP language is inheritance. Classes form a hierarchy, in which classes lower in the hierarchy inherit default information from their superclasses, possibly from more than one superclass in languages such as C++ that support multiple inheritance. Since Maple does not use classes it does not support class inheritance.

Incidentally, Maple is a dynamically typed language, unlike C++ and Java, which are statically typed. The difference is that in a statically typed language, the data type of every variable must be declared (statically) before it can be used, whereas in a dynamically typed language only the value of a variable has a type, which can be tested (dynamically) when the program is run. A class definition is essentially a user-defined type definition to be used in a (static) variable type declaration. Since Maple does not use variable type declarations it is not surprising that it does not use class definitions either.

15.2 Block matrices

As subject matter for OOP let us consider block or partitioned matrices. These arise in linear algebra and are not particularly well supported by default in Maple. They are described in most standard linear algebra textbooks; for example, they appear in several places in *Linear Algebra* by Seymour Lipschutz [21]. The idea is this. Take a (reasonably large) matrix and *partition* it by (imagining) drawing one or more horizontal and/or vertical lines between some of the rows and columns. Then the matrix can be regarded as a smaller matrix whose elements are themselves matrices; e.g., a 3×3 matrix could be partitioned as a 2×2 block matrix, which we can do explicitly in Maple like this:

```
> with(linalg):

   Warning, the protected names norm and trace have been redefined
   and unprotected

> M := matrix(3,3):  MM := evalm(M):

> BM := matrix(
      [[submatrix(M,1..2,1..2),submatrix(M,1..2,3..3)],
       [submatrix(M,3..3,1..2),submatrix(M,3..3,3..3)]]):

> eval(MM) = eval(BM);
```

$$\begin{bmatrix} M_{1,1} & M_{1,2} & M_{1,3} \\ M_{2,1} & M_{2,2} & M_{2,3} \\ M_{3,1} & M_{3,2} & M_{3,3} \end{bmatrix} = \begin{bmatrix} \begin{bmatrix} M_{1,1} & M_{1,2} \\ M_{2,1} & M_{2,2} \end{bmatrix} & \begin{bmatrix} M_{1,3} \\ M_{2,3} \end{bmatrix} \\ \begin{bmatrix} M_{3,1} & M_{3,2} \end{bmatrix} & \begin{bmatrix} M_{3,3} \end{bmatrix} \end{bmatrix}$$

It will be useful to keep the unblocked and blocked forms of this matrix available for later use, which is why I have assigned them to the variables `MM` and `BM`, respectively. (The use of `evalm` above is essentially just a trick to generate a symbolic matrix with elements named by a specific identifier, here `M`.)

What makes block matrices useful is that computations can be performed as if the blocks were normal matrix elements, provided that matrix algebra is used for the elements as well as the matrix itself. This is obvious for addition, subtraction, and scalar multiplication, less obvious for other operations. There is, however, an extra constraint, which is that formally both the block matrix and its blocks must be conformable (i.e., the right size and shape) for the operation to be performed. However, Maple does not support block matrix algebra. The only explicit support for block matrices provided by Maple is the function `linalg[blockmatrix]`, which creates an unblocked matrix from its blocks, and `LinearAlgebra[JordanBlockMatrix]`, which creates an unblocked matrix from its Jordan blocks. Hence, block matrix support is a good candidate for a new object oriented implementation. The essential information about a block matrix is its elements and its partitioning, which is needed to check conformability, but the way the matrix is actually stored and the way operations are actually performed should be irrelevant.

I will use only table-based matrices in this chapter, and I will use the table-based `linalg` package as appropriate. This is slightly more convenient for present purposes than rtable-based matrices and the rtable-based `LinearAlgebra` package. It would be entirely possible to use rtable-based as well as, or instead of, table-based matrices, but that would just add unnecessary complication and is left as an exercise for the reader.

15.3 Modules as records

The simplest kind of object contains only data and is just a user-defined data type, which is often called a *record* or *structure*. We can represent the partitioning of a matrix by lists of the numbers of rows and columns in each partition, which suggests the following block matrix representation:

```
> B := module()
      export elements, rowp, colp;
  end module;
```

$B :=$ **module()** **export** *element*, *rowp*, *colp*; **end module**

One way to put data into this object is by simply assigning to its fields or exported variables by using the syntax *module*`:-`*member* to access them.

Then to make block matrix B represent the explicit block matrix BM defined above we could set its partitioning information like this:

```
> B:-rowp := [2,1];  B:-colp := [2,1];
```

$$rowp := [2, 1]$$
$$colp := [2, 1]$$

In order to develop a concrete implementation, let us represent the matrix elements as a single unblocked matrix. Then we can assign all the matrix elements at once as follows. (As we will see later, the advantage of this internal representation is that it makes block matrix algebra trivial.)

```
> B:-elements := eval(MM);
```

$$element := \begin{bmatrix} M_{1,1} & M_{1,2} & M_{1,3} \\ M_{2,1} & M_{2,2} & M_{2,3} \\ M_{3,1} & M_{3,2} & M_{3,3} \end{bmatrix}$$

Modules have the same evaluation rules as do procedures and tables (i.e., last name evaluation), and there is an obvious analogy between accessing the members of a module and the elements of a table. Hence, explicit evaluation is required to see the module to which a variable evaluates:

```
> B;
```

$$B$$

```
> eval(B);
```

module() **export** *element*, *rowp*, *colp*; **end module**

This is essentially all that one ever sees; evaluation of a module does not show its internal values, only its data structure, which constitutes its "signature". But a module does not need to be explicitly evaluated in order to access its exported (public) members:

```
> B:-rowp;
```

$$[2, 1]$$

However, explicit evaluation is again required to see the matrix elements assigned to the exported variable `element`, because in our present representation it is an array and arrays, like modules, obey last name evaluation:

```
> B:-elements;
```

$$element$$

```
> eval(B:-elements);
```

$$\begin{bmatrix} M_{1,1} & M_{1,2} & M_{1,3} \\ M_{2,1} & M_{2,2} & M_{2,3} \\ M_{3,1} & M_{3,2} & M_{3,3} \end{bmatrix}$$

Before we continue with this example let us take a more formal look at Maple modules.

15.4 Modules in more detail

A Maple module combines aspects of both a data structure and a control structure: it can contain both local variable declarations and code that is executed immediately when the module is defined, which may include assignments to variables and procedure definitions. When using a module to represent an object, this code is used to initialize the information in the object. Variables can be declared `local`, just as in a procedure, which makes them completely private to the module. Alternatively, they can be declared `export`, which makes them local but public. If such variables evaluate to procedures then they correspond, respectively, to private or public methods; otherwise, they correspond respectively to private or public fields or properties. Any global variables used should be explicitly declared `global`, just as in a procedure definition. (In OOP parlance, local and exported variables correspond to instance members, and global variables are the nearest equivalent to class members.)

The full syntax of a module definition is this:

```
module()
    local variables;
    export variables;
    global variables;
    options ...;
    description ...;
    body_statements
end module
```

All the sections are optional, but any body statements must come last. Note that the keyword `module` must (in the anonymous modules that we are considering here) be followed immediately by empty parentheses, which are part of the module definition syntax; however, modules cannot take parameters in the way that procedures can. There are many parallels between modules and procedures; the principal differences are the lack of parameters, the `export` declaration, and the fact that module body statements are executed when the module is defined. Hence, there is no mechanism for "calling" or "applying" a module after it has been defined and a variable to which a module has been assigned can never be followed by parentheses. A `return` statement in the body of a module terminates execution of the body at that point, just as in a procedure, although a return statement in a module cannot be followed by a value to be returned, since modules do not return values. Apart from `export`, declarations in modules are very similar to declarations in procedures, and I will not pursue the details further here.

However, it is the declarations that identify a module, in the sense that once a module has been defined only the declarations (and the values assigned within the module to any variables) remain as part of the module data structure; the module body code is transient and once it has been executed when the module is defined it no longer plays any role and so in some sense no longer exists:

```
> module() local a; export b; global c;
     option foo; description "A module!";
     return  # redundant module body!of
  end module;
```

module()
local a;
export b;
global c;
option *foo*;
description “A module!”;
end module

In the above example, the `return` statement is executed when the module is defined and then no longer remains as part of the module data structure created by executing the module definition. This is much different from procedure definitions, which preserve the procedure body code unevaluated until the procedure is called or applied. Moreover, there is currently no facility within the Maple interface to control the amount of information about a module that is displayed.

Modules can be used quite freely as expressions and a module body can contain definitions of other modules. Modules obey exactly the same lexical scoping rules as do procedures, and it is this lexical scoping that is the key to most of the clever things that one can do with modules in Maple!

15.5 Module constructors

A typical OOP language will provide an elegant syntax for constructing an object, typically consisting of the keyword *new* followed by a call of a constructor function, the name of which is the same as the name of the class of object to be constructed. The constructor function is used to initialize the object, and it can be overloaded to allow the object to be initialized differently depending on the nature of its arguments. Some languages, such as C++, also require explicit destruction of objects once they are no longer required, whereas others, such as Java, handle this by automatic garbage collection. Maple is like Java in this — and many other — respects, and modules that are no longer in use (i.e., which can no longer be accessed in any way) are automatically destroyed by garbage collection when necessary to recover memory.

Maple does not provide any elegant syntax for constructing objects, but the idea of using constructor functions is still a good one, as too is the idea of overloading them. Indeed, it would be unusual to define a Maple module intended to represent an object or record explicitly; usually such a module would be defined within a procedure, which may be thought of as a module constructor, and usually this procedure will rely on lexical scoping to initialize the module. Here is a very simple constructor for our block matrix representation:

```
> BlockMatrix := proc(Elements::matrix,
                      Rowp::list(nonnegint),
                      Colp::list(nonnegint))
    module()
       export elements, rowp, colp;
       description "Block matrix";
       rowp := Rowp;
       colp := Colp;
       elements := Elements
    end module
  end proc;
```

BlockMatrix := **proc**(*element*::*matrix*, *Rowp*::list(*nonnegint*), *Colp*::list(*nonnegint*))
module()
export *element*, *rowp*, *colp*;
description “Block matrix”;
rowp := *Rowp* ; *colp* := *Colp* ; *element* := *element*
end module
end proc

One thing that we gain by using a constructor function to initialize the module is type checking of the values assigned to the module exports. There is no way to do this if they are assigned directly. The arguments of the constructor function play the role that you might have expected to be played by arguments taken by the module definition itself, and this use of lexical scoping provides a way to write parameterized modules. Any aspect of the module can be parameterized in this way; it is not restricted to initialization of exports.

Note that `rowp` and `Rowp` are completely different identifiers in Maple, as they must be for this code to work. We could have defined the module first and then assigned to its exported members, as we did interactively before, but it is more elegant to use assignments in the module body together with lexical scoping to perform the initialization as part of the module definition. You may be wondering why Maple has displayed the body of this module. The reason is that the module has not yet been defined, so what we are seeing is actually the body of the procedure. If we now execute the procedure we see only the signature of the module:

```
> B := BlockMatrix(eval(MM), [2,1], [2,1]);
```

$B :=$ **module**()
export *element*, *rowp*, *colp*;
description "Block matrix";

end module

although the module really has been initialized; e.g., its row partition is

```
> B:-rowp;
```

$$[2, 1]$$

I have added a brief description string to the block matrix module that could be useful to identify modules representing different classes of object (as opposed to different objects) in a large program.

We can make this constructor function much more flexible by overloading it, so that it behaves differently depending on the number and type of arguments with which it is called. (In Maple, this overloading must be handled dynamically by writing a polymorphic function, whereas in OOP languages is is handled statically by the compiler.) Let us allow either full initialization or no initialization, but not partial initialization. Unfortunately, lexical scoping does not apply to the special identifiers `args` and `nargs`, so now it is more elegant to define the module first and then assign to its exported members. It is no longer necessary to use distinct identifiers for the procedure arguments and the module exports, although it is perhaps clearer so I will continue to do so. It is also a good idea to include a consistency check on the initialization, i.e., to check that the partitioning is compatible with the matrix of elements. We can do this by checking that the row dimension of the `element` matrix is equal to the sum of the row dimensions of the blocks in the partition, and similarly for the column dimensions. It is easiest to use the `linalg` package to extract the dimensions of the `element` matrix; note that within procedures it is best to use the full forms of the names of package members rather than to rely on the use of `with`. In this implementation it is necessary to assign the module to a variable that is declared to be local to the constructor procedure. Note that, since modules obey last-name evaluation, it is necessary to apply eval to this local variable in order to return the module rather than the local variable to which it was assigned.

```
> BlockMatrix := proc(Elements::matrix,
                      Rowp::list(nonnegint),
                      Colp::list(nonnegint))
     local m;

     m := module()
        export elements, rowp, colp;
        description "Block matrix";
     end module;

     if nargs = 3 then
        if linalg[rowdim](Elements) <> `+`(op(Rowp)) or
           linalg[coldim](Elements) <> `+`(op(Colp)) then
           error "Incompatible initialization data!"
        end if;
        m:-elements := Elements;
        m:-rowp := Rowp;
        m:-colp := Colp
     elif nargs > 0 then
        error "Must initialize fully or not at all!"
     end if;
     eval(m)
  end proc:
```

Here are a few tests of the new overloaded constructor:

```
> B := BlockMatrix(eval(MM), [2,1], [2,1]);
```

$B :=$ **module**()
export *element*, *rowp*, *colp*;
description "Block matrix";
end module

```
> B:-rowp;  eval(B:-elements);
```

$$[2, 1]$$

$$\begin{bmatrix} M_{1,1} & M_{1,2} & M_{1,3} \\ M_{2,1} & M_{2,2} & M_{2,3} \\ M_{3,1} & M_{3,2} & M_{3,3} \end{bmatrix}$$

```
> B := BlockMatrix();
```

$B :=$ **module**()
export *element*, *rowp*, *colp*;
description "Block matrix";

end module

```
> B:-rowp;
```

rowp

```
> BlockMatrix(eval(MM));
```

```
Error, (in BlockMatrix) Must initialize fully or not at all!
```

```
> BlockMatrix([2,1], [2,1]);
```

```
Error, BlockMatrix expects its 1st argument, Elements, to be of
type matrix, but received [2, 1]
```

15.6 Modules as objects

So far, we have considered only a simple object that contains only data (fields or properties) and so just plays the role of a record or structure. But objects can contain procedures (methods), which are usually intimately related to the data in the object. So let us add a method called `get` to our block matrix object that outputs the data in block matrix form. The basic idea is to generalize the code that was used early in this chapter to display a partitioned matrix, i.e., to convert a matrix into a matrix of matrices according to some specified partitioning. The `get` method is a procedure that *must* be defined *inside* the module that represents a block matrix, and the variable `get` is added to the module exports, as follows. No other change to the constructor procedure `BlockMatrix` is required.

Note that procedure `get` accesses the module exports as variables global to the procedure, which relies on lexical scoping. The procedure `ranges` is local to the module and converts a partition list into a list of ranges. (I could have made procedure `ranges` local to procedure `get`, but it turns out to be useful in other methods.) A double `seq` iterates over the row and column range sequences to build a list of lists of sub-matrices of the `element` matrix. Each sub-matrix is constructed by calling `linalg[submatrix]`, and the list of lists is converted to a matrix by applying the standard Maple function `matrix`.

```
> m := module()
    local ranges;
    export elements, rowp, colp, get;
```

```
    description "Block matrix";

    ranges := proc(P)  # P::list(posint)
       # Convert a partition (list of submatrix dimensions)
       # to a list of matrix sub-ranges.
       local beg, Rs, r, nxt;
       beg := 1;  Rs := NULL;
       for r in P do
          nxt := beg + r;
          Rs := Rs, beg..nxt-1;
          beg := nxt
       end do;
       [Rs]
    end proc;

    get := proc()
       # Return a matrix of matrices
       local rows, cols;
        if not assigned(elements) then
          error "Block matrix not initialized!"
       end if;
       rows := ranges(rowp);
       cols := ranges(colp);
       matrix([seq([seq(linalg[submatrix](elements,i,j),
          j=cols)],i=rows)])
    end proc;

 end module:
```

Each time the code is changed it is necessary to reconstruct any block matrix representations, so that everything remains consistent. Having done that, the new get method can be tested. Note that it is called with no arguments, since all the necessary information is contained within the (new) block matrix object assigned to the variable B:

```
> B := BlockMatrix(eval(MM), [2,1], [2,1]):
```

```
> B:-get();
```

$$\begin{bmatrix} \begin{bmatrix} M_{1,1} & M_{1,2} \\ M_{2,1} & M_{2,2} \\ M_{3,1} & M_{3,2} \end{bmatrix} & \begin{bmatrix} M_{1,3} \\ M_{2,3} \\ M_{3,3} \end{bmatrix} \end{bmatrix}$$

```
> eval(BM);
```

$$\begin{bmatrix} \begin{bmatrix} M_{1,1} & M_{1,2} \\ M_{2,1} & M_{2,2} \\ M_{3,1} & M_{3,2} \end{bmatrix} & \begin{bmatrix} M_{1,3} \\ M_{2,3} \\ M_{3,3} \end{bmatrix} \end{bmatrix}$$

```
> B := BlockMatrix():
```

```
> B:-get();
```

```
Error, (in get) Block matrix not initialized!
```

15.7 Data encapsulation

In the same way that we can use methods to get data from objects, we could use methods to set the data. A model in which the data in an object is completely private and is accessed only via public methods has the advantage that the implementation is entirely hidden and that the access methods can perform various checks that cannot be performed if direct access to the data is allowed. To convert our block matrix implementation to this model, it is just necessary to write access methods for the partitioning data, which is trivial. I will call these methods `getRowp` and `getColp`, following the established capitalization conventions of OOP [6, page 190].

We also need to convert the constructor function into a method, which I will call `set`, that can be used both to set (initialize) and to reset the data. We still need a simple constructor function, but most of its work will now be done by calling the `set` method. We can usefully overload the `set` method (and constructor function) further. It should accept a list of lists, which is a common representation for a matrix and is acceptable as input to the `matrix` function, and it should accept the input matrix in either unblocked or explicitly blocked form. It would also be convenient to be able to construct symbolic block matrices by simply specifying an identifier to be subscripted to represent the elements. So, finally, let us allow a block matrix to be initialized by either

- a matrix or a list of lists of algebraic values or an identifier, plus partitioning data, or
- a matrix or a list of lists, the elements of which are *matrices* of algebraic values.

In the second case, explicit partitioning data are optional, but if provided then they must agree with the actual partitioning of the data. We can distinguish among these options by performing on `element`, the first argument of the constructor function, a more specific type test than that used for basic validation in the procedure header. Note that the type `'matrix(matrix)'` *must* be enclosed in uneval quotes, because `matrix` is also the name of a procedure.

In the first case, allowing a list of lists is a minor generalization and we can just apply the function `matrix` to convert it to a matrix. (Incidentally, the online help states that the function `matrix` is in the `linalg` package, but in fact it is available in the main library!) Also, we will now be more careful when the initializer is a matrix and use a copy rather than the matrix itself. Otherwise, once we start assigning to block matrix elements, we run the risk of inadvertently changing the initializer matrix, which is potentially very confusing!

In the second case, the data structure containing the elements implies the block partitioning data, which we need to extract and use. For this, I use a short (local) procedure called `unblock`. This unblocks the list of lists — often abbreviated to listlist — data structure by calling `linalg[blockmatrix]` with suitably constructed arguments and returns an *expression sequence* consisting of the unblocked matrix and the row and column partitions. This expression sequence can then conveniently be used in multiple-assignment statements. A matrix of matrices is first converted to a list of lists of matrices, because this is closer to the list of matrices required by `linalg[blockmatrix]` as its first argument.

The following constructor function allows maximum flexibility in the number of arguments provided. If there are no arguments then it just defines the data structure module and returns immediately. Otherwise, it calls the `set` method, which processes all the arguments that are provided and then checks that the module is fully initialized by using the standard Maple `assigned` function to check that all the module exports are assigned. It constructs a block matrix object in which the implementation is entirely hidden and the data can be accessed only via public methods.

```
> restart;


> BlockMatrix := proc(
        Elements::{matrix,list(list),symbol},
        Rowp::list(nonnegint), Colp::list(nonnegint))
     local m;

     m := module()
        local ranges, elements, rowp, colp;
        export get, getRowp, getColp, set;
        description "Block matrix";

        ranges := proc(P)  # P::list(posint)
           # Convert a partition (list of submatrix dimensions)
           # to a list of matrix sub-ranges.
           local beg, Rs, r, nxt;
           beg := 1;  Rs := NULL;
           for r in P do
```

```
        nxt := beg + r;
        Rs := Rs, beg..nxt-1;
        beg := nxt
    end do;
    [Rs]
end proc;

get := proc()
    # Return a matrix of matrices
    local rows, cols;
     if not assigned(elements) then
       error "Block matrix not initialized!"
    end if;
    rows := ranges(rowp);
    cols := ranges(colp);
    matrix([seq([seq(linalg[submatrix](elements,i,j),
       j=cols)],i=rows)])
end proc;

getRowp := () ->
    # Return the row partition list
    if not assigned(rowp) then
       error "Block matrix not initialized!"
    else rowp end if;

getColp := () ->
    # Return the column partition list
    if not assigned(colp) then
       error "Block matrix not initialized!"
    else colp end if;

set := proc(Elements::{matrix,list(list),symbol},
            Rowp::list(nonnegint),
            Colp::list(nonnegint))
    # Set the block matrix data flexibly

    local unblock;

    unblock := proc(LLM)
       # LLM is a list of lists of matrices
       # Return: elements, rowp, colp
       linalg[blockmatrix](                       # elements
          nops(LLM), nops(LLM[1]), # rows, cols
          map(op, LLM)),  # flat list of matrices
       map(linalg[rowdim], map2(op,1,LLM)), # rowp
```

```
      map(linalg[coldim], LLM[1])          # colp
   end proc;

   unassign('rowp', 'colp');

   if Elements::'matrix(matrix)' then
      elements, rowp, colp :=
         unblock(convert(Elements,listlist))
   elif Elements::'list(list(matrix))' then
      elements, rowp, colp :=
         unblock(Elements)
   elif Elements::'matrix' then
      elements := copy(Elements)
   elif Elements::'list(list)' then
      elements := matrix(Elements)
   else  # Elements::symbol
      if nargs < 3 then
         error "Must specify dimensions for symbolic or
zero block matrix!"
      end if;
      elements := matrix(`+`(op(Rowp)),
         `+`(op(Colp)), (i,j)->Elements[i,j]);
      rowp := Rowp;  colp := Colp;
      return  # nothing
   end if;

   if nargs > 1 then
      if assigned(rowp) then
         if rowp <> Rowp then
            error "Incompatible initialization data!"
         end if
      elif linalg[rowdim](Elements)<>`+`(op(Rowp)) then
         error "Incompatible initialization data!"
      else
         rowp := Rowp
      end if
   end if;

   if nargs > 2 then
      if assigned(colp) then
         if colp <> Colp then
            error "Incompatible initialization data!"
         end if
      elif linalg[coldim](Elements)<>`+`(op(Colp)) then
         error "Incompatible initialization data!"
```

```
            else
               colp := Colp
            end if
         end if;

         if not(assigned(elements) and
                  assigned(rowp) and assigned(colp)) then
            error "Must initialize fully or not at all!"
         end if

      end proc;

   end module;

   if nargs > 0 then m:-set(args) end if;
   eval(m)
end proc:
```

Now we need to test it. First, let us construct a *symbolic* block matrix B that is initialized during construction:

```
> B := BlockMatrix(M, [2,1], [2,1]);
```

$B :=$ **module**()
local *ranges*, *element*, *rowp*, *colp*;
export *get*, *getRowp*, *getColp*, *set*;
description "Block matrix";

end module

We can access all the data about this block matrix via methods of B:

```
> B:-get();
```

$$\left[\begin{bmatrix} M_{1,1} & M_{1,2} \\ M_{2,1} & M_{2,2} \\ M_{3,1} & M_{3,2} \end{bmatrix} \quad \begin{bmatrix} M_{1,3} \\ M_{2,3} \\ M_{3,3} \end{bmatrix}\right]$$

```
> B:-getRowp();
```

$$[2, 1]$$

```
> B:-getColp();
```

$$[2, 1]$$

Now let us construct an initialized block matrix:

```
> B := BlockMatrix();
```

$B :=$ **module**()
local *ranges*, *element*, *rowp*, *colp*;
export *get*, *getRowp*, *getColp*, *set*;
description "Block matrix";
end module

```
> B:-get();
Error, (in get) Block matrix not initialized!

> B:-getRowp();
Error, (in getRowp) Block matrix not initialized!

> B:-getColp();
Error, (in getColp) Block matrix not initialized!
```

We can reset (or in this case set) the block matrix by calling its `set` method, e.g.,

```
> B:-set(M, [2,1], [2,1]);

> B:-get();
```

$$\left[\begin{bmatrix} M_{1,1} & M_{1,2} \\ M_{2,1} & M_{2,2} \\ M_{3,1} & M_{3,2} \end{bmatrix} \begin{bmatrix} M_{1,3} \\ M_{2,3} \\ M_{3,3} \end{bmatrix} \right]$$

```
> B:-getRowp();
```

$$[2, 1]$$

```
> B:-getColp();
```

$$[2, 1]$$

Finally, let us test some of the input validation:

```
> BlockMatrix(M);
Error, (in set) Must specify dimensions for symbolic or zero
block matrix!
```

```
> BlockMatrix([2,1], [2,1]);

   Error, BlockMatrix expects its 1st argument, Elements, to be of
   type {matrix, list(list), symbol}, but received [2, 1]
```

15.8 Accessing block matrix elements

Having provided `get` and `set` methods to get and set the whole of a block matrix, we should also provide methods to get and set individual elements. Here is the relevant part of the module. The methods `getElements` and `getElement` check that the block matrix elements have been assigned before attempting to return them, but this example code does not perform any other data validation. The method `getElements` takes no arguments and simply returns all the elements as a single unblocked matrix, which we will need later, whereas `getElement` returns the single element referenced by its single argument. Let us implement the convention that a block matrix element can be referenced by a list consisting of a row and column index, in that order, where each index can be either a positive integer representing the index within the unblocked matrix, or a list of positive integers representing the index of the block and the index within the block, in that order. The second (blocked) representation is converted to the first (unblocked) representation by the local procedure `unblockIndex`.

The method `setElement` takes a single argument consisting of an equation with a block matrix element reference on the left and its intended (algebraic) value on the right and sets the element; the method `setElements` takes an arbitrary number of such equations and sets all the specified elements to their specified values, by applying `setElement` to each argument. I will illustrate the use of `setElements` in the next section and the use of `getElements` later.

```
> m := module()
     local ranges, elements, rowp, colp, unblockIndex;
     export get, getRowp, getColp, set,
            getElements, getElement, setElements, setElement;
     description "Block matrix";

     unblockIndex := (index,partition) ->
        if index::list then  # unblock
           `+`(op(1..index[1]-1,partition),index[2])
        else index end if;

     getElements := () ->
        # Return all the elements as a (flat) matrix
        if not assigned(elements) then
           error "Block matrix not initialized!"
        else elements end if;
```

```
    getElement := proc(a::[{posint,[posint,posint]},
                           {posint,[posint,posint]}])
       # Return one element
       if not assigned(elements) then
          error "Block matrix not initialized!"
       end if;
       elements[unblockIndex(a[1],rowp),
                unblockIndex(a[2],colp)]
    end proc;

    setElements := proc()
       # Set one or more elements
       local arg;
       for arg in args do setElement(arg) end do;
       return  # nothing
    end proc;

    setElement := proc(a::[{posint,[posint,posint]},
                           {posint,[posint,posint]}]=algebraic)
       # Set one element
       elements[unblockIndex(lhs(a)[1],rowp),
                unblockIndex(lhs(a)[2],colp)] := rhs(a)
    end proc;

 end module:
```

15.9 Square and inverse block matrices

A matrix is *square* if its row and column dimensions are equal; a block matrix is square if its row and column partitions are equal. A square matrix is *invertible* if it is non-singular, and so is a square block matrix. Let us add a method that returns the inverse of a block matrix, provided it is square and non-singular. This is the first method we have written that returns an object of the same class, namely a block matrix module. Here is the relevant part of the new module definition:

```
> m := module()
    local elements, rowp, colp;
    export get, getRowp, getColp, set, inverse;
    description "Block matrix";

    inverse := proc()
       if rowp <> colp then
```

```
        error "Only square block matrices are invertible!"
      end if;
      BlockMatrix(linalg[:-inverse](elements), rowp, colp)
    end proc;

  end module:
```

It is convenient to use `linalg[inverse]` to compute the elements of the inverse block matrix, but note that there is a name clash between the `inverse` function in the `linalg` package and the local (exported) identifier `inverse` in our module; this clash is resolved by preceding the global identifier with the module member selection operator `:-` preceded by no module.

At about this point in the development, the code begins to get too big to maintain comfortably in a Maple worksheet. I therefore exported it to a text file called `BlockMatrices.txt` (the reason for the name will become clearer later) and used a normal text editor to maintain the file. Text can easily be transferred between Maple worksheets and other applications, such as a text editor. One easy way to do this is via the clipboard (i.e., by using "cut-and-paste"). But a more convenient way to import a text file of Maple program code into Maple is by using the `read` statement, as follows, which assumes that the file to be read is in the current directory. If you keep all the files related to a project in the same directory, and start Maple from that directory (e.g., by simply opening one of the worksheet files), then this directory should be the current directory within Maple. Otherwise, you may need to use a full pathname for the file to be read.

From now on, I will display only changes to the code, and not repeatedly display code that has not changed. However, each time the code is changed the file must be re-read to re-execute the definition of the whole body of code, and any block matrix representations must also be reconstructed, so that everything remains consistent. Let us use a simple *diagonal* block matrix to test this facility (and the `setElements` method):

```
> restart;  read "BlockMatrices.txt":
```

```
> B := BlockMatrix(M, [2,1], [2,1]):
  B:-setElements([[2,1],[1,1]]=0,[[2,1],[1,2]]=0);
  B:-setElements([[1,1],[2,1]]=0,[[1,2],[2,1]]=0);
  B:-get();
```

$$\begin{bmatrix} \begin{bmatrix} M_{1,1} & M_{1,2} \\ M_{2,1} & M_{2,2} \end{bmatrix} & \begin{bmatrix} 0 \\ 0 \end{bmatrix} \\ \begin{bmatrix} 0 & 0 \end{bmatrix} & \begin{bmatrix} M_{3,3} \end{bmatrix} \end{bmatrix}$$

```
> B:-inverse():-get();
```

$$\left[\begin{array}{cc} \left[\begin{array}{cc} \frac{M_{2,2}}{\%1} & -\frac{M_{1,2}}{\%1} \\ -\frac{M_{2,1}}{\%1} & \frac{M_{1,1}}{\%1} \end{array}\right] & \left[\begin{array}{c} 0 \\ 0 \end{array}\right] \\ \left[\begin{array}{cc} 0 & 0 \end{array}\right] & \left[\begin{array}{c} \frac{1}{M_{3,3}} \end{array}\right] \end{array}\right]$$

$$\%1 := M_{1,1}\, M_{2,2} - M_{1,2}\, M_{2,1}$$

This result is sufficiently simple that we can see that it is obviously correct, based on the well-known property of block diagonal matrices that the blocks essentially behave independently. It also illustrates composition of the module member selection operator `:-`, which is left-associative, so that `B:-inverse():-get()` is parsed as `(B:-inverse()):-get()`, which is what we want.

15.10 Modules and types

To make flexible use of user-defined data types it is necessary to be able to apply type tests to them. A module is just a data structure and its primitive type is `module`:

```
> whattype(module() end module);
```

module

However, to use the identifier `module` in a type test it *must* be enclosed in name (backward) quotes as follows, because it is a Maple keyword:

```
> type(module() end module, `module`);
```

true

(In passing, it is worth recalling that, although not always *necessary*, it is generally good practice also to enclose type expressions in uneval [forward] quotes, *except* when specifying types in procedure headers, as explained in Chapter 10.)

The type of a module can be qualified by some or all of its exports to give a special structured type, which is called an *interface* because it is only the exports of a module that can be accessed from outside the module and so define its interface. We might consider the essential block matrix exports to be `get`, `getRowp`, `getColp`, `set`, in which case we could define the *BlockMatrix type* to be

```
> `type/BlockMatrix` :=
    'module'(get, getRowp, getColp, set):
```

A general module will not be of this particular structured type:

```
> type(module() end module, BlockMatrix);
```

$$false$$

but a `BlockMatrix` module will be, provided it exports at least the specified variables. (It may export more variables, making it effectively a subtype of this particular BlockMatrix type.)

```
> type(BlockMatrix(), BlockMatrix);
```

$$true$$

Note that, although last name evaluation applies to modules, the type system automatically allows for this:

```
> z := BlockMatrix():  z;
```

$$z$$

```
> type(z, BlockMatrix);
```

$$true$$

The exports of a module are usually named to indicate the function of the export rather than the function of the module as a whole, so there is in principle a small risk that two modules could have some common exports. A more reliable way to test for `BlockMatrix` modules might be to arrange that they export a field called (say) `BlockMatrix`, the purpose of which is purely to act as a *tag* to indicate the data type of the module. The use of *tagged* data types is the conventional approach in older languages that do not explicitly support user-defined data types. For example, in (some dialects of) a language such as LISP, one would represent a block matrix as a list, the first element of which is a tag used to identify the data type, or equivalently as an inert function data structure, where the name of the function serves as the tag. The fields of the data structure would be the subsequent list elements or function arguments, which would normally be accessed via special functions (like our access methods). We could have used this approach in Maple, and I probably would have done before the introduction of modules, but modules have several advantages, one of which is the special `:-` syntax for accessing module members.

15.11 Block matrix multiplication

A matrix, and hence a block matrix, can be multiplied by a scalar or by a conformable matrix (or vector). It is straightforward to implement multiplication by a scalar as a method that takes the scalar as its argument, and in this context "scalar" means a value with the Maple type `algebraic`. I will call the method `scalarmul`, and it will just be an interface to the equivalent function in the `linalg` package.

Two block matrices are (formally) conformable for multiplication if the column partition of the first block matrix is equal to the row partition of the second. If we want to implement block matrix multiplication as a method of one of the block matrices then that method must take the other block matrix as an argument; let us choose that the argument represents the second matrix in the product. This asymmetry seems a little strange, but matrix multiplication does not commute (generally) and so is not a symmetric operation. (The method asymmetry is stranger for symmetric operations, such as equality testing!) This is the point at which we need the access method `getElements` written earlier; it gives the unblocked form of a block matrix, which can be manipulated directly by the `linalg` package. Without this method, there would be no way to access the unblocked form of one block matrix from within the module defining another block matrix. I will call the block matrix multiplication method `multiply`, and again it will just be an interface to the equivalent function in the `linalg` package. (I will not implement multiplication by a vector, which is left as an exercise.)

Here are the two new methods:

```
> scalarmul := (k::algebraic) ->
     # Multiply by a scalar k
     linalg[:-scalarmul](elements, k);
     BlockMatrix(%, rowp, colp);

  multiply := (M::BlockMatrix) ->
     # Multiply by a block matrix M
     if colp = M:-getRowp() then
        linalg[:-multiply](elements, M:-getElements());
        BlockMatrix(simplify(%), rowp, M:-getColp())
     else
        error
           "Block matrices not conformable for multiplication!"
     end if;
```

It is not obvious *a priori*, but experimenting with this code shows that, except in trivial cases, it is usually necessary to simplify the result of multiplying two matrices, so I have applied `simplify` to the matrix returned by `linalg[:-multiply]` in the `multiply` method. I have split the value returned by the `multiply` method into two expressions coupled by using the `%` operator. This is not necessary, but makes the code a little easier to read. It also illustrates the use of `%` within a procedure to save the need to use a local variable. (`%` is an environment variable, which means that in effect its use within a procedure is local to that procedure.)

Here are some simple tests of the two new methods:

```
> B := BlockMatrix(M, [2,1], [2,1]):
```

```
> B:-scalarmul(k):-get();
```

$$\left[\begin{bmatrix} k\,M_{1,1} & k\,M_{1,2} \\ k\,M_{2,1} & k\,M_{2,2} \\ k\,M_{3,1} & k\,M_{3,2} \end{bmatrix} \quad \begin{bmatrix} k\,M_{1,3} \\ k\,M_{2,3} \\ k\,M_{3,3} \end{bmatrix}\right]$$

```
> BI := B:-inverse():
```

```
> B:-multiply(BI):-get();
```

$$\left[\begin{bmatrix} 1 & 0 \\ 0 & 1 \\ 0 & 0 \end{bmatrix} \quad \begin{bmatrix} 0 \\ 0 \\ 1 \end{bmatrix}\right]$$

15.12 Limitations of modules as objects

Before moving on from the representation of objects as Maple modules, let me conclude the story so far by noting a few limitations of the approach.

It is not possible to customize the display (i.e., the default pretty-printing) of a module at present. (It is possible to redefine the print function, but that has no effect on implicit display.) It is one advantage of using an inert function representation for user-defined data types that the Maple `print` command can be customized to print inert functions in any desired way. Let us hope that future versions of Maple will allow such customized display of modules.

As I remarked earlier, Maple does not provide any straightforward mechanism for module inheritance, which is an important feature of OOP. The best that is possible is to reassign some or all of the exports of a module, but this does not allow the set of exported identifiers to be changed, and it gives no way to affect any of the local or internal behaviour of the module.

The object-oriented approach illustrated here is probably not very efficient, because every instance of an object implemented as a Maple module contains the definition of all the methods in the module. To avoid this, the actual code for the methods should be defined once, with pointers to the correct data. It would not be difficult to do this in Maple: the methods could be procedures that are local to an enclosing (package) module and the object data could be passed to them as arguments. A user of such a package would not see any difference and there should be a potentially huge saving in memory by avoiding multiple almost-identical copies of all the methods. However, this is a technical detail, which I will not pursue because it would take us away from the spirit of OOP.

15.13 Operator overloading

You will probably agree that the syntax we have used so far for block matrix multiplication is cumbersome. Wouldn't it be nice if we could use the elegant dot syntax that Maple provides for non-commuting multiplication? To do so would require us to *overload* the dot operator, i.e., to make it serve more purposes than it currently does. Operator overloading is a feature of OOP that is supported by some OOP languages, such as C++, but not by others, such as Java. In fact, many Maple operators are already overloaded; for example, operators such as `+`, `-`, and `.` can already be used with both scalar and rtable quantities. These (and most other) operators are implemented as Maple procedures:

```
> restart;
```

```
> print(`.`):
```

$$\mathbf{proc}() \ldots \mathbf{end\ proc}$$

so the identifier `` `.` `` can be reassigned a new procedure to perform whatever operation we want. However, to do so carelessly could cause havoc, so all such important Maple identifiers are *protected* to prevent accidental reassignment:

```
> `.` := foo;
```

```
Error, attempting to assign to `.` which is protected
```

They can be unprotected, reassigned, and then optionally re-protected, but this would be a dangerous way to proceed. Modules come to the rescue again. We simply define new versions of any operators that we want to overload as module exports. This allows the new versions to be used temporarily or locally, which means that we can easily undo the overloading. This is very convenient for program development. Let us work toward the goal of overloading `.` to represent block matrix multiplication in small steps.

Let us first establish the behaviour we want for the dot operator in terms of the operator `&.`, which we can use freely. It is essentially this:

```
> `&.` := (A::BlockMatrix,B::BlockMatrix) -> A:-multiply(B):
```

The Maple parser automatically allows both `.` and `&.` to be used as infix (as well as prefix) operators:

```
> read "BlockMatrices.txt":
```

```
> B := BlockMatrix(M, [2,1], [2,1]):
```

```
> (B &. B:-inverse()):-get();
```

$$\begin{bmatrix} \begin{bmatrix} 1 & 0 \\ 0 & 1 \\ 0 & 0 \end{bmatrix} & \begin{bmatrix} 0 \\ 0 \\ 1 \end{bmatrix} \end{bmatrix}$$

But if we really want to overload the `.` operator then we should preserve its previous behaviour, which requires more complicated type testing:

```
> `&.` := proc(A,B)
     if A::'BlockMatrix' then
        if B::'BlockMatrix' then
           A:-multiply(B)
        else error
           "Can multiply block matrices only with each other!"
        end if
     elif B::'BlockMatrix' then
        error
           "Can multiply block matrices only with each other!"
     else
        A . B
     end if
  end proc:
```

Here are some simple tests of multiplying a block matrix by another block matrix and a regular matrix (which is invalid), and also of multiplication not involving block matrices to show that the default behaviour has been preserved:

```
> (B &. B:-inverse()):-get();
```

$$\begin{bmatrix} \begin{bmatrix} 1 & 0 \\ 0 & 1 \\ 0 & 0 \end{bmatrix} & \begin{bmatrix} 0 \\ 0 \\ 1 \end{bmatrix} \end{bmatrix}$$

```
> B &. matrix(3,3);

  Error, (in &.) Can multiply block matrices only with each other!
```

```
> a &. b;
```

$$a \;.\; b$$

Now let us implement this operator with the name `.` within a module and export it. (This new module, which I have called `BlockMatrices`, represents the start of a package for manipulating block matrices, which we will develop further in the rest of this chapter. The use of modules as packages is conceptually completely different from their use to represent objects.) Let us also overload the dot operator further to support multiplication of a block matrix by a scalar. Note that it must be enclosed in name (backward) quotes except when it is actually used as an infix operator, and that when we fall back to the global definition of dot we must precede it with the null module selection operator, which means that we must use function application rather than operator syntax. If we do not force the global version of `.` to be used then we get infinite recursion when this branch of the operator definition is executed.

```
> BlockMatrices := module()
     export `.`;

     `.` := proc(A,B)
        if A::'BlockMatrix' then
           if B::'BlockMatrix' then
              return A:-multiply(B)
           elif B::'algebraic' then
              return A:-scalarmul(B)
           end if
        elif B::'BlockMatrix' then
           if A::'algebraic' then
              return B:-scalarmul(A)
           end if
        else
           return :-`.`(A,B)
        end if;
        error "Can multiply block matrices only with each other
  and scalars!"
     end proc;

  end module;
```

$$BlockMatrices := \textbf{module}()\,\textbf{export}\,`.`;\ \textbf{end module}$$

If this implementation of the procedure detects argument types that it can multiply then it explicitly returns their product by applying either the appropriate block matrix method if one argument is a block matrix or by applying the standard `.` operator if neither argument is a block matrix. If no explicit return statement is executed then an error has been caused by trying to multiply a block matrix by an invalid type (i.e., not a block matrix or scalar). Coding the procedure this way means that only one error statement is needed, which avoids the repetition of the same error message that was necessary in the first implementation.

We can invoke the dot operator in the `BlockMatrices` module by using module member selection syntax, which is useful for initial testing, for example for testing the new support for multiplication of a block matrix by a scalar:

```
> BlockMatrices:-`.`(3,B):-get();
```

$$\left[\begin{bmatrix} 3\,M_{1,1} & 3\,M_{1,2} \\ 3\,M_{2,1} & 3\,M_{2,2} \\ 3\,M_{3,1} & 3\,M_{3,2} \end{bmatrix} \quad \begin{bmatrix} 3\,M_{1,3} \\ 3\,M_{2,3} \\ 3\,M_{3,3} \end{bmatrix}\right]$$

but the more elegant local syntax that I referred to above employs the `use` statement, as follows:

```
> use BlockMatrices in
      (3 . B):-get();
      (B . B:-inverse()):-get():
      a . b;
      B . matrix(3,3)
  end use;
```

$$\left[\begin{bmatrix} 3\,M_{1,1} & 3\,M_{1,2} \\ 3\,M_{2,1} & 3\,M_{2,2} \\ 3\,M_{3,1} & 3\,M_{3,2} \end{bmatrix} \quad \begin{bmatrix} 3\,M_{1,3} \\ 3\,M_{2,3} \\ 3\,M_{3,3} \end{bmatrix}\right]$$

$$\left[\begin{bmatrix} 1 & 0 \\ 0 & 1 \\ 0 & 0 \end{bmatrix} \quad \begin{bmatrix} 0 \\ 0 \\ 1 \end{bmatrix}\right]$$

$$a\;.\;b$$

```
Error, (in .) Can multiply block matrices only with each other
and scalars!
```

Among other things, the `use` statement has the same effect locally that calling `with` has globally, which is particularly convenient for testing; the effect of `with` cannot be undone other than by executing a `restart`! To be precise, `use` rebinds identifiers locally, and the identifiers can be either the exports of a module or they can be the left sides of one or more equations, the right sides of which define their bindings, which we will also use very soon. However, it is not possible to display the output from *some* of the statements but not others (e.g., the second in the example above) executed within a `use` statement; either all or none are displayed, as in a `do`-loop.

Using modules for operator overloading is most useful when several operators are overloaded by the same module. Having overloaded the dot operator to include block matrix multiplication, we would probably also want to overload `^`, `+`, and `-`. I will leave these as exercises and instead, and somewhat unconventionally, overload `/` to perform "block matrix division", partly because we have already developed the necessary support and partly because this exercise turns out to be quite instructive! By "block matrix division" I mean that if A and B are block matrices and k is any scalar then $A/B = A \cdot B^{-1}$, $A/k = (1/k)\,A$, and $k/B = k\,B^{-1}$. Note that, although `-` is regularly used as a unary prefix operator, Maple does not allow `/` to be used in input as a *unary* operator:

```
> restart;
```

```
> /x;
```

```
Error, `/` unexpected
```

even if it is enclosed in backquotes:

```
> `/`(x);
```

```
Error, (in /) / uses a 2nd argument, y, which is missing
```

although it can be used as a *binary* prefix operator:

```
> `/`(x,y);
```

$$\frac{x}{y}$$

Overloading `.` and `+` (and `*`) as binary operators is straightforward. However, in order to overload `-` and `/` it is necessary to overload them as unary operators (even though `/` cannot be used in input as a unary operator). This is essentially because there is only one internal representation (`PROD`) for (repeated) products and quotients, and only one internal representation (`SUM`) for (repeated) sums and differences (see Appendix A2 of the *Maple 6 Programming Guide* [22]), so the Maple parser generates output that uses `-` and `/` *only* as unary operators. To see exactly what is happening, let us take advantage of a `use` statement to temporarily rebind these operators as inert functions in some test expressions:

```
> use `+`=Plus, `-`=Minus, `*`=Times, `/`=Divide in
    a+b, a-b, -b, a*b, a/b, 1/b
  end use;
```

$$\mathrm{Plus}(a, b), \mathrm{Plus}(a, \mathrm{Minus}(b)), \mathrm{Minus}(b), \mathrm{Times}(a, b),$$
$$\mathrm{Times}(a, \mathrm{Divide}(b)), \mathrm{Times}(1, \mathrm{Divide}(b))$$

We see that `Minus` and `Divide` are only ever called with a single argument, which is why `-` and `/` must be overloaded as unary operators. Moreover, `Divide` always appears as the second argument of `Times` and never alone (unlike `Minus`); i.e., the fact that `/` can only be used in input as a binary operator manifests itself in a binary `Times` function on output. Hence, we discover that, in order to overload `/`, we are forced also to overload `*`. But the Maple `*` operator is commutative; i.e., the simplifier may swap the operands around:

```
> a*b = b*a;
```

$$a\,b = a\,b$$

Hence, we should allow `*` to be used *explicitly* only in commutative contexts; i.e., we should disallow `A*B` and allow only `A*k`, `A/B => A*(/B)`, `k*B` and `k/B => k*(/B)`, where `A` and `B` are block matrices and `k` is any scalar as before. This means that we must be able to detect when an operand of `*` is a request to invert a block matrix. One way to do this is for the form `/B` to remain symbolic and then to be interpreted, if appropriate, by `*` as block matrix inversion. This is achieved as follows.

When the argument `A` of `/` is a block matrix then the `/` function returns the inert function `BlockMatrixInverse(A)`, where `BlockMatrixInverse` is declared to be a local variable; otherwise, `/` returns the value of `1/A` computed by calling the global function `/` *as a binary function*. The function `*` detects an argument of the form `/B` by testing for the type

```
specfunc(anything, BlockMatrixInverse)
```

which represents the specific function named `BlockMatrixInverse` applied to an argument of any type — since we already know that it must be a block matrix it would be inefficient to test it again. When `*` accepts an argument of the form `/B`, it extracts `B` by applying `op` to the argument.

One final change that we need to make to the whole package is to replace expressions of the form `a&b` by `:-`&`(a,b)` where `&` represents any overloaded operator, but we want to use the standard version in order to avoid infinite recursion. This is illustrated below for `*` and `/`:

```
> BlockMatrices := module()
    local BlockMatrixInverse;
    export `*`, `/`;

    `/` := proc(A)
       if A::'BlockMatrix' then
          BlockMatrixInverse(A)  # inert!
       else
          :-`/`(1,A)  # binary!
       end if
    end proc;

    `*` := proc(A,B)
       # Disallow BM*BM.
       # Allow only BM*k, BM/BM=BM*(/BM), k*BM, k/BM=k*(/BM).
       if A::'BlockMatrix' then
          if B::'BlockMatrix' then
             error "Must use . to multiply block matrices with
 each other!"
```

```
         elif B::specfunc(anything,BlockMatrixInverse) then
            A:-multiply(op(B):-inverse())
         elif B::'algebraic' then
            A:-scalarmul(B)
         else
            error "Can only multiply/divide block matrices and
scalars by each other!"
         end if
      elif B::'BlockMatrix' then
         if A::'algebraic' then
            B:-scalarmul(A)
         else
            error "Can use * only to multiply block matrices
by scalars!"
         end if
      elif B::specfunc(anything,BlockMatrixInverse) then
         if A::'algebraic' then
            op(B):-inverse():-scalarmul(A)
         else
            error "Can only divide block matrices and scalars
by each other!"
         end if
      else
         :-`*`(A,B)
      end if
   end proc;

end module:
```

First, let us test the overloading of * and /:

```
> restart;  read "BlockMatrices.txt":

> B := BlockMatrix(M, [2,1], [2,1]):

> use BlockMatrices in
     (2*B):-get();
     (B*2):-get();
     (B*B):-get()  # should be invalid!
  end use;
```

$$\left[\begin{bmatrix} 2M_{1,1} & 2M_{1,2} \\ 2M_{2,1} & 2M_{2,2} \\ 2M_{3,1} & 2M_{3,2} \end{bmatrix} \quad \begin{bmatrix} 2M_{1,3} \\ 2M_{2,3} \\ 2M_{3,3} \end{bmatrix}\right]$$

$$\left[\begin{array}{cc} \left[\begin{array}{cc} 2\,M_{1,1} & 2\,M_{1,2} \\ 2\,M_{2,1} & 2\,M_{2,2} \end{array}\right] & \left[\begin{array}{c} 2\,M_{1,3} \\ 2\,M_{2,3} \end{array}\right] \\ \left[\begin{array}{cc} 2\,M_{3,1} & 2\,M_{3,2} \end{array}\right] & \left[\begin{array}{c} 2\,M_{3,3} \end{array}\right] \end{array}\right]$$

```
Error, (in *) Must use . to multiply block matrices with each
other!
```

```
> use BlockMatrices in
     (B/k):-get();
     (B/B):-get()
  end use;
```

$$\left[\begin{array}{cc} \left[\begin{array}{cc} \frac{M_{1,1}}{k} & \frac{M_{1,2}}{k} \\ \frac{M_{2,1}}{k} & \frac{M_{2,2}}{k} \end{array}\right] & \left[\begin{array}{c} \frac{M_{1,3}}{k} \\ \frac{M_{2,3}}{k} \end{array}\right] \\ \left[\begin{array}{cc} \frac{M_{3,1}}{k} & \frac{M_{3,2}}{k} \end{array}\right] & \left[\begin{array}{c} \frac{M_{3,3}}{k} \end{array}\right] \end{array}\right]$$

$$\left[\begin{array}{cc} \left[\begin{array}{cc} 1 & 0 \\ 0 & 1 \end{array}\right] & \left[\begin{array}{c} 0 \\ 0 \end{array}\right] \\ \left[\begin{array}{cc} 0 & 0 \end{array}\right] & \left[\begin{array}{c} 1 \end{array}\right] \end{array}\right]$$

But, because the inverse of the general 3×3 matrix is complicated, let us use a simpler block-diagonal matrix for the final test:

```
> B:-setElements([[2,1],[1,1]]=0,[[2,1],[1,2]]=0);
  B:-setElements([[1,1],[2,1]]=0,[[1,2],[2,1]]=0);
  use BlockMatrices in
     (k/B):-get()
  end use;
```

$$\left[\begin{array}{cc} \left[\begin{array}{cc} \frac{k\,M_{2,2}}{\%1} & -\frac{k\,M_{1,2}}{\%1} \\ -\frac{k\,M_{2,1}}{\%1} & \frac{k\,M_{1,1}}{\%1} \end{array}\right] & \left[\begin{array}{c} 0 \\ 0 \end{array}\right] \\ \left[\begin{array}{cc} 0 & 0 \end{array}\right] & \left[\begin{array}{c} \frac{k}{M_{3,3}} \end{array}\right] \end{array}\right]$$

$$\%1 := M_{1,1}\,M_{2,2} - M_{1,2}\,M_{2,1}$$

15.14 Using a private Maple library

Libraries are also called repositories in Maple 6. Each Maple library or repository resides in a separate directory (or folder in GUI parlance) and is referred to by the name of that directory.

If you have write access to your Maple installation directory (e.g., on a single-user system) then you could, in principle, add your own code to the standard Maple library, which is the *default* value of the Maple variable `libname`. For example, on my standard single-user Windows installation it is this:

```
> libname;
```

"C:\\PROGRAM FILES\\MAPLE 6/lib"

(The double backslash is the external representation for a single backslash, which is the Windows equivalent of the single forward slash used by UNIX and conventionally by Maple to separate file pathname components; it is normally better to use forward slashes on all platforms when explicitly *entering* file paths.)

However, I strongly recommend that you do not write to the standard Maple library, because it is quite easy to corrupt it and so prevent Maple from operating correctly. The only way to recover is to reinstall either Maple or at least the damaged library. (You can do the latter fairly easily if, before you do anything that might write to the library, you save copies of the files in the standard Maple library directory.) But from now on I will assume that you take my advice and use only a private Maple library. In that case, it is probably also wise to change the access permissions on the standard Maple library files to read-only, to avoid any chance of accidentally corrupting the standard library.

When you start to use a private Maple library for the first time you must explicitly create a new directory (folder) to contain the library. It can be anywhere in your accessible file system; since I have write access to my Maple installation directory I will create a subdirectory of the standard Maple library directory called `FJW`; i.e., the name of my private library as a Maple string will be

```
> mylibname := cat(libname, "/FJW");
```

mylibname := "C:\\PROGRAM FILES\\MAPLE 6/lib/FJW"

Probably the easiest way to create the library directory is to call the Maple `mkdir` function, e.g.,

```
> mkdir(mylibname);
```

This function returns nothing (`NULL`) unless an error occurs.

Next, you need to use `march`, which is the Maple archive manager, to create an empty archive (library). The `march` command has a number of capabilities that are described in the online help. I will not explore them in detail, although you will need to use most of them if you start making serious use of private Maple libraries. Note that `march` is built into Maple 6 (it no longer exists as an external program, although some of the online help describing libraries seems to be out of date), so the following *Maple function-call* will create an empty archive in my new private library directory:

```
> march('create', mylibname, 100);
```

The final argument in this invocation of `march` is the approximate number of distinct objects to be stored in the library; 100 should be large enough for a small private library. The effect of this command is to create the two files `maple.lib` and `maple.ind` in the directory named by the second argument (`mylibname`). (When certain types of information are saved by Maple to this new library another file named `maple.rep` will also be created.)

Now Maple needs to be told explicitly where to find the private library. The value of `libname` can, in general, be a *sequence* of directories (represented as strings). Maple will search the directories in this sequence in turn to find a library to which it can save information or from which it can retrieve information. Provided any private library is the first directory assigned to `libname` (and it contains a valid writable Maple library), Maple will preferentially save to and retrieve from the private library. This also means that one can redefine standard library code by saving a replacement to a private library (which can be useful but must clearly be used with care). Hence, to use my new private library, I need to initialize Maple as follows:

```
> libname := mylibname, libname;
```

libname := "C:\\PROGRAM FILES\\MAPLE 6/lib/FJW",
"C:\\PROGRAM FILES\\MAPLE 6/lib"

Provided `libname` is set correctly, the easiest way to save information to a library is to use the Maple command `savelib`, which can be used to save any information that has the form of a value assigned to a variable. (But beware that trying to use it to save information in any other form is likely to corrupt the library.) The `savelib` function takes as its arguments the *names* of the variables to which the information to be saved has been assigned as values, so it is necessary to ensure that the arguments evaluate to the variable names and not to their values. I can now test my new private library by saving some trivial data, e.g.,

```
> foo := bar;
```

$$foo := bar$$

```
> savelib('foo');
```

Note that the unevaluation quotes here are essential.

The `march` function can also be used to list the contents of a library, e.g.,

```
> march('list', mylibname);
```

$$[[\text{"foo.m"}, [2001, 5, 8, 14, 37, 48], 1, 19]]$$

To understand this, note that the `savelib` command first saves the value assigned to a variable (e.g., `foo`) to a Maple binary file with the same name as the variable and extension `.m` (e.g., `foo.m`) and then archives the binary file, which is why the information must always be referenced in `march` via its binary file name (as a string, e.g., `"foo.m"`). The numbers shown in `march` *list* output are the date and time of creation of the archive member in the form of a list, followed by location and size information. For full details see the online help.

Now, to test that this information has indeed been saved to the library and can be automatically retrieved, we need to restart Maple:

```
> restart;
```

and set `libname` correctly:

```
> libname := cat(libname, "/FJW"), libname;
```

libname := "C:\\PROGRAM FILES\\MAPLE 6/lib/FJW",
"C:\\PROGRAM FILES\\MAPLE 6/lib"

```
> foo;
```

$$bar$$

So it works!

Another option of the `march` function allows archive members to be deleted from a library, so let me now delete my trivial test data:

```
> march('delete', libname[1], "foo.m");
```

and check that my private library is now empty:

```
> march('list', libname[1]);
```

$$[]$$

Probably the easiest and most reliable way to delete an *entire* library is to delete all the files in the directory (and start again to create a new library if necessary).

For a private library to be really useful it needs to be selected *automatically* for both saving and retrieving, which can be achieved using a Maple initialization file. No initialization file need exist, but if one does then it should contain normal Maple code, which is executed before Maple accepts any other input. Any output from the initialization code will appear at the start of each Maple session, so usually you will want to suppress output from such code. In order to use the private Maple library described above, I could create an initialization file containing the following assignment:

```
libname := cat(libname, "/FJW"), libname:
```

The names and possible locations of Maple initialization files are very platform dependent; see the online help for the `maple` command for details. It is possible to have more than one initialization file and Maple will use different initialization files depending on exactly how it is started. So it would be possible to have a default initialization file, which could be over-ridden for specific projects. On my single-user Windows system the default initialization file I use is

```
C:\Program Files\Maple 6\Users\maple.ini
```

but I could put this file in several other places. One reason why I choose to put it in this particular directory is that the file `maple6.ini` that Maple uses to store user options under Windows already lives in this directory. (But note that these two files serve very different purposes and must not be confused.)

Another way to use a private library would be to use the `-b` option in the command used to start Maple; again, see the online help for the `maple` command for details. This could be automated by editing the shell script, alias, or function used to start Maple on UNIX or the shortcut used on Windows.

15.15 Modules as packages

We have developed a lot of code in this chapter to support block matrices and the exercises invite you to develop some more. As I hinted earlier, such a collection of related facilities should be encapsulated in a Maple package, and the way to implement a package in Maple 6 is as a module. Let us call the package `BlockMatrices`. All of the code that has been developed in this chapter should now be moved into a module assigned to the variable `BlockMatrices`. Any public variables (and their values) that we want to be visible outside the package must be either exported or declared global; any private variables should be declared local. The constructor function for `BlockMatrix` objects and the operators (and any functions) that this package will overload need to be exported. However, type definitions must be global and do not work as module exports. (We might also declare the name `BlockMatrices` to be *protected* like a Maple keyword, although I will not do so.)

Once the `BlockMatrices` module definition has been executed, the package can be used just like any standard Maple package. The package exports can be used locally either via their full names as module members of the form `BlockMatrices:-`*export* (or `BlockMatrices[`*export*`]`) or via their short names within a `use BlockMatrices` statement, or they can be used globally via their short names after calling `with(BlockMatrices)`.

In practice, the code for this module is sufficiently large that I prefer to maintain it outside of Maple as a text file called `BlockMatrices.txt`. Then the statement:

```
> read "BlockMatrices.txt":
```

reads the file into Maple and executes the code in it, thereby executing the `BlockMatrices` module definition. This is convenient while developing the package, but once it is developed it is more convenient to make it available as a binary archive member in the Maple library, thereby giving it *exactly* the same status as a standard Maple package. But in order to make this work there is one further change that must be made.

The code in the body of a module is executed when the module is defined, which is when the module definition is either executed in a worksheet or read into Maple from a file of source code. Now suppose that the module is saved to a library. When it is later loaded back into Maple the module body is not re-executed; all that happens is that the values assigned to any variables, including module exports and local variables, are re-established. Thus, the original execution of the module body assigns values to exports and local and global variables, but when the module is loaded from a library only the values of the exports and local variables are re-established. In particular, any assignments to global variables made in the body of the module are lost, which in our case means the type definitions are lost. It is essential that assignments to global variables are executed when the module is *either* defined *or* loaded, and there are several ways to achieve this.

There are two further options specific to modules, which have the form:

`options load =` ***`pname1`***`, unload =` ***`pname2`***

The values ***pname1***, ***pname2*** on the right of the equations should be the names, which must to be either local or exported, of procedures defined in the module body. These procedures are called, with no arguments, when the module is respectively loaded from a library or unloaded. A module is *loaded* when it is referenced by accessing any of its exports or by naming it in a `with` call or `use` statement; it is *unloaded* either by garbage collection once it is no longer accessible or when Maple terminates. We can ensure that global variables are initialized by performing their assignments in a procedure assigned to a local variable and named in a `load` option. For testing, we may also want the initialization to be performed when the module is first defined, in which case we can simply include an explicit execution of the load procedure in the module body, since executing the code that defines a module does not count as loading it.

(Alternatively, a procedure assigned to a package module export named `init` is also automatically called with no arguments when the package is loaded, but such a procedure is accessible outside the package — the `load` option allows the initialization procedure to be private, which is usually preferable.)

By convention, modules that are used as packages include the option `package` among their options, although it is not clear what effect, if any, this has at present. Here is the skeleton of the entire `BlockMatrices` package, without repeating the code developed earlier:

```
> BlockMatrices := module()
     local init, BlockMatrixInverse;
     export BlockMatrix, `.`, `/`, `*`;
     option package, load=init;
     description "Prototype support for block matrices";

     init := proc()
        # Types MUST be global!
        global `type/BlockMatrix`;
        `type/BlockMatrix` :=
           '`module`(get, getRowp, getColp, set)'
     end proc;

     init();  # run also when module defined

     # Constructor function:
     BlockMatrix := proc(...)
     end proc;

     # Overload arithmetic operators:

     `.` := proc(A,B)
     end proc;

     `/` := proc(A)
     end proc;

     `*` := proc(A,B)
     end proc;

  end module:
```

The package can now be saved to a Maple library, as described in the previous section:

```
> libname := [libname][-1]:  # Just a precaution!
  libname := cat(libname, "/FJW"), libname;
```

libname := "C:\\PROGRAM FILES\\MAPLE 6/lib/FJW", "C:\\PROGRAM FILES\\MAPLE 6/lib"

```
> savelib(BlockMatrices);
```

Then, to check that it has really worked, we just need to restart Maple and try to use the new package as we would any other Maple package, e.g.,

```
> restart;
```

```
> libname := cat(libname, "/FJW"), libname:
```

```
> with(BlockMatrices);
```

$$[*, +, -, ., /, \mathit{BlockMatrix}, \mathit{evalb}]$$

```
> `type/BlockMatrix`;
```

$$\mathrm{module}(\mathit{get}, \mathit{getRowp}, \mathit{getColp}, \mathit{set})$$

```
> BlockMatrix(M, [2,1], [2,1]):-get();
```

$$\begin{bmatrix} \begin{bmatrix} M_{1,1} & M_{1,2} \\ M_{2,1} & M_{2,2} \\ M_{3,1} & M_{3,2} \end{bmatrix} & \begin{bmatrix} M_{1,3} \\ M_{2,3} \\ M_{3,3} \end{bmatrix} \end{bmatrix}$$

There is one curious feature associated with storing a module in a library, namely that it generates a lot of archive members with numeric names, e.g.,

```
> march('list', libname[1]);
```

[[":-3.m", %1, 549, 129], [":-8.m", %1, 1962, 201],
[":-4.m", %1, 678, 274], [":-9.m", %1, 2163, 14],
[":-5.m", %1, 952, 170], [":-6.m", %1, 1122, 567],
[":-1.m", %1, 2177, 3813],
["BlockMatrices.m", %1, 20, 330], [":-2.m", %1, 350, 199],
[":-7.m", %1, 1689, 273]]
%1 := [2001, 5, 8, 14, 38, 7]

These numeric archive members are used to store module members, but the actual numbers are of no particular consequence. However, they are not deleted automatically when the module is deleted from the archive:

```
> march('delete', libname[1], "BlockMatrices.m");
```

```
> march('list', libname[1]);
```

[[":-3.m", %1, 549, 129], [":-8.m", %1, 1962, 201],
[":-4.m", %1, 678, 274], [":-9.m", %1, 2163, 14],
[":-5.m", %1, 952, 170], [":-6.m", %1, 1122, 567],
[":-1.m", %1, 2177, 3813], [":-2.m", %1, 350, 199],
[":-7.m", %1, 1689, 273]]
%1 := [2001, 5, 8, 14, 38, 7]

They can be deleted by explicitly "garbage collecting" the library:

```
> march('gc', libname[1]);
```

9, 0, 5640

```
> march('list', libname[1]);
```

[]

This is done automatically when a module is updated, i.e., a module is saved to a library that already contains a copy of that module. The wasted space can be finally recovered by packing the archive:

```
> march('pack', libname[1]);
```

This may be necessary occasionally in a library that is frequently updated.

15.16 Adding online help

To add online help for the code you add to your private library (or anything else), the easiest way is just to prepare a normal Maple worksheet to provide the help. It is probably a good idea to copy the style of the standard Maple help worksheets, but this is not obligatory. Then, with the new help worksheet current, simply select the item *Save to Database...* from the Help menu. A dialogue box will appear, showing the current default *Database* and a list of *Writable Databases in 'libname'*. For example, on my standard Windows installation the default help database is this:

```
c:\PROGRAM FILES\MAPLE 6\lib\maple.hdb
```

To create a private help database in my private library directory (which *must* be included in the directory sequence assigned to `libname` for the new help database to be accessible), I just need to change the pathname in the *Database* box to

```
c:\PROGRAM FILES\MAPLE 6\lib\FJW\maple.hdb
```

When the help worksheet is saved it will create the new database file if necessary, which must always be named `maple.hdb`. Once a new help database has been created (and provided its directory is included in the directory sequence assigned to `libname`) it will appear in the list of *Writable Databases in 'libname'*, from which any available database can be selected to be the current database.

To save a help worksheet it is also necessary to provide an entry in the *Topic* box; the *Parent* and *Aliases* entries are optional. The entries in these boxes must not contains any spaces. When adding help for the block matrices package a suitable topic might be `BlockMatrices`. The *Parent* entry shows which help topic will be selected when the "up arrow" button on the help toolbar is pressed. If you want to provide a parent topic then an easy way to determine how to specify it is to select the desired page in the help browser; the topic entry appears in square brackets in the window title bar (but should be entered without the surrounding square brackets). A suitable parent page for block matrices might be the page entitled "Linear Algebra Computations in Maple", which is identified by the topic `LinearAlgebra,General,linalgebragen`. The *Aliases* entry lists other keywords that will lead to the same help page, and several aliases can be given, separated by commas and/or spaces. Suitable aliases for block matrices might be `PartitionedMatrices, Blocks`.

It is also possible to add online help without using the Maple GUI by calling the Maple function `makehelp`, which can be used to add either a Maple worksheet file or a plain text file as help. Further control over help databases than is available via the GUI is possible by *printing* a call of the (inert) Maple function `INTERFACE_HELP`, which simply passes information from the Maple kernel to the GUI, in the same way that printing a plot structure does. For further details of both facilities see the online help.

Epilogue

We have now covered most, but not quite all, features of modules. See the online help or the *Maple 6 Programming Guide* [22] for the few remaining features and some different examples. The complete source code file `BlockMatrices.txt` for the package developed in this chapter, which includes the solutions to the exercises, is available from the book web site given in the preface.

15.17 Exercises

1. As a convenience, overload the block matrix constructor further to accept as first argument one of the integers 0 or 1 to indicate, respectively, a zero or unit block matrix. In both cases, the dimensions must be specified. Test the facility.

2. A square block matrix is called *block diagonal* if its non-diagonal blocks are all zero matrices; it is called *block upper/lower triangular* if the blocks below/above the diagonal are all zero matrices. Add Boolean-valued methods called `isDiagonal`, `isUpperTriangular` and `isLowerTriangular` to the block matrix object to test these conditions. You can use ideas similar to those used in the `get` method to check that the appropriate sub-matrices are zero.

3. Implement and test a block matrix addition method, `matadd`, modelled on the implementation of block matrix multiplication. Java objects provide a method called *clone* (provided they are cloneable), which returns a copy of the object. Implement a `clone` method for block matrices by calling the standard Maple `copy` function that copies tables and arrays. Implement and test for block matrices the following methods based (except for `isunit`) on those in the `linalg` package: `equal`, `iszero`, `isunit`, `transpose`, `exponential`. (Only try to take the exponential of *simple* matrices!)

4. Overload the operators `+`, `-` to accept block matrix operands. (This then gives the full set of operations needed for the mathematical structure called a *ring*, namely `+`, `-`, `*`, `iszero`, and `isunit`.) Convince yourself that `evalb(A=B)` does not give a reliable comparison for block matrices (e.g., by comparing `2*B` with `B+B`). Hence, overload `evalb` to accept an equation involving either block matrices or 0 or 1, where the latter two values are used as a convenient shorthand (*cf.* exercise 1) to test equality with, respectively, a (conformable) zero or unit block matrix.

5. Add support for "partitioned vectors". Re-implement block matrices using rtables and the `LinearAlgebra` package for the internal representation. Consider other internal representations. (No solution provided.)

Bibliography

[1] A. V. Aho, J. E. Hopcroft, and J. D. Ullman, *Data Structures and Algorithms*, Addison-Wesley (1983).

[2] I. Bratko, *Prolog Programming for Artificial Intelligence*, Addison-Wesley (1988).

[3] F. P. Brooks, Jr., *The Mythical Man Month*, Addison-Wesley (1975), Anniversary Edition (1995).

[4] D. Cox, J. Little, and D. O'Shea, *Ideals, Varieties, and Algorithms*, Springer-Verlag (1992).

[5] P. Drijvers, "White-Box/Black-Box Revisited", *The International DERIVE Journal*, Vol. 2, No. 1, 3–14 (1995).

[6] D. Flanagan, *Java in a Nutshell: A Desktop Quick Reference*, O'Reilly (3rd ed., 1999).

[7] J. von zur Gathen and J. Gerhard, *Modern Computer Algebra*, Cambridge University Press (1999).

[8] K. O. Geddes, S. R. Czapor, and G. Labahn, *Algorithms for Computer Algebra*, Kluwer (1992).

[9] M. Goosens, F. Mittelbach, and A. Samarin, *The LaTeX Companion*, Addison-Wesley (1994).

[10] M. Goosens and S. Rahtz, *The LaTeX Web Companion*, Addison-Wesley Longman (1999).

[11] K. M. Heal, M. L. Hansen, and K. M. Rickard, *Maple 6 Learning Guide*, Waterloo Maple, Inc. (2000).

[12] A. Heck, *Introduction to Maple*, Springer-Verlag (1997).

[13] H. Jeffreys and B. Jeffreys, *Methods of Mathematical Physics*, Cambridge University Press (3rd ed., 1992).

[14] B. W. Kernighan and R. Pike, *The Practice of Programming*, Addison-Wesley (1999).

[15] B. W. Kernighan and P. J. Plauger, *Software Tools*, Addison-Wesley (1976).

[16] B. W. Kernighan and P. J. Plauger, *The Elements of Programming Style*, McGraw-Hill (1978).

[17] B. W. Kernighan and D. M. Ritchie, *The C Programming Language*, Prentice-Hall (2nd ed., 1988).

[18] D. E. Knuth, *The Art of Computer Programming*, Vol. 1, *Fundamental Algorithms*, Addison-Wesley (3rd ed., 1997).

[19] D. E. Knuth, *The Art of Computer Programming*, Vol. 2, *Seminumerical Algorithms*, Addison-Wesley (3rd ed., 1998).

[20] L. Lamport, *LaTeX: A Document Preparation System*, Addison-Wesley (2nd ed., 1994).

[21] S. Lipschutz, *Linear Algebra*, Schaum's Outline series, McGraw-Hill (2nd ed., 1991).

[22] M. B. Monagan, K. O. Geddes, K. M. Heal, G. Labahn, S. M. Vorkoetter, and J. McCarron, *Maple 6 Programming Guide*, Waterloo Maple, Inc. (2000).

[23] W. H. Press, B. P. Flannery, S. A. Teukolsky, and W. T. Vetterling, *Numerical Recipes: The Art of Scientific Computing*, Cambridge University Press (1986).

[24] D. Redfern, *The Maple Handbook*, Springer-Verlag (1996).

[25] J. Stoer and R. Bulirsch, *Introduction to Numerical Analysis*, Springer-Verlag (1983).

[26] B. Stroustrup, *The C++ Programming Language*, Addison-Wesley (2nd ed., 1995).

[27] F. Vivaldi, *Experimental Mathematics with Maple*, Chapman & Hall / CRC Press (2001).

[28] P. H. Winston and B. K. P. Horn, *LISP*, Addison-Wesley (3rd ed., 1988).

Index

!, 31, 244, 245, 268, 271, 318–319, 405
!=, 31
" (double quote), 27–28, 39–41, 68–69
  in identifier, 39
  in string, 39
#, 16, 26, 36
' (forward quote), 25, 29, 36, 42, 44, 49, 56, 65, 76, 94, 135, 182, 201, 211, 229, 230, 235, 236, 277, 283, 303, 306, 320, 322, 349, 373, 378, 382, 384, 412, 419, 422, 430, 457, 471, 479, 480, 485
  in identifier, 26, 39, 371
  in string, 39
( ), 28–29, 220, 221, 450
*, 30, 33, 34, 49, 51, 52, 133, 134, 163, 206, 214, 249, 269, 270, 274–275, 307–310, 336, 339–341, 404, 470–477, 483
`*`
  function, 58, 66, 306
  Maple type, 300, 389
**, 30, 274–275
`**` (Maple type), 300
⊞, 16
+, 30, 49, 308, 336, 339, 341, 470–477
`+`
  function, 58, 66, 306, 309, 311, 345, 382, 389, 454
  Maple type, 300, 307, 382
`+`({integer,symbol}) (Maple type), 309
`+`({symbol,integer}) (Maple type), 309
`+`(symbol) (Maple type), 307
, (comma)
  column constructor, 161, 162, 168, 181, 183
  operator, 40, 55, 64, 268, 298
⊟, 16
-, 30, 63, 470–477, *see also* ->, :-
->, 25, 48, 50, 75, 96, 107, 120, 126, 158, 165, 183, 194, 199–201, 220, 239, 242, 243, 248, 249, 251, 270–273, 281, 286, 288, 296, 297, 302, 316, 322, 331–333, 340, 358, 382, 389, 391, 394, 407, 414, 458, 463, 468, 470
. (dot), 32–33, 309–310, 470–477, 483
`.` (Maple type), 300
`.`(symbol) (Maple type), 309
.., 37, 40–41, 56, 58, 62, 64–70, 77, 79, 89, 91, 95, 98, 107–109, 113, 145–147, 165, 171, 181, 182, 213, 215, 228, 247, 290, 297, 308, 332, 334, 336, 338, 340, 348, 376, 391, 430, 431, 442, 443, 447, 455, 463
`..` (Maple type), 300
/

- in identifier, 141, 146–147, 238, 321, 322, 333, 384, 466
- in pathname, 123, 343, 433, 434, 478, 481, 483
- operator, 29, 30, 32, 34, 49, 51, 77, 84, 91, 98, 135, 136, 138, 140, 155, 165, 179, 183, 227, 232, 243, 248, 257, 319, 345, 372, 373, 390, 406, 419, 420, 470–477, 483

`/` (function), 310, 474

: (colon), 35, 195, 205, 222, 351, 402, 404

:-, 432, 446, 448, 465–467, 475, 481, 484

::, 233, 234, 246, 248, 251, 255, 262, 271, 272, 285, 288, 289, 301–303, 310–313, 332, 334, 336, 338–341, 358, 372–374, 376, 378, 382, 389, 391, 393, 394, 421, 429, 437, 442, 452, 458, 463, 468, 470, 471, 475

:=, 24, 25, 42–46, 54–56, 61, 65, 86, 88, 141, 143, 149, 155, 220, 223, 270

; (semicolon), 35, 195, 205, 222

<, *see also* <...>, <>
- in UNIX shell, 347
- operator, 31, 69, 187, 193, 199, 206, 210–212, 248, 286, 314, 320, 378, 379, 419, 420

`<`
- in convert, 333
- Maple type, 300

<...> (rtable constructor), 161, 168, 181, 183

<=, 187, 207, 232, 234, 316, 349, 382, 430

`<=`
- in convert, 333
- Maple type, 300

<>, 187, 272, 339, 341, 372–374, 379, 382, 454

`<>`
- in convert, 333
- Maple type, 300

=, *see also* :=, <=, <>, >=
- in interface, 348, 350
- in kernelopts, 385
- in module options, 482
- operator, 31, 32, 46–49, 54–56, 61–62, 64–67, 77, 79, 80, 82, 85, 89–91, 93, 95, 107, 110, 113, 116, 118, 142, 143, 148, 152, 166, 167, 170, 187, 194, 214, 215, 237, 240, 279, 308, 330, 331, 378, 416, 474
  - non-scalar, 170

`=`
- in convert, 333
- Maple type, 300

>, *see also* ->, <...>, <>
- operator, 31, 50, 187, 193, 199, 201, 204, 210, 223, 225, 228, 235, 254, 316, 376, 378, 379, 394, 437
- prompt, 12, 16, 349, 361

>=, 187, 319, 372, 376, 382

?, 12, 41

?: (ternary operator), 204

@, 32, 274

@@, 32, 203, 274, 276

[constant,constant] (Maple type), 314

[integer,integer,integer] (Maple type), 307

[integer,integer] (Maple type), 307

[numeric,numeric] (Maple type), 314

[ ], 29–30, 171, 214, 345, 410, 463
- for list, 48, 55, 57, 59–64, 69, 79–82, 84, 88, 90, 92, 93, 95, 98, 107, 110, 111,

113–114, 149, 155, 165, 167, 170, 176, 182, 192, 243, 251, 257, 258, 272, 273, 286, 310, 311, 330, 331, 334, 431, 435, 447, 449
for selection, 69, 136, 141, 170, 172, 179, 206, 214, 228, 232, 248, 278, 285, 290, 333, 338, 339, 341, 345, 356, 406, 432, 442, 454, 480, 483
#, 36
$, 33, 40, 41, 64–70, 276
%
ditto, 19, 28, 32, 34, 38, 47, 98, 171, 222, 268, 468
in format, 134, 232, 318, 435–437, 441
&
in UNIX shell, 347
inert operator, 34, 153, 268–294
&! (user operator), 271, 272
&* (Maple operator), 34, 269, 308
&+ (Maple operator), 269, 307, 308
&++ (user operator), 272
&- (user operator), 281, 282
&. (user operator), 470, 471
&/ (user operator), 271
&< (user operator), 314
&^ (Maple operator), 154, 155, 269
&div (user operator), 270
&member (user operator), 283
&where (Maple operator), 277
_ (underscore)
in autosave filenames, 18
in identifiers, 35, 134, 135, 138, 143–144
\, 38, 69, 128, 281, 285, 347, 427, 428, 436, 437
{algebraic, procedure} (Maple type), 378
{list,set,‘+‘,‘*‘,function} (Maple type), 336
{list,set} (Maple type), 316
{matrix,list(list),symbol} (Maple type), 458
{name,procedure} (Maple type), 332
{set,list} (Maple type), 283, 302
{ }, 30, 279–286
for alternative types, 298–301, 304, 308, 322, 384
for set, 49, 55, 57, 113, 391, 421, 431
for statement grouping, 195
in solve, 142
^ (circumflex)
e^x, 53–54
operator, 30, 64, 133, 134, 152, 154–155, 164, 169, 174, 206, 242, 470–477
‘^‘ (Maple type), 300
‘ (backward quote), 26–28, 38, 39, 58, 66, 136, 155, 167, 188, 192, 194, 204, 238, 257, 269–271, 279, 280, 283, 300, 304, 306, 308–311, 314, 316, 320–322, 333, 336, 345, 371, 382, 384, 391, 403–405, 419, 420, 466, 470, 472–475
in identifier, 39
in string, 39
| (row constructor), 161, 162, 168, 183, 249
| |, 33, 37, 40, 56, 252, 412
~ (tilde)
character, 35, 68
in assume, 36
in spreadsheet, 403, 412

2 (Maple type), 318

Abelian groups, 213, *see also* additive groups
abnormal, 351
about

assumption, 36
help, 12
information, 356, 357
abs, 190, 199, 288, 320, 379
absolute
cell reference, 402
error, 137, 377
file pathname, 343
value, 190, 199, 288, 320, 379
abstract
function, 6, 74
linear algebra, 159, 171
acceleration, 413
accented letters, 70, 426
accumulating sum, 340, 341
accuracy, 135–137, 161, 166, 169, 177, 179, 184, 343, 377, 419
action of key or mouse, 13, 15
active, 7
functions, 52, 58, 146, 208, 301, 306
mappings, 7
operators, 267, 270, 280, 314
ADD (user proc), 247, 248
add, 40, 66–67, 222, 224, 342, 345, 382
AddArgs (user proc), 311
addition, *see* +, add, sum
of lists, 272, 435
of matrices and vectors, 163
of sequence elements, 58, 247
of vectors, 338
repeated, 67
type testing, 308
additive groups, 128, *see also* Abelian groups
address in memory, 58, 311
addword (user proc), 438
Aho, A. V., 5, 436
algebra
block matrix, 447–477
Boolean, 191–194
computer, 4, 5
linear, 159–184
matrix, 337–342
of operators, 32
on relations, 188
set, 284
type, 299
vector, 337–342
algebraic
equations, 142–146, 150–151, 154–155
numbers, 128, 130, 151–154
operators, 30
programming, 369–397
types, 300
algebraic (Maple type), 300, 468
Algol, 195, 197, 445, 446
algorithms, 5
bisection, 377, 418
Buchberger's, 389, 391
chord, 419
digit extraction, 289
Euclidean GCD, 287, 373
generalized polynomial division, 381–397
Horner's, 289, 290
in program design, 261
interval halving, 377, 418
Koch snowflake, 257
Markov chain, 440
multivariate polynomial, 381–397
numerical failure in, 408
power sets, 284–286
prime factorization, 207, 232
recursive power, 248
ring, 286
squarefree factorization, 371
text formatting, 437
text processing, 425
tracing of, 356
unit matrix construction, 215
univariate polynomial, 369–381
word count, 429
alias
command shell, 481

for help topic, 486
alias, 151–154
alignment
of plots, 109, 158
of text in plots, 93, 125, 202, 259
allvalues, 143
α, 152
alpha software release, 355
Alt key, 17, 18, 38
alternative
execution, *see* elif, else
types, 298
ambiguous
evalf, 30
dot, 33, 164
file location, 433
juxtaposition, 51
line numbers, 361
newline, 428
terminology, 6, 273
variable concept, 43
ampersand, 34, 268, *see also* &
amplitude of pendulum, 412–422
analysis, numerical, 5, 139, 150, 160
And (type operator), 304–305
'and' (Maple type), 300
and (Boolean operator), 191–194
And(fraction,negative) (Maple type), 304
And(integer,Not(prime)) (Maple type), 305
angle
brackets, 161
of pendulum, 413
of rotation, 257
parameter, 95
plotting, 94
viewing, 111, 112
angular
acceleration, 413
frequency, 414
animation, 89–100, 112, 119–121, 255–259
export, 101
Animation menu, 90
annotation
mark-up, 426
of plots, 84, 93–94
of program code, 36
anonymous
modules, 450
procedures, 25, 242, 244, 248, 331, 332
ANSI, 138
antihermitian, 170, 171
antisymmetric, 170
anything (Maple type), 299, 475
apostrophe, 25, 430
appending
lists, 272
sequences, 59
appendto, 343
applying functions, 7, 28, 29, 31, 32, 45, 48, 51, 53, 58–61, 63, 64, 66, 140, 143, 144, 189, 220, 225, 227, 237, 273–275, 296, 330, 334, 335, 378, 384, 412, 451, 453, 472
approximate
eigenvalues, 172–176
evaluation, 140–142
integration, 146–147
linear algebra, 160–184
roots of equations, 142–146, 377
roots of linear equations, 167–169, 180–184
solution of ODEs, 148–150
approximation, floating-point, 32, 132–150, 160–184, 190, 206, 316, 407, 415–422
arbitrary
constants, 148
precision arithmetic, 138–141
arc (plotting), 94
archive, 478–485
area under curve, 147

`argc`, `argv`, 227
`args`, 227–232, 247, 248, 251, 270, 272, 282, 290, 310, 311, 434, 453
arguments, 7, 57–60, 65, 220–222, 224–229, 236, 239, 241–242, 247, 248, 268, 273, 275, 278, 290, 301–304, 306, 310–313, 315, 318, 323, 330, 331, 336, 347, 388, 452, 453, 467
arithmetic
    arbitrary precision, 140–141
    floating-point, 140, 141, 177
    for linear algebra, 166
    IEEE, 138
    series, 404, 405
arity, 268
`Array` (Maple type), 300
`array`, 334
    Maple type, 300, 334
arrays, 8, 61, 63, 164, 176, 192, 212, 213, 215, 243, 256, 297, 334, 337–342, 382, 410, 449
    indexing, 29
    mapping over, 330
    of plots, 91–93, 158
arrow
    plot (`gradplot`), 117
    syntax, 25, 50, 220, 221, 239, 270
artificial intelligence, 4, 193
ASCII, xvi, 24, 41, 68–70, 94, 343, 426–428, 430, 431
`ASSERT`, 382, 384
`assertlevel` (kernel option), 385
`assign`, 55, 143
assignment, 24–25, 42–46
`assume`, 35, 189
at, 32, *see also* `@`
augmented matrix, 181
AutoCAD, 121
AutoSave, 18
Awk, 436, 438, 440, 443
`axes`, 78, 83, 93, 111
axis
    labelling, 74, 75, 77, 79, 115
    scaling, 80, 83, 93, 111

background
    execution, 346, 347
    references, 3
    to animation, 92, 98, 120
backquote, 26, *see also* `‘`
backslash, 38, 343, *see also* \
Backspace key, 14, 400
backward
    quote, 26, *see also* `‘`
    slash, 38, 343, *see also* \
ball (animation), 119–121
balloon help, 13
banded matrices, 161
bar, vertical, 37, *see also* `|`
base
    of logarithm, 29
    of natural logarithms, 53, 152
    of number, 132, 133, 288, 290
    of recursion, 244–248, 285, 287, 318, 363, 405, 406
Basic, Visual, 260
basis, Gröbner, 381–397
batch processing, 346
Bauldry, W. C., 256
`begin ... end`, 195
beta software release, 355
`BINARY`, 428
binary
    files, 11, 36, 427, 428, 480, 482
    numbers, 289, 426
    operators, 188, 191, 267, 268, 270, 271, 274, 279–281, 474
    trees, 253
binding, 24, 35, 270, 301, 303, 473, 474
bisection algorithm, 377, 418
bitmap, xvi, xvii, 9, 101
bivariate polynomials, 386
black, *see* colour

blank
  character, 426
  line, 436
block
  matrix, 161, 447–477
    inverse, 464–466, 475
    multiplication, 467, 470
    square, 464–466
  structure, 224, 445
`BlockMatrices` (user package), 481–485
`BlockMatrix`
  user proc, 452, 454, 458
  user type, 466–468, 470
`blockmatrix`, 448, 458
blue, *see* colour
body, 7
  of loop, 205, 208–212, 387
  of module, 46, 241, 450, 451, 453, 482
  of procedure, 46, 48, 50, 58, 199, 221, 222, 225, 231, 237–242, 248, 273, 312, 352, 451
Boole, G., 191
Boolean, 31, 50, 85, 188–190, 201, 205, 282, 302–304, 314, 322, 335, 380, 385, 402, 418
  algebra, 191–194
  inverse, 191
  types, 300
`boolean` (Maple type), 300
border of spreadsheet, 401, 403
bouncing ball, 119–121
bound
  of array, 301, 339, 341
  of loop, 209
  on polynomial zeros, 377
`BOX`, 83
`BOXED`, 78, 83, 84, 96
braces, 30, *see also* { }, set
brackets, 28, *see also* `( )`, `[ ]`, { }
branch
  complex, 138, 408
  of tree, 253, 305
Bratko, I., 4
`break`, 211, 349, 387, 435, 443
break, line, 16, 39, 427, 428, 434, 436
breakpoint, 361–363
Brooks, F. P., Jr., 443
browser
  help, 13, 486
  rtable, 165
  virtual reality, 112, 122, 123
  web, 9, 100, 101, 121, 122
Buchberger's algorithm, 389, 391
Buchberger, B., 389
bugs, 354–364
builtin procedures, 238, 311, 332, 335, 336
Bulirsch, R., 5, 160
button
  context bar, 52, 90, 98, 112, 400, 404, 405, 409
  mouse, xv, 13–15, 112, 164
  toolbar, 10, 11, 13, 16, 18, 19, 69, 357, 486
`by`, 209
byte, 426, 427, 434
`bytes`, 69, 70, 431

C, C++, xiv, 3–5, 24, 31, 38, 68, 70, 135, 139, 193, 195, 197, 204, 211, 224, 227, 238, 268, 295, 296, 299, 311, 329, 343, 346, 427, 428, 431, 436, 437, 440, 441, 443, 445–447, 452, 470
call, procedure, 219, 222, 226, 227, 229, 232, 239, 244, 245, 247, 248, 250, 253–255
canonical, 49, 57, 188, 308, 427, 433
capital letters, 24, 52–53, 58, 63, 135, 153, 154, 160, 166, 457
cardinal numbers, 128

cardinality, 129, 286
`Cardioid` (user proc), 100
cardioid, 94–100
carriage return, 69, 427
Cartesian, 80, 81, 96, 254
cartoon, *see* animation
case
  base, 244–247, 285, 287, 363, 405, 406
  of letters, 52–53, 68, 82, 85
  statement, 197
  study, 94, 119, 255, 412
`cat`, 37, 68, 261, 350, 438, 478
`catch`, 351–354
catenation, 37, 162, 262, 350
`ceil`, 206
cell, spreadsheet, 399–422
`cfrac`, 132
CGI, 9
chain
  Markov, 440
  rule, 276
characters, 6, 7, 26, 41, 68–70, 262
  blank, 426
  control, 69, 343, 426, 427, 430, 431
  counting, 428
  encoding, 68–70, 94, 426–431
  escape, 38, 343
  European, 70, 426
  graphic, 68, 69, 426, 427, 430
  invisible, 16, 69
  Latin, 426, 427
  linefeed, 427
  national, 426
  newline, 39, 356, 427, 434–437, 441
  non-ASCII, xvi, 427
  NUL, 431
  positions, 427
  printable, 68
  punctuation, 430
  sets, 68, 70, 427
  space, 12, 26, 33, 41, 51, 52, 68, 69, 161, 164, 209, 222, 232, 269, 343, 347, 426, 428–430, 436, 437, 486
  tab, 69, 428–430
  word, 429, 430
chord algorithm, 419
`CIRCLE`, 83, 96
circle, 32, 79–81, 94, 95, 98, 113, 253, 274
class
  continuity, 202–204
  equivalence, 130
  of object, 446, 447, 450, 451, 453, 464
clearing variables, 18, 25, 44, 76, 240
click, mouse, 12–19, 69, 87, 101, 112, 121, 164, 165, 400–402, 404–406, 409, 433
clipboard, 14, 15, 346, 465
`cmaple`, 347
code, program, 7, 10, 16, 17, 19, 36, 65, 87, 116, 136, 139, 159, 187, 198, 204, 205, 208, 210, 238, 322, 332, 340, 346, 347, 354, 355, 357, 360, 361, 412, 441, 443, 445, 447, 450, 465, 469, 478–480, 482, 485
coefficient matrix, 161, 181
coefficients, polynomial, 129, 288–290, 369, 381, 382, 384, 389, 393, 397
`coldim`, 339, 454
collating sequences, 68
collection
  garbage, 42, 452, 482, 485
  of terms, 49
  of values, 55
colon, 19, 24, 35, 86, 205, 222, 351, 352, 359, 402, *see also* `:`, `::`, `:-`, `:=`
color, colour, 78, 83, 84, 91
`Column`, 179, 182
column
  of array, 212–216

of matrix, 161–184, 339, 341, 447–469
of spreadsheet, 150, 400–418
combination
linear, 180, 181, 381, 382, 396
of plots, 90–93, 105, 112, 114, 118, 253
of types, 304, 322
comma, 12, 40, 60, *see also* ,
Command key, 14, 15
command
debugger, 361, 363
line, 8–10, 12, 329, 346, 348, 400
shell, 347, 481
comment, 36, 211, 223, 240, 346, 355, 401
commutativity, 32–34, 40, 163, 169, 249, 269, 278, 308–310, 468, 474, 475
comparing, 68, 170, 187, 190, 194, 210, 314, 316
compatibility
of Maple versions, 19, 27, 28, 30, 33, 195, 205
of matrices, 161, 341, 453
with UNIX, 343
compiled code, 116, 141, 197, 295, 296, 329, 453
`Complex`, 131
`complex` (Maple type), 299
complex numbers, 127, 128, 130, 131, 138, 139, 145, 170–172, 174, 175
random, 171
composition of functions, 32, 267, 274–275
compound statement, 445
computer algebra, 4, 5
concatenation, 33, 37, 40, 68, 162, 262, 350
condition
ill, 160, 176–180
number, 160, 176, 177, 179
conditional
evaluation, 204
execution, 187, 193–216, 221, 233, 241, 302, 351, 378, 390
conditions on ODE, 148, 279, 414–416
conformability of matrices, 161–163, 183, 341, 448, 467, 468
conjugate, 171
constant, 27, 28, 40, 45, 53, 85, 134, 140, 148, 152, 172, 181, 190, 194, 257, 314–316, 318, 345, 401
`constant` (Maple type), 299, 316, 378
`CONSTRAINED`, 83
constructor, 131, 161, 162, 166, 301, 334, 451–455, 457, 458, 481
`cont`, 361
contagious floating point, 140
context menu, 14, 17, 87, 112, 121, 164, 165, 400, 404, 409
continued fractions, 132
continuing
execution, 351
loop, 211
continuity class, 202–204
`CONTOUR`, 111
contours, 114–119
control
characters, 69, 343, 426, 427, 430, 431
structures, 35, 61, 87, 187–216, 222, 357, 450
transfer of, 351
variables, 65, 67, 222, 224, 226
convergence, 203, 406, 419, 421
conversion of Maple format, 14, 19
`convert`, 69, 70, 160, 251, 252, 276, 277, 332–334, 336, 344, 382, 431, 458

coordinates, 79–81, 87, 93, 95, 96, 107–109, 111, 113, 254, 256, 345, 401
copy, 458
copying
    arguments, 225, 226
    spreadsheet cells, 399, 402–404
    via GUI, 11, 14, 15, 87, 165, 346
correcting errors, 52, 354–364
cos, 39, 78, 79, 81, 108
cosh, 92
Cosmo Player, 122, 123
CountDown (user proc), 225, 226
counter, 208, 209
counting
    characters, 428
    components, 59
    lines, 428
    words, 428
Cox, D., 381
CR (carriage return), 427
crand (user proc), 171
crash, 18
create
    archive, 478
    directory, 434, 478
    file, 434
    help, 485, 486
CreateSpreadsheet, 411, 412
critical points, 116
CROSS, 83
Ctrl key, 14, 17, 69
cube roots, 408
cubic, 150
curly
    brackets, 30, *see also* { }, set
    partial derivative, 148, 278
currency symbols, 33
curve, 79, 80, 91, 106, 113–114, 117–119, 147
CVS, 10
cycloid, 95

D, 32, 203, 256, 267, 275–279, 414
D2N (user proc), 290
data structures, 450
data-type, *see* type
database, help, 485, 486
datatype, 166, 167
date of archive, 480
DBG> prompt, 361
debug, 359–360
debugging, 221, 354–364, 417
    procedures, 359
decimal, 32, 68, 131–134, 137, 164, 178, 190, 288, 289, 320, 436
    repeating, 133
decision, 187, 197, 419
declaration, 35, 67, 136, 154, 222, 224–226, 233, 241, 246, 295, 310–313, 445, 447, 450, 451, 453, 475, 481
decomposition, singular value, 161, 178, 182–184
default, 7
    array index range, 213
    directory, 343, 433, 478
    I/O mode, 428
    number system, 131, 133, 135, 140, 145, 155, 166, 171, 178
    plotting, 77, 78, 80, 82, 83, 85, 87, 90–93, 107, 111–114, 116
    spreadsheet names, 400
definite integration, 146–147, 407–410
degree of polynomial, 130, 152, 156, 286–289, 382, 384
degrees (angular), 95, 111
Delete key, 14, 15, 18, 400
deleting
    archive members, 480, 484, 485
    elements, 62, 64
    library, 480
delimiters

control structure, 35, 198, 205, 351
group, xvi, 16, 400
section, 16
word, 436, 437, 441
DELTA (user proc), 241
denary, *see* decimal
denom, 318, 322, 323
denominator, 50, 131, 322, 389
dense matrices, 171
dependence of variables, 153–154
derivatives, 32, 33, 66, 148, 150, 153, 201–204, 208, 275–280, 330, 370, 371, 373
repeated, 278
description, 221, 450
Det, 155
Determinant, 168
determinant, 155, 160, 168
diagonal, 163, 174, 179, 213, 465, 466
diagonal, 170
DiagonalMatrix, 174, 183, 184
DIAMOND, 83
dice (user proc), 349, 350, 352, 353
die (user proc), 348
Diff, 208
diff, 33, 66, 148, 153, 201, 202, 223, 238, 267, 276, 278, 279, 330, 372
differential equations, 148–150, 153, 279, 356, 413, 414
differentiation, 32, 33, 66, 148, 150, 153, 201–204, 208, 275–280, 330, 370, 371, 373
Digits, 135–137, 140, 141, 178, 190, 224, 239, 343, 344
digits, 6, 19, 68, 131, 132, 134, 135, 137, 140, 154, 166, 168, 178, 190, 288–290, 320–321, 343, 344, 426, 429, 430
Diophantine, 146
diphthongs, 427
directory, 347, 478
creating, 434, 478
current, 343, 433, 465
library, 477–479
separators, 40, 342, 433
discont, 85, 202, 203
discontinuities, 84–85, 138, 198, 199, 202–204, 241
discrete, 40, 41, 68, 89, 107, 114, 119, 128, 129, 191, 240, 345
disjoint types, 298, 299
display, 88–95, 98–100, 250, 254
display3d, 112, 120
displaying
assignments, 213
floats, 134
information, 357, 359, 360, 363
modules, 451, 453, 469
plots, 85–88, 250–253
procedures, 237–238
rtables, 165
tables, 213, 215, 400
ditto, 19, 28, 34, 225, 268, *see also* %, "
DIV (user proc), 227, 228
Div (user proc), 310
divide, 207, 232–235, 381, 384, 389, 396, *see also* /
divide, 382, 384
DivideAll (user proc), 232
DivideIfPossible (user proc), 230, 231
divisibility, 207, 233, 382, 384, 387, 389
division, 128, 155, 207, 227, 230, 233, 235, 248, 288, 370, 372, 381, 396, 397, *see also* /, multiplication
block matrix, 473
Euclidean, 270, 286–290, 381
generalized polynomial, 381–397
do, 65, 204–211

dollar, 33, *see also* $
dolls, Russian, 187
domain, 57, 107, 111, 114, 117, 118, 138, 191, 198, 203, 240, 274, 369, 393
`Domains` (Maple package), 447
dot, 32–33, 40–41, 83, 163, 164, 248, 470, 472, 473, *see also* `.`, `..`
double
    at, 32
    backslash, 342, 478
    bar, vertical, 37
    click, 12, 13, 121, 165
    colon, 35, 301, 311
    ditto, 34
    dot, 40
    loop, 212–216, 341
    precision, 141
    quote, 19, 27–28, 39, 342, 347, 349, 401, *see also* `"`
    roots, 150
    vertical bar, 37
Drijvers, P., 95
`dsolve`, 148–150, 279, 356, 414–416, 421
DVI file, 100
`dvips`, 100
DXF, 121, 122
dynamic types, 447

e, 53–54, 152
`e`, `E`, 133
e-mail, 14, 256, 436
`e^x`, 53–54
Edison, T. A., 354
Edit menu, 11, 14, 16, 17, 19, 357
editing, xvi, xvii, 9–11, 14, 16, 17, 19, 84, 112, 346, 357, 426, 428, 433, 436, 446, 465, 481
efficiency, 61, 63–65, 67, 92, 131, 134, 147, 170, 171, 197, 213, 231, 233, 235, 239, 248, 304, 311, 318, 319, 336, 341, 378, 381, 387, 389–391, 402, 405, 430, 441, 443, 469, 475
`Eigenvalues`, 172
eigenvalues, 159, 160, 172–176, 178, 179
`Eigenvectors`, 172, 175
eigenvectors, 160, 172–176
    independent, 174
    orthogonal, 175
elements, 29, 57–65, 143, 165, 170–172, 174, 176, 179, 188, 210, 213, 215, 228, 242, 260, 279–282, 285, 286, 298, 301, 306–308, 310, 311, 330, 331, 335, 341, 391, 393, 401, 442, 447–449, 463
    repeated, 58
`elif`, 196–201, 248
elimination, 57, 389
    ordering, 388
ellipse, 81, 118
ellipsoid, 109, 118
elliptic
    functions, 412
    integrals, 415
`else`, 195–201
Emacs, xvii, 9–11, 15, 428
empty
    archive, 478
    catch, 352
    collection, 260
    identifier, 236
    loop body, 209, 210, 387
    parentheses, 450
    selection, 69
    sequence, 41, 63–64, 142
    set, 284, 285
    statement, 35
    string, 262, 430
    table, 442
Encapsulated PostScript, xvi, 100
Encarta, Microsoft, 6

encoding of characters, 68–70, 94, 426–431
end, 35, 195, 222, 351
  do, 204–211
  if, 194–201
  module, 450
  proc, 221, 222
  try, 351
end-points, 40, 41, 98, 378
English, 68, 191, 208, 427
Enter key, 15, 17, *see also Return* key
entry point, 357
environment
  Maple, xvi, 178, 223, 255, 352, 361
  programming, 2, 3, 10, 15, 356, 412, 425, 427
  variables, 136, 166, 178, 224, 268, 468
epicycloid, 95
EPS, 100, 101
Equal, 170
equal, 24, 31, 54, *see also* =
Equation Editor, Microsoft, xvi, 9
equations, 24–25, 31, 32, 47, 48, 53–55, 61, 74, 77, 79, 80, 82, 90, 110, 118, 152, 237, 330, 333, 334, 369, 370, 377, 388, 413, 414, 416, 473, 482
  algebraic, 150–151, 154–155
  differential, 148–150, 153, 279, 356, 413, 414
  linear, 160, 167–169, 180, 183
  matrix, 174, 181
  solving, 142–146
equation (Maple type), 300
equilateral, 256, 258
equilibrium, 422
equivalence class, 130
error, 168, 231, 232, 251, 271
errors
  assertion, 385
  checking, 298, 301–302
  correcting, 354–364
  execution, 231, 232, 321, 384
  floating-point, 137
  handling, 351–354
  in this book, xv
  messages, 34, 231, 232, 313, 318, 351–353, 384, 385
  numerical, 132, 134, 135, 137, 147, 160–184, 190, 344, 377
  plotting, 75, 76, 115
  programming, 51–55, 205, 222
  spreadsheet, 401
  stack overflow, 229, 246
  syntax, 51, 193
escape character, 38, 343
escaping, 39, 69, 168, 192, 195, 269, 279, 299, 349, 403, 429, 431, 436
estimating
  errors, 174
  roots, 419
Euclidean
  division, 270, 286–290, 381
  GCD algorithm, 287, 373
  norm, 167
European characters, 70, 426
eval, 25, 45–49, 53, 148, 192, 215, 230, 237, 239, 240, 243, 273, 276, 277, 283, 297, 312, 334, 338, 339, 380, 384, 414, 447, 449, 453, 454
evala, 153
evalb, 31, 55, 69, 148, 170, 189, 190, 242, 273, 286, 303, 314, 378
evalc, 131
evalf, 30, 32, 34, 53, 136, 140–141, 144, 152, 176, 190, 206, 314, 378, 409, 410, 412
  /Int, 146–147
evalhf, 76, 77, 140–142, 377–380
evalm, 155, 256, 257, 269, 340, 447, 448

evaln, 312
Maple type, 311, 312
evaln(numeric) (Maple type), 312
EvaluateSpreadsheet, 411, 412
evaluation, 7, 25–27, 32, 42–49, 58, 64–66, 76, 87, 88, 141, 146, 147, 153, 162, 188, 190, 192, 193, 201, 227, 229, 230, 237, 243, 277, 290, 306, 312, 322, 377, 380, 401, 403, 408, 409, 411, 416, 418, 449, 450, 479
argument, 296, 297, 311, 313
Boolean, 31, 189, 191
complex, 131, 138
conditional, 204
last name, 164, 243, 453, 467
matrix, 449
numerical, 30, 136, 140–142, 147, 190, 224, 289, 314, 377–379, 406, 408–410, 412, 415
repeated, 45
even (Maple type), 299
exact, 132, 133, 138, 139, 142, 169, 177, 179, 184, 316, 407–409, 415
roots, 142–144
exact (in convert), 344
Excel, Microsoft, 400, 403
exceptions, 351, 352, 363
exclamation, 31, *see also* !
execution, 3, 7, 9, 17, 19, 25, 34, 35, 48, 58, 349
background, 346, 347
block, 445
conditional, 187, 193–216, 221, 233, 241, 302, 351, 378, 390
efficiency, 386
errors, 231, 232, 321, 384
groups, xv, xvi, 7, 15–17, 400, 411
loop, 234, 244
module, 445, 450, 451, 481, 482
of exception code, 351, 352
path, 347
procedure, 219, 220, 223–225, 228, 229, 231, 239, 240, 244, 245, 301, 303, 352, 355, 443, 453
program, 295, 296, 329, 351, 357, 361, 363, 385, 412, 465, 482
repeated, 205, 207, 210
shell, 347
single step, 361
tracing, 359
exit point, 357
exp, 53–54, 148, 152, 407
expand, 50, 51, 67, 189, 236, 238, 249, 330, 373, 382, 384, 390
expanding, 50, 65, 68, 249, 372
RootOf, 144
''(...), 236
strings, 262
Explorer
Internet, 122
Maple, 101
Windows, 121, 433, 434
exponent, 133–135, 137, 320
exponential, 53–54, 149
exponentiation, 269, 274
export
format, 100–101, 121–123
from module, 431, 448–483
of plot, xv, xvi, 87, 92, 100–101, 121–123
of worksheet, xv, xvii, 10, 100–101, 121–123, 346, 426, 446, 465
export, 241, 448, 450
Export menu, 100, 101, 121, 122
expressions, 6, 7, 14, 24, 25, 28, 29, 31, 32, 42, 44, 46–50, 54,

55, 59, 61, 64, 270, 273, 275, 391, 441, 451
arithmetic, 140, 236
Boolean, 187, 189, 193–194, 201, 205, 418
complex, 131
conditional, 204, 337
constant, 345
data-type, 305, 322, 466
displaying, 205, 221, 237, 359
in spreadsheets, 401
matrix, 167
plotting, 74–76, 78, 82, 107, 113, 118
recalling, 34
sequences, 29, 40, 55, 57, 142, 146, 172, 349, 382, 391, 429, 441, 458
set, 284
zeros of, 142, 377
exprseq, 297

`Factor`, 153, 154
`factor`, 153, 154, 370, 371, 373
`Factorial` (user proc), 245, 246, 318, 358
factorial, 31, 244–247, 268, 271, 318–320, 358–364, 405, *see also* !
`Factorial1` (user proc), 319
factorization, 153, 207, 232–236, 289, 290
squarefree, 369–373, 377
factors
common, 131
integer, 207, 236
numerical, 309
repeated, 235, 370–372
squarefree, 370–372, 377
`FAIL`, 190
failure
computer, 18
numerical, 408
of `add`, 66, 67
of `fsolve`, 418
of `int`, 146, 147
of `plot`, 74
of `try`, 352
of algorithms, 429
of large powers, 155
program, 232
`false`, 85, 190–194, 283, 382, 402, 429, 438
Maple type, 300
`fi`, 195
Fibonacci, 405, 406
field, 128, 370
finite, 127, 130, 155
gradient, 117–118
of coefficients, 267, 286, 288, 369, 381, 382, 389, 393
of object, 446, 448, 450, 455, 467
vector, 117
figures, 9
bitmap, xvi, xvii
geometrical, 83, 94
in this book, xvi
Lissajous, 90
significant, 30, 132, 135, 137, 141, 344
file
autosave, 18
binary, 11, 36, 480
export, xvi, 100–101, 121–123
help, 486
hierarchy, 238
initialization, 480, 481
input, 343–344, 346–347, 433–444
library, 478, 479
Maple text, 10, 36, 446, 465, 482, 486
names, 27
output, 342–343, 346–347
pathnames, 40, 465, 478
plot, 85, 92
text, 11, 36, 426–428, 433–444
worksheet, 11, 19, 486
File menu, 10, 11, 346

filling
  spreadsheet, 150, 404–406, 409
  text lines, 425, 436
`finally`, 351
finite number systems, 130
`FirstPrime` (user proc), 231
fixed
  point, 133, 134, 137
  precision, 139, 141
Flanagan, D., 4, 139, 457
flattening, 315, 316
`Float`, 133
`float`, 166, 320
  Maple type, 299, 320
floating-point, 32, 33, 40, 132–150, 160–184, 190, 206, 233, 257, 298, 314, 316, 320, 343, 344, 377–379, 406, 407, 409, 415–422
`floor`, 206
`fmt`
  UNIX, 436
  user proc, 437, 438
font, xvi, 24, 83
  symbol, 68, 93, 94
`font`, 93, 94
`for`, 65, 67, 204–216, 222
`forget`, 240
format
  export, 100–101, 121–123
  input, 346
  output, 86, 134, 235, 236
  paragraph, 87
  plot, 100
  source code, 16
  specification, 437, 441
  strings, 134, 349, 436
  UNIX, 427
  worksheet, 426
Format menu, 400
formatting text, 436
formula in spreadsheet, 399, 401–403
FORTRAN, 5, 135, 139
`frac`, 206
`fracinteger` (user type), 323
fractal, 219, 255–259
fractions, 128, 143, 206, 298, 304, 323
  continued, 132
  decimal, 32
`fraction` (Maple type), 299, 318, 322
`FRAME`, 83
frame
  around plot, 83, 87, 112
  of animation, xvi, 90–92, 98, 101, 258
  sequences, 90, 120
  wire, 107
`frames`, 90
free
  software, xvii, 9, 10, 101, 122
  variables, 74–76, 78, 82, 89, 90, 141, 143, 145, 225
`from`, 209
`fscanf`, 437, 438, 441, 442
`fsolve`, 41, 142, 145–146, 377, 409, 417–418
`function` (Maple type), 300, 302, 310, 332
functions, 6, 25, 28, 29, 32, 34, 51, 52, 58–59, 63, 64, 73, 79, 105, 107, 114, 140, 148–150, 153, 191, 219, 220
  abstract, 6, 74
  argument indexing, 59
  argument selection, 336
  arguments, 55, 57, 60, 61
  composition, 32, 267, 274–275
  critical points, 116
  discontinuous, 84–85
  exponential, 53–54, 149
  identity, 149
  inert, 52, 53, 57–58
  piecewise, 194, 198–204

Gamma function, 318, 408
GAP, 43

garbage collection, 42, 452, 482, 485
Gathen, J. von zur, 5
Gaussian numbers, 130, 171, 370
GB (user proc), 394
GBreduce (user proc), 393
GCD, 287, 288, 370, 373, 381
gcd, 371–373
Geddes, K. O., 5
GeneralizedPolynomialDivision (user proc), 382, 387
GenerateEquations, 168
GenerateMatrix, 168
generating strings, 260
generators of ideals, 381, 388, 389, 396
generic structured types, 306–307, 309–311, 314
geometrical plotting, 73, 80, 83, 93, 94, 105, 256
Gerhard, J., 5
get (method), 455–477
getColp, 457–477
GetDim (user proc), 338, 339
getElement (method), 463–468
getElements (method), 463–468
getRowp, 457–477
GhostScript, xvii, 101
GIF, 100, 101, 121
global, 35, 44–46, 67, 131, 135, 136, 239, 243, 321, 388, 389, 430, 431, 438, 450, 455, 465, 472, 473, 475, 481, 482
  scope, 222–225, 241–242
  variables, 35, 44, 136, 239, 321, 388, 450, 455, 482
global, 221, 222, 450
GNU, xvii, 9, 428
golden ratio, 406
Goosens, M., 100, 427
graded lex order, 392, 396
gradient
  fields, 117–118
  vectors, 117
gradplot, 117–118
graphic characters, 68, 69, 426, 427, 430
graphical
  investigation, 94
  user interface, 8
graphics, xiv, xv, 141, 257
graphs, 114, 116, 117, 418, 419
  plotting, 73–77, 106–107
gravity, 120, 413
greater, *see* >, >=
greatest common divisors, 287–288, 370
Greek, 93, 94, 191
green, *see* colour
grid, 107, 111, 401
grid, 117
Groebner (Maple package), 381, 393
Gröbner bases, 381–397
Gröbner, W., 389
GroebnerBasis (user proc), 391
groups
  delimiters, 400
  discussion, 6
  execution, 7, 15–17, 205, 343
  mathematical, 128, 213
GUI, 8–11, 13–15, 17, 18, 116, 343, 402, 429, 433, 477, 486

half-integer, 318, 319, 322
halfinteger (user type), 322
halving, interval, 376–381, 418
handles, resizing, 87, 112, 401
handling errors, 351–354
hardware floating point, 76, 138, 141–142, 166, 177, 178, 377, 379
harmonic motion, 414
has, 373, 389
hash, 36, *see also* #
  tables, 301, 441
.hdb, 485
Heal, K. M., 4
Heaviside, 198

Heck, A., 4
help, 2, 10–13, 485–486
    balloon, 13
Help menu, 12, 13, 485
HELVETICA, 125
hermitian, 170, 171, 174, 178
hermitian, 171
HermitianTranspose, 171, 176
hexadecimal, 289
HIDDEN, 111
hidden lines, 111
Hilbert, D., 165
HilbertMatrix, 176–179
horizontal
    alignment, 400
    axis, 74, 79, 80
    coordinates, 79, 81, 93, 345, 401
    plot range, 77, 95
    tabulator, 428
Horn, B. K. P., 4
Horner's algorithm, 289, 290
HTML, 100, 426

I, 130, 131
ideals, 381, 388, 389, 396–397
identical (Maple type), 384
identifiers, 6, 7, 26–28, 36
    _ in, 35
    \ in, 38
    arbitrary, 26, 27, 192, 215
    array, 192
    as strings, 45
    backquotes in, 39
    e, 53, 152
    empty, 236
    font for, xvi
    I, 131
    joining, 37
    juxtaposing, 51
    keyword, 195
    local, 338
    newlines in, 39
    operator, 34, 58
    pi, 53
    protected, 131, 243, 470, 481
    rebinding, 473
    special, 227–230, 244, 248, 453
    spreadsheet, 411
    type, 301, 322
    upper case, 52
identity
    function, 149
    matrix, 168, 215
identity, 170
IdentityMatrix, 168, 177, 178
IEEE, 138, 139
'if' function, 204, 316, 382, 387, 391, 419, 420
if, 194–201, 303
ifac (user proc), 232, 234, 235
ifactor, 207, 235–236
IGCD (user proc), 288
ill condition, 160, 176–180
Im, 58
imaginary, 129, 131, 138, 168
    part, 131, 171
imaginaryunit, 131
implicit
    declaration, 224, 226
    definition, 114, 118–119
implicitplot, 116, 118–119
importing
    into spreadsheet, 411
    into worksheet, 465
in, 210–211
inc (user proc), 312, 313
incomplete Gamma, 408
inconsistent system, 181, 182
increment, 43, 208, 209, 212, 312, 404
.ind, 479
indefinite integration, 208
indentation, 10, 198, 208, 222
independent
    eigenvectors, 174
    variables, 75, 153, 223, 224
indeterminates, 43, 45, 129, 289, 295
indets, 373, 374, 378

`indexed` (Maple type), 300, 301
indexing, 7, 29, 61, 63, 88, 136, 140, 213, 214, 228, 232, 278, 298, 301, 339, 341, 345, 441
  negative, 63, 248
inequalities, 9, 31, 76, 142, 188
inert, 7, 34, 52, 53, 57–58, 67, 85, 105, 146, 147, 166, 188, 208, 236, 268–294, 303, 308, 408, 467, 469, 474, 475, 486
infinite
  fields, 131
  loops, 18, 204, 205, 208, 442
  numbers, 129, 135, 137–139, 180
  period, 422
  recursion, 229, 244, 246–247, 472, 475
  series, 129
  sets, 128, 130
  structures, 256
`infinity`, 137, 210, 223
infix, 24, 128, 154, 268, 270, 274, 279–281, 283, 470, 472
`infolevel`, 141, 356–357, 417
inheritance, 447, 469
`.ini`, 481
`init`, 482
initial conditions, 414–416
initialization, 43, 64, 225, 320, 431
  array, 192, 213, 215
  file, 480, 481
  loop, 16, 205, 206, 208, 210, 316, 429, 441
  Maple, 479
  matrix, 169
  module, 450–453, 457, 458
  package, 482, 483
  table, 442
input, 8–10, 13, 14, 16, 85, 152, 221, 260, 268, 346, 347, 431, 474
  advanced, xiv
  examples, xvi
  file, 10, 343–344, 346–347, 433–444
  spreadsheet, 401
  standard, 347, 428
  tabular, 400
  test, 355
  validation, 301–302, 351, 352, 462
`insequence`, 91, 92, 98
Insert menu, 8, 16, 400
`Int`, 52, 53, 146–147, 408
`int`, 41, 52, 67, 201, 208, 331, 332, 407
integer, 40, 41, 66, 68, 128, 129, 139
  conversion, 344
  division, 270, 286–290
  factorial, 244–247
  factorization, 207, 232–236
  Gaussian, 130, 370
  largest, 131
  modular, 130, 154–155
  multiplication table, 213, 404
  output, 436
  part, 32, 33, 206
  power, 205, 248–250
  random, 348, 443
  range, 207–209
  solution, 146
`integer`
  in `readdata`, 343, 344
  Maple type, 248, 299
`integer &+ symbol` (Maple type), 309
integration, 41, 52, 198, 201, 204, 331
  definite, 146–147, 407–410
  indefinite, 208
interface
  external function, xiv
  method, 468
  object, 446
  procedure, 320, 377
  spreadsheet, 399, 412

to Excel, 400
type, 466
user, 1, 8–12, 85, 131, 237, 346, 348, 451
worksheet, 19, 53, 436
`interface`, 85, 86, 131, 165, 237, 240, 348, 350
`INTERFACE_HELP`, 486
Internet, 122
discussion groups, 6
Explorer, Microsoft, 122
`intersect`, 280, 281, 284
intersection, 280–282
interval
autosave, 18
halving, 376–381, 418
isolating, 146, 369, 376, 377
range, 41
representation, 132
root, 145
`into`, 361
invariant, loop, 384–386
inverse
block matrix, 464–466, 475
Boolean, 191
functional, 415, 421
matrix, 160, 168, 169, 174, 175, 177, 179, 183
invisible characters, 16, 69
`iquo`, 155, 207, 235, 270, 286, 287, 289, 320
`irem`, 130, 207, 235, 248, 270, 286–289, 320
Iris, 85, 116
irrational, 128, 131–133, 139
irreducible, 153, 369, 370, 372
`is`, 189
ISO, 122, 426
`Isolate` (user proc), 376
isolating
code, 103
intervals, 146, 369, 376, 377
`isolve`, 146
`IsOrthogonal`, 175
`isprime`, 204, 209, 231
`IsUnitary`, 175
`ITALIC`, 103
italic, 53
iteration, 64, 67, 68, 204, 207–211, 244, 247, 248, 379, 384, 386, 420, 455

Java, 3, 4, 24, 139, 193, 195, 197, 204, 295, 296, 351, 427, 440, 443, 445–447, 452, 470
Jeffreys, H. and B., 413
joining
execution groups, 16
identifiers, 37
lines, 436
lists, 60, 272
sequences, 59
strings, 37, 262
`JordanBlockMatrix`, 448
justification (alignment), 343, 400
juxtaposition, 51–52

kernel, 9, 10, 17, 18, 85, 141, 238, 431, 486
`kernelopts`, 385
Kernighan, B., 4, 354, 355, 425, 427, 428, 436
keyboard, 11, 13, 15, 18, 19, 23, 24, 26, 33, 38, 68, 69, 93, 94, 347, 348, 401, 429
keyword, 27, 82, 168, 192, 195–197, 204, 205, 208, 222, 227, 231, 301, 351, 352, 450, 451, 466, 481, 486
Knuth, D. E., 5, 376, 397
Koch, H. von, 255

labelling
axes, 74, 75, 77, 79, 115
plots, 79, 83, 202
spreadsheets, 401
sub-expressions, xvi
tables, 214, 215
`labels`, 125

Λ, 172
λ, 172
Lamport, L, xv
LaTeX, xv–xvii, 100, 426, 443
    Companion, 100
    Web Companion, 427
Latin characters, 426, 427
`lc` (user proc), 394
lcm, 389
`lcm`, 390
`lcoeff`, 384
leading
    coefficient, 288, 384, 393
    diagonal, 213
    monomial, 382, 389
    term, 382, 384, 386, 388, 389, 394
    zero, 289
least
    common multiple, 389
    squares, 184
`LeastSquares`, 184
`legend`, 83, 84, 251
length
    of line, 425, 436, 437
    of sequence, 248, 311
    of vector, 167, 172, 175
`length`, 428, 429, 438
less, *see* `<`, `<=`
letters, 6, 23, 26, 33, 35, 52–53, 68, 70, 93, 94, 260, 262, 299, 426–428, 430
    `e`, `E`, 133
lexical scope, 224, 241–242, 380, 388, 438, 445, 451–453, 455
lexicographic, 261, 384, 388, 392, 396
LF (line feed), 427
`lhs`, 173, 188, 376
`.lib`, 479
`lib` directory, 478–480, 485
`libname`, 478–481, 483–486
library, 10, 36, 73, 105, 139, 159, 166, 226, 227, 237, 238, 351, 355, 425, 436, 437, 446, 458, 477–485
`limit`, 241
limits, 66, 67
`linalg` (Maple package), 159, 166, 337, 339, 359, 445–469
`LINE`, 83
line
    blank, 436
    breaking, 15–16, 39, 82, 84, 427, 428, 434, 436
    count, 428
    end, 356, 427–429, 434
    feed, 69, 427
    hidden, 111
    length, 425, 436, 437
    numbers, 361
    oriented I/O, 346
    style, 91
linear
    algebra, 159–184, 255, 337, 447
    combinations, 180, 181, 381, 382, 396
    equations, 160, 167–169
    printing, 86
`LinearAlgebra` (Maple package), 89, 159–161, 166–184, 337, 339, 448
`LinearSolve`, 167, 168, 180–182
linefeed character, 427
`linestyle`, 83, 84, 251
Linux, xv, 8, 122
Lipschutz, S, 447
LISP, 4, 193, 224, 295, 311, 316, 329, 330, 445, 467
Lissajous figures, 90
`list` (Maple type), 272, 300, 306
`list(algebraic)` (Maple type), 311
`list(constant)` (Maple type), 314
`list(integer)` (Maple type), 306
`list(list(matrix))` (Maple type), 458

list(list) (Maple type), 458
list(name) (Maple type), 306, 382, 391
list(nonnegint) (Maple type), 452, 458
list(numeric) (Maple type), 314
list(polynom) (Maple type), 374, 376, 382
listlist, 458
listprocedure, 149, 416, 421
lists, 55, 57, 58, 62–64, 69, 223, 262
- adding, 272, 435
- as vectors, 257
- constructing, 29, 40, 64–66
- dsolve output, 149, 416
- in evalhf, 141
- in plotting, 78, 79, 81, 84, 91–93, 96, 107, 110, 112, 113, 116
- in substitution, 48
- indexing, 29, 48, 59, 64
- input into, 343
- iterating over, 210
- mapping over, 330, 331
- membership, 282
- of archive, 480
- of divisors, 382, 387
- of equations, 142
- of indices, 463
- of letters, 261
- of lists, 455
- of partitions, 448
- of plots, 86
- operating on, 59–63
- selecting from, 336
- tagged, 467

listset (user type), 322
load, 482, 483
local, 46, 67, 136, 349, 388, 469
- arrays, 338
- binding, 473
- context, 103
- control variables, 65
- evaluation, 45–46
- procedures, 320, 379, 458, 463, 469
- reassignment, 136, 178
- scope, 222–227, 241–243, 431
- substitution, 47
- variables, 45–46, 244, 248, 340, 341, 361, 386, 445, 450, 468, 475

local, 100, 103, 125, 136, 221, 222, 228, 450
locus, 95, 98
log, 29, 140, 193, 206
logarithm, 29, 408
logic, 187–206
logical (Maple type), 300
loop, 16, 37, 43, 61, 65, 187, 194, 204–216, 222, 231, 235, 244, 330, 337
- errors, 353
- interactive, 349
- invariant, 384–386
- read, 435
- repeat, 349
- summation, 342

lower case letters, 52–53
lprint, 25, 28, 38, 85, 86, 232, 239, 268, 318, 343, 349
lt (user proc), 389

Macintosh, xv, 8, 10, 13–15, 17, 101, 112, 122, 164, 400, 427, 434
macro, 152
macro, LISP, 311
magic, 1, 5, 122
magnification, worksheet, 19
makehelp, 486
mangler, Markov, 437, 441
mantissa, 133, 135, 137, 320
MAP (user proc), 332, 334
Map, 183
map, 155, 188, 242, 251, 272, 316, 329–332, 334–336, 374, 375, 387, 458
map2, 330–332, 458

Maple Explorer, 101
mappings, 7, 24–25, 273
  plotting, 75
  simple, 7
march, 478–480, 484, 485
Markov
  chain algorithm, 440
  mangler, 437, 441
markov (user proc), 442
mass, 413
Mathematica, 301
MathML, 100
MatMul (user proc), 341
Matrix, 160, 166
  Maple type, 300
matrix, 155, 159–184, 192, 212, 213, 215, 339, 407, 410, 411, 445
  addition, 163
  algebra, 337–342
  banded, 161
  block, 161, 447–477
  coefficient, 161
  inverse, 160, 168, 169, 174, 175, 177, 179, 183
  multiplication, 341
  partitioned, 447–477
  random, 166
  rotation, 257
  sparse, 161, 171, 172
  special, 159, 161, 169–172
  square, 164, 168, 172, 179, 248
  triangular, 161
  tridiagonal, 161
  zero, 163
matrix, 155, 257
  Maple type, 300, 341, 452
matrix(matrix) (Maple type), 458
MatrixInverse, 177, 178
MAX (user proc), 223, 228, 316
max, 210
McCarthy, J., 193
mdeg, 382
member, 282, 283, 382, 431
membership, 282–283, 396–397
menu, 8
  Animation, 90
  context, 14, 17, 87, 112, 121, 164, 165, 400, 404, 409
  Edit, 11, 14, 16, 17, 19, 357
  Export, 100, 101, 121, 122
  File, 10, 11, 346
  Format, 400
  Help, 12, 13, 485
  Insert, 8, 16, 400
  Options, 14, 18, 86, 87, 250
  Spreadsheet, 400, 401, 404–406, 409
  Start (Windows), 94, 433
  View, 16, 19, 69
  Window, 17
messages, 349
  error, 34, 231, 232, 313, 318, 351–353, 384, 385
methods, 446–477
Microsoft
  Encarta, 6
  Equation Editor, xvi, 9
  Excel, 400, 403
  Internet Explorer, 122
  Paint, 9
  Windows, xv, 8, 11, 40, 68, 141
  Word, 9, 11, 16, 260, 436
    Picture, xvi
middle mouse-button, 15
minimal polynomial, 130, 152
minus, *see* -
minus (set), 128, 281, 282, 284, 285, 431
mkdir, 478
mod, 154, 155, 269
mode, file access, 427, 428
mods, 155
Modula, 195, 197, 445, 446
modular integers, 130, 154–155
'module' (Maple type), 301, 466
module, 45, 46, 87, 103, 136, 178, 224, 412, 431, 450

modules, 45, 46, 87–89, 103, 136, 137, 178, 222, 224, 241–243, 282, 297, 388, 412, 431, 432, 445–477, 481–485
  body, 46, 241
  types, 466–467
Monagan, M. B., xiv, 4, 378, 474, 486
monic, 130, 288, 393
monomial, 382, 389, 397
  leading, 389
  ordering, 384, 388, 389
`monord` (user type), 382, 384, 391
mouse, xv, 8, 11–19, 69, 87, 90, 101, 112, 121, 164, 165, 400–402, 404–406, 409, 433
moving
  spreadsheet cells, 399, 402–404
  via GUI, 11, 14
`.mpl`, 347
`msolve`, 146, 154
μ, 345
`mul`, 40, 66–67, 222, 224
multidegree, 382
multiple
  assignments, 55–56, 458
  derivatives, 33, 66
  factors, 370
  inheritance, 447
  integration, 208
  least common, 389
  plots, 78–80, 83, 84, 110
  roots, 150–151, 370
  solutions, 142
  substitutions, 48–49
multiplication, *see* `*`, `&*`, `.`, `mul`, `product`
  block matrix, 467, 470
  matrix, 163, 341
  non-commuting, 32, 269
  of sequences, 58
  repeated, 67
  tables, 213, 404
  type testing, 308
multiplicity, 150, 235, 377
multivariate, 117
  polynomial algorithms, 381–397
mushroom, 122
`.mws`, 18, 19, 426, 433

n-ary, 268, 270, 272, 282, 304
`N2D` (user proc), 289, 320
NAG, 139, 159
`name` (Maple type), 300–302, 313, 372
names, 6, 24, 25, 42
  alternative, 151, 152, 299
  argument, 278
  as axis labels, 74, 77
  assigned, 42, 43
  capitalized, 58, 153
  clearing, 25
  conflict of, 359, 465
  dynamic, 37
  empty, 236
  evaluation to, 311, 312
  functional forms as, 240
  generated, 180
  help database, 486
  in assignments, 42
  indexed, 29
  library, 477, 478
  library member, 480
  numeric, 484
  of environment variables, 225
  operator, 268–270
  package member, 88, 95, 112, 166, 255, 453
  primed, 26
  procedure, 220, 229, 237, 244, 248, 273, 333, 384, 388
  rtable data-type, 160
  scope of, 223, 225, 242
  special, 301
  spreadsheet, 400
  sub-procedure, 238

testing for, 313
to save, 479
UNIX, 428, 436
NaN, 138
`nargs`, 125, 227–229, 232, 247, 248, 271, 290, 311, 372, 378, 382, 453
national characters, 426
natural numbers, 128
Navigator, Netscape, 122
negating, 191
negative
indexing, 63, 248
numbers, 128
`negative` (Maple type), 299
`negint` (Maple type), 299
nested types, 295, 304–307
nesting, 25, 187, 197, 206, 212, 361
Netscape Navigator, 122
neutral, *see* inert
new types, 321–323
newline, 39, 82, 349, 356, 427, 434–437, 441
`next`, 211, 212, 361
`nextprime`, 207, 232, 233
nomenclature, 58, 273
`NONE`, 83
`nonnegative` (Maple type), 299
`nonnegint` (Maple type), 246, 289, 299, 318, 358
`nonposint` (Maple type), 299
`nonpositive` (Maple type), 299
`nops`, 59, 63, 272, 286, 316, 333, 345, 373, 382, 393, 442, 458
`Norm`, 167, 173, 177, 183, 184
norm, 167
`NORMAL`, 83
`Not` (type operator), 304–305
`` `not` `` (Maple type), 300
`not` (Boolean operator), 191–194
NUL (ASCII), 70, 431
`NULL`, 57, 63–65, 142, 207, 211, 225, 226, 229, 232, 261, 316, 334, 336, 337, 349, 378, 391, 478
null
module, 472
set, 284
value, 34
nullary, 225, 267, 268
number
condition, 160, 176, 177, 179
exponential (e), 53
of elements, 286
systems, 128
theory, 233
numbers, 128, 298, 401
algebraic, 128, 130, 151–154
cardinal, 128
complex, 127, 128, 130, 131, 138, 139, 145, 170–172, 174, 175
finite, 130
floating-point, 409
fractional, 128
Gaussian, 130, 171, 370
in products, 33
infinite, 129, 135, 137–139, 180
irrational, 128, 131–133, 139
line, 361
Maple, 131
natural, 128
negative, 128
random, 343–345
complex, 171
rational, 128, 133, 143, 369
real, 127, 128, 132–150
statement, 361
`numer`, 37
numerator, 37, 131, 389
`numeric` (Maple type), 299, 313, 316, 374
numerical
analysis, 5, 139, 150, 160
failure, 408
types, 299

Numerical Algorithms Group (NAG), 139, 159
numtheory (Maple package), 132, 233, 234
NZeros (user proc), 376

Object Linking and Embedding, 8
objects, 445–477, 481–485
  classes of, 446, 447, 450, 451, 453, 464
oblate, 109, 120
octal, 289
od, 205
odd (Maple type), 299
odd multiplicity, 377
ODE, 148, 279, 415, 416, 421
OLE, 8
ω, 414
OO, 446
OOP, 446, 447, 450, 451, 453, 457, 469, 470
op, 48, 59–63, 69, 170, 188, 239, 243, 272, 286, 316, 320, 332, 334, 336, 339, 341, 345, 373, 374, 378, 382, 391, 419, 435, 437, 454, 458, 463, 475
operands, 32, 58, 59, 61, 63, 65, 188, 190, 193, 243, 267, 268, 270, 308–310, 333, 334, 338, 349, 474, 475
operators, 267–283
  algebra of, 32
  Maple
    &*, 34, 269, 308
    &+, 269, 307, 308
    &^, 154, 155, 269
    &where, 277
  overloading, 470–477, 481
  user
    &++, 272
    &-, 281, 282
    &/, 271
    <, 314
    &div, 270
    &member, 283
    &!, 271, 272
    &., 470, 471
Option key, 14, 17
options, 7
  2D plot, 82–85, 87, 90, 91, 93, 255
  3D plot, 110–116
  command-line, 481
  interface, 165
  march, 480
  module, 432, 482, 483
  procedure, 238–241, 319, 360
  shape, 169
  user, 481
options, 221, 222, 450
Options menu, 14, 18, 86, 87, 250
Or (type operator), 304–305
'or' (Maple type), 300
or (Boolean operator), 191–194, 211
Or(symbol,string) (Maple type), 304
Order, 239
order
  canonical, 49, 57, 58, 308
  elimination, 388
  graded lexicographic, 392, 396
  lexicographic, 261, 384, 388, 392, 396
  monomial, 382, 384, 388, 389
  of derivative, 208
  of power series, 150
  relation, 190
  reverse, 63, 248, 289
  sort, 384
orientation, 111, 113–115
orientation of plot, 112, 114, 115
orthogonal, 113
  eigenvectors, 175
  matrices, 175, 179
outfrom, 362
output, 8, 9, 12–14, 16, 19, 52, 85, 346
  advanced, xiv

buffering, 434
context menu, 164
debugging, 355–364
device, 83
dsolve, 149, 416
examples, xvi
file, 342–343, 346–347
float, 133–134
format, 86, 235, 236
formatted, 436
from initialization, 481
from modules, 137, 224
from procedures, 195, 220, 221, 250–259
integer, 436
labels, xvi
march, 480
message, 349
plot, 85, 100
spreadsheet, 401, 412
strings, 436
tabular, 400
within use, 473
output, 149, 182, 416, 421
outputoptions, 166, 167, 170, 171
overflow
numerical, 134–135, 138
stack, 229, 246, 380
overhead
ASSERT, 385
float conversion, 141, 379
GUI, 8
overloading, 106, 161, 273, 298, 313, 446, 451–454, 457, 470–477, 481

pack, 485
package
Maple
Domains, 447
Groebner, 381, 393
linalg, 159, 166, 337, 339, 359, 445–469
LinearAlgebra, 89, 159–161, 166–184, 337, 339, 448
numtheory, 132, 233, 234
plots, 88–100, 112–121, 158, 202, 254, 255, 258, 259, 345
plottools, 122, 123
Spread, 410–412
member names, 453
user
BlockMatrices, 481–485
text_processing, 431–433
package (module option), 431, 432
packages, xiv, 88–89, 255, 445, 469, 471, 481–486
packing archive members, 485
Paint, Microsoft, 9
Pairs (user proc), 391
pairs
of backslashes, 38
of coordinates, 345
of dots, 40
of points, 191
of symbols, 28, 31
sets of, 391
parametric plotting, 79–81, 90, 94, 107–110, 113–114, 118
parentheses, 28–29, *see also* ( )
parse, 350, 352
parsing, 268, 270, 350, 352, 466, 470, 474
partial
derivatives, 30, 148, 278
factorization, 289, 290
partitioned matrices, 447–477
Pascal, 5, 195, 197, 445, 446
pasting via GUI, 11, 14, 15, 18, 87, 465
PATCH, 111, 117
PATCHCONTOUR, 111, 114, 126
PATCHNOGRID, 111
path, execution, 347
pathnames, 40, 342, 343, 347, 433, 465, 478, 485

pattern
  matching, 301, 425
  types, 305, 307, 308, 318
PDF, xvii
pendulum, 412–422
percent, 34, *see also* %
period, 32, *see also* .
Perl, 204, 347, 425, 440, 441, 443
Persistence of Vision, 121, 122
phase shift, 90
ϕ, 108, 109, 111
`Pi`, 25, 34, 53
π, 25, 34, 53
`piecewise`, 201–202, 204
piecewise functions, 194, 198–204
Pike, R., 4, 354, 355, 427
plain text, 346, 486
plane, 106, 114, 120
  curves, 105, 116
platform, xv, 8, 9, 11, 13, 15, 17, 40, 85, 101, 122, 164, 400, 427, 429, 434, 478, 481
Plauger, P. J., 4, 425
`plex`, 382, 384, 390
`PLOT`, 82, 85, 105
`plot`, 41, 73–101, 106, 107, 110, 113, 145, 250, 254
  failure, 74
`PLOT3D`, 85, 105
`plot3d`, 106–123, 250
plots, xvi, 8, 9, 14, 27, 68, 400, 426, 486
  combining, 90–93
  multiple, 78–79, 110
  superimposing, 78–79, 90–93, 110, 118, 251
`plots` (Maple package), 88–100, 112–121, 158, 202, 254, 255, 258, 259, 345
`plotsetup`, 100
plotting, 73–123, 141, 145, 147, 149, 151, 198, 200, 202, 203, 227, 250–259, 345, 408, 416, 422
  expressions, 75
  graphs, 73–77, 106–107
  mappings, 75
  points, 81–82
`plottools` (Maple package), 122, 123
plus, *see* +
`POINT`, 83, 96, 114
point, 32, *see also* .
  branch, 408
  decimal, 32, 68, 131–134, 137, 164, 178, 190, 288, 289, 320, 436
  entry, 357
  exit, 357
  fixed, 133, 134, 137
  floating, 32, 33, 40, 132–150, 160–184, 190, 206, 233, 257, 298, 314, 316, 320, 343, 344, 377–379, 406, 407, 409, 415–422
  mapping, 241
  plot, 81–82
polar coordinates, 80, 95, 96, 108, 109, 111
polygon plotting, 81
polymorphism, 298, 313, 453
`polynom` (Maple type), 300, 372, 373, 382
polynomial, 42–44, 49, 50, 128–130, 144–146, 150, 152, 153, 248–250, 286–288, 300, 355
  algorithms, multivariate, 381–397
  algorithms, univariate, 369–381
  coefficients, 288–290
  division, generalized, 381–397
  minimal, 130, 152
  primitive, 152
  variables, 289
`PolyPow` (user proc), 249
pop-up menu, 14, 17, 409
`posint` (Maple type), 234, 299, 338
`positive` (Maple type), 299, 378

postfix, 32, 245, 268
PostScript, xvi, 100
POV, 121, 122
Pow (user proc), 248
power, *see* ^, exponent
  fractional, 143
  inert, 154
  integer, 129, 205, 235, 248–250
  large, 155
  of 10, 133
  of matrix, 164
  of variable, 382
  plotting, 251
  reduced, 371, 372
  series, 150
  set, 284–286, 391
  with map, 242
PowerPlots (user proc), 252
PowerSet (user proc), 285
PowersPlot (user proc), 251
precision, 30, 135–142, 161, 166, 168, 177, 190, 224, 379, 409, 420
predicate, 302
prefix, 268, 270, 470, 474
prepending, 59, 260, 261
Press, W. H., 5, 160
pretty-printing, 52, 232, 239, 268, 274, 278, 386, 412, 469
prevprime, 209
prime, 370
  factorization, 207, 232–236
  numbers, 130, 131, 155, 209, 231, 305
prime (Maple type), 299
primed names, 26
primitive
  polynomials, 152
  types, 296–299, 305, 332, 466
print, 28, 38, 45, 87–89, 134, 137, 192, 195, 213, 214, 220, 221, 232, 237, 238, 250, 252, 343, 348, 349, 360, 469, 470
printable characters, 68
printf, 38, 134, 342, 346, 348, 349, 435–438, 442
printing, linear, 86
PrintIt (user proc), 220
printlevel, 245, 246, 286, 357–359, 417, 418
printline (user proc), 438
proc, 35, 100, 125, 190, 220–223
procedural type definition, 308, 322, 323
procedure (Maple type), 50, 298, 301, 336, 378
procedures, 24–25, 36, 38, 45, 50, 67, 87, 136, 149, 178, 194, 219–263, 267, 270, 297, 298
  anonymous, 25, 242, 244, 248, 331, 332
  body, 46, 48, 50, 58, 199, 221, 222, 225, 231, 237–242, 248, 273, 312, 352, 451
  builtin, 238, 311, 332, 335, 336
  call stack, 363
  calling, 219, 222, 226, 227, 229, 232, 239, 244, 245, 247, 248, 250, 253–255
  debugging, 359
  displaying, 237–238
  execution, 219, 220, 223–225, 228, 229, 231, 239, 240, 244, 245, 301, 303, 352, 355, 443, 453
  interface, 320, 377
  local, 320, 379, 438, 458, 463, 469
  names, 220, 229, 237, 244, 248, 273, 384, 388
  numerical, 416, 418
  options, 238–241, 319, 360
  output from, 195, 220, 221, 250–259
  polymorphic, 313
  recursive, 285, 289, 290

special argument types, 311–313
special names, 333
subsidiary, 338, 378, 379
tracing, 359
user
ADD, 247, 248
AddArgs, 311
addword, 438
BlockMatrix, 452, 454, 458
Cardioid, 100
CountDown, 225, 226
crand, 171
D2N, 290
DELTA, 241
dice, 349, 350, 352, 353
die, 348
DIV, 227, 228
Div, 310
DivideAll, 232
DivideIfPossible, 230, 231
Factorial, 245, 246, 318, 358
Factorial1, 319
FirstPrime, 231
fmt, 437, 438
GB, 394
GBreduce, 393
GeneralizedPolynomialDivision, 382, 387
GetDim, 338, 339
GroebnerBasis, 391
ifac, 232, 234, 235
IGCD, 288
inc, 312, 313
Isolate, 376
lc, 394
lt, 389
MAP, 332, 334
markov, 442
MatMul, 341
MAX, 223, 228, 316
N2D, 289, 320
NZeros, 376
Pairs, 391
PolyPow, 249
Pow, 248
PowerPlots, 252
PowerSet, 285
PowersPlot, 251
PrintIt, 220
printline, 438
PSeries, 243
R2P, 242
ReturnIt, 220
ScaProd, 339, 340
SELECT, 336
Side, 257
Snowflake, 258
solveh, 377–380
solveh1, 379, 380
Spoly, 390
SqFree, 372
sqr, 220, 239, 240
Sturm, 373
SubTree, 254
SubWords, 261
termcoeff, 389
Theta0, 421
Tree, 255
unblock, 458
unblockIndex, 463
Var, 374
VecAdd, 338
wc, 435
wc_line, 429
Words, 261, 262
writefile, 434, 437, 438, 440
procname, 227, 229–231, 244, 248, 283, 314, 316
PROD, 474
Product, 58
product, 40, 66–67
products, 32
block matrix, 467
commutative, 310
expanding, 249
iterating over, 210

mapping over, 330
matrix, 341–342
numerical factors in, 309
of factors, 370
of powers, 382
of repeated factors, 371
polynomial, 355, 369, 372
repeated, 53, 66, 474
scalar, 339–340
selecting from, 336
programming types, 301
prolate, 109
Prolog, 4, 316
prompt
debugger, 361
Maple, 16, 346, 349–353
shell, 347
`prompt`, 350, 352
protected identifiers, 131, 243, 470, 481
`PSeries` (user proc), 243
pseudo
Boolean, 190
types, 311
variables, 225
public, 446, 449, 450, 457, 458, 481
punctuation, 23, 426, 429, 430, 441

quadratic, 150
quadrature, 146
question mark, 12, 41, *see also* ?
queue, 245
`quiet`, 348, 349
`quit`, 347, 362–364
`quo`, 286, 287, 381, 382
quotes, 6, 7, 19, 24, 25, *see also* ", ', `
quotient, 207, 235, 286, 287, 289, 300, 304, 382, 384, 396, 474
sets, 130
quoting types, 303

`R2P` (user proc), 242
radial lines, 94
radians, 95, 111
radicals, 143, 144
radius, 79, 81, 95, 97, 108
radix, 132, 133, 138, 288, 290
raise
exception, 351, 352, 363
to power, 30, 164, 242, 372
`rand`, 171, 181, 286, 343, 344, 348, 442, 443
random, 440, 441
complex numbers, 171
integers, 286, 348, 443
matrices, 166–168, 170, 171, 180, 181
numbers, 343–345
vectors, 166
`RandomMatrix`, 166, 167, 170, 171, 180, 181
`RandomVector`, 166, 167, 180
`randpoly`, 287
range, 37, 40, 41, 62, 65, 67, 68, 82, 90, 95, 132, 145
cell, 402–419
`fsolve`, 145
in selection, 248
iterating over, 68, 207–209
parameter, 108, 113
plot, 77–78, 107
`range` (Maple type), 300
`range(numeric)` (Maple type), 376
ratio, golden, 406
rational
approximations, 132
arithmetic, 177
expressions, 50
numbers, 127, 128, 133, 143, 295, 318–320, 369
`rational`
in convert, 344
Maple type, 299
`ratpoly` (Maple type), 300
RCS, 10
`Re`, 58
`READ`, 428

`readdata`, 343, 344, 346
`readline`, 346, 348, 350, 435, 438
`readstat`, 349
real, 170, 175
  cube roots, 408
  numbers, 41, 127, 128, 132–150, 318
  part, 131, 171
  values, 131
  zeros, 373, 375–377
`real`, 35
`realcons` (Maple type), 299
reciprocal, 227
records, 448–450
rectangular tables, 159, 301
recurrence relations, 150
recursion, 219–263
  infinite, 246–247
recursive procedures, 285, 289, 290
red, *see* colour
Redfern, D., 4
redo, 11, 15
REDUCE, 301
reducible, 370, *see also* irreducible
regions
  text, xv, 36
  worksheet, 17
`relation` (Maple type), 300
relations, 31, 187–190, 193, 282, 283, 314–315, 330, 331, 378
  mapping over, 330
  recurrence, 150
  selecting from, 336
relative error, 137
`rem`, 286–288, 373, 381, 382
remainder, 130, 270, 286, 287, 289, 381, 382, 391, 393, 396
  sequences, 373
`remember`, 239, 240, 319
remember tables, 238–241
`remove`, 330, 335–337, 389
removing
  elements, 62, 330, 335–337
  output, 19, 357
  prompts, 346
repeat loops, 211, 349
repeated
  addition, 67
  decimal, 133
  derivatives, 278
  evaluation, 45
  execution, 205, 207, 210
  factors, 235, 370–372
  multiplication, 67
  products, 53, 66, 474
  roots, 370
  self-composition, 32, 274
  set elements, 58
  sums, 53, 66, 474
repetition, 33, 204, 244
repository, 477
reserved variables, 135
residual, 167
  vector, 167, 180
resize, 112, 401, 404
`restart`, xvi, 18, 95, 104, 158, 348, 413, 458, 465, 470, 473, 474, 476, 480, 484
Return key, 15–17, 39, 82
return, 57, 61, 62, 87, 137, 220–222, 225, 226, 229–232, 242–243, 427
  carriage, 69
`return`, 231, 234, 235, 249, 257, 362, 450, 451, 458, 472
`ReturnIt` (user proc), 220
reversing, 63, 245
rewriting, 49, 425
`rhs`, 33, 173, 188, 339, 341, 376
Ritchie, D. M., 4, 428
`RootOf`, 35, 143–144, 152
roots, 128, 143, 145, 152, 178
  approximate, 145–146
  cube, 408
  double, 150
  estimating, 419
  exact, 142–144
  integer, 146
  intervals, 145

multiple, 150–151, 370
of equations, 142–146, 377
of linear equations, 167–169, 180–184
repeated, 370
triple, 150
rotation
matrices, 257
of 3D plot, 112, 121
`round`, 206, 286
round brackets, 28, *see also* `( )`
row
of array, 212–216
of matrix, 161–184, 339, 341, 447–469
of spreadsheet, 400–418
`rowdim`, 339, 454
`rsolve`, 150
`rtable`, 165
Maple type, 300
rtables, 159, 160, 164, 165, 176, 248, 256, 301, 337, 446, 448, 470
browsing, 165
displaying, 165
`rtablesize`, 165
RTF, 101, 426
Russian dolls, 187

`savelib`, 479, 480, 484
saving
code, 36
help, 485, 486
plots, 101, 121
to library, 479, 480, 482, 483, 485
worksheets, 9–11, 17–19, 90, 100, 106
scalar, 470, 474, 475
as matrix, 163
multiplication, 163, 448, 467, 472
product, 32, 339–340
`scalar`, 170
`scalarmul`, 468, 469, 472, 475
`scaling`, 79–81, 83, 93, 95, 108, 109, 111, 113, 118, 119
`scanf`, 346
`ScaProd` (user proc), 339, 340
scientific, 8, 127, 133, 134, 139
scope
lexical, 380, 388, 438, 445, 451–453, 455
local, 431
of variables, 222–227, 241–244
searching
help, 12, 384
library, 479
`SELECT` (user proc), 336
`select`, 330, 335–337
selecting, 32, 61–63, 68, 69, 89, 248
elements, 330, 335–337
module members, 465, 466, 472, 473
`selectremove`, 330, 336, 337
self-composition, 32, 274
semantics, 243, 268, 311
semicolon, 19, 35, 195, 205, 221, 222, 250, 359, *see also* `;`
sentinels, 211, 316
separators, 35
`seq`, 33, 40, 64–67, 69, 91, 98, 100, 120, 149, 176, 202, 204, 222, 224, 241, 258, 261, 262, 272, 311, 316, 332, 334, 345, 349, 391, 455
sequences, 7, 55, 57, 62, 63, 66, 146, 172, 349, 382, 391, 393, 440, 441, 458
adding, 58, 247
animation, 90, 91, 120
appending, 59
argument, 28, 57, 220, 227, 228, 231, 232, 273
constructing, 40, 64–67
digit, 289, 320–321
directory, 479, 485, 486
empty, 41, 63–64, 142
from ranges, 40, 41
from strings, 69

in concatenation, 37
in evalf, 30
in spreadsheets, 405–406
indexing, 29, 59
iterating over, 68, 210
joining, 59
length, 248, 311
mapping over, 330
maximizing, 210
multiplying, 58
of exact roots, 144
of plots, 86, 98
of solutions, 142
operating on, 59–63
operator, 33
remainder, 373
return, 429
Sturm, 373–377
type testing, 310–311
sequential substitutions, 49
series
arithmetic, 404, 405
infinite, 129
power, 150
spreadsheet, 399, 404
Taylor, 150, 151, 156, 203
server
parallel, 10, 17
web, 9, 347
X Window, 15
set
Maple type, 285, 300, 391
method, 457, 458, 462
set(numeric) (Maple type), 251
set(polynom) (Maple type), 391
SetCellFormula, 412
setElement (method), 463
setElements (method), 463, 465
SetMatrix, 411
setoptions, 91, 93, 95, 103, 112, 255, 259
setoptions3d, 112, 119
sets, 38, 55, 57, 58, 60, 62, 63, 223, 261, 430, 431
algebra of, 284
character, 68, 70, 427
constructing, 40, 64–66
empty, 284, 285
in plotting, 78, 79, 91–93, 110, 112
in solve, 142
in substitution, 48
in type testing, 310
indexing, 29, 59, 143
infinite, 128, 130
iterating over, 210
mapping over, 330
minus, 128
notation, 9
null, 284
of equations, 142
of pairs, 391
of polynomials, 381
of types, 298, 304, 308
operating on, 59–63
operations on, 279–283
power, 284–286, 391
quotient, 130
selecting from, 336
shake, 132
shape, 169, 170
sharp, 36, *see also* #
Shift key, 15, 16, 39, 82, 429
shift, phase, 90
shortcuts
keyboard, 11, 14, 15, 19
Windows, 343, 433, 481
showstat, 360, 361
showstop, 363
Side (user proc), 257
σ, 345
sign (arithmetic), 138, 150, 288, 373, 374, 376, 379
signature of module, 449, 453
significant figures, 30, 132, 135, 137, 141, 344
signum, 137, 374
simple mappings, 7
simplification, 8, 49–51, 57–58, 189, 191, 194, 201, 236,

238, 277, 280, 284, 304, 307, 381, 468
simplify, 50, 51, 54, 152, 153, 189, 237, 238, 284, 318, 468
simultaneous substitutions, 48, 49
sin, 25, 32, 39, 41, 47, 50–53, 74, 75, 77–79, 81, 108
singular
  matrices, 161, 168, 178, 180, 464
  value decomposition, 161, 178, 182–184
SingularValues, 179, 182
sinh, 92
size
  of archive, 480
  of file, 434
  of plot, 87, 112
  of spreadsheet, 400
  of stack, 246
slash, 12, 342, 478, *see also* /, \
  backward, 38, 343
  forward, 40, 343
Snowflake (user proc), 258
snowflake, 255–259
solve, 31, 35, 38, 55, 80, 132, 142–143, 188, 419
solveh (user proc), 377–380
solveh1 (user proc), 379, 380
solving
  differential equations, 148–150
  equations, 142–146
  linear equations, 167–169
sort, 261, 262, 286, 382, 384, 389
sorting, 58, 260, 261, 286, 384
source code, 10, 36, 141, 332, 346, 347, 355, 360, 361, 482, 486
space
  character, 12, 26, 33, 41, 51, 52, 68, 69, 161, 164, 209, 222, 232, 269, 343, 347, 426, 428–430, 436, 437, 441, 443, 486
  curves, 105, 113–114, 118
spacecurve, 113–114, 125
sparse, 172
sparse matrices, 161, 171, 172
specfunc (Maple type), 475
special
  identifiers, 227–230, 244, 248, 453
  matrices, 159, 161, 169–172
  names, 301
  procedure argument types, 311–313
  semantics, 311
  symbols, 23–41, 68, 70, 83, 192, 238, 268, 269, 299, 301, 321, 427
specific structured types, 307–311, 314, 318
spelling, 52, 53, 82, 260, 280, 381
spheres, 108–110, 120
spherical coordinates, 108, 109, 111
spheroids, 108–110, 120
spinner, 112
Spoly (user proc), 390
Spread (Maple package), 410–412
Spreadsheet menu, 400, 401, 404–406, 409
spreadsheets, xiv, xvi, 8, 14, 36, 150, 399–422
SqFree (user proc), 372
sqr (user proc), 220, 239, 240
sqrt, 140, 257, 258, 319, 345, 414
square
  block matrices, 464–466
  boxes, 69
  brackets, 29, *see also* [ ]
  matrices, 164, 168, 172, 179, 248
squarefree, 373, 377
  factorization, 369–373, 377
squashing, 81, 83, 109, 120
sscanf, 438

stack, 244–246
- matrices, 161, 162, 174
- overflow, 229, 246, 380
- procedure call, 363
- size, 246

Start menu (Windows), 94, 433
statements, xvi, 8, 16, 19, 34, 35, 86, 164, 188, 194–197, 204, 205, 207, 208, 210, 211, 221, 222, 231, 239, 349–352, 355, 357–360, 445, 450, 473
- compound, 445
- numbering, 361

static
- animation background, 92, 120
- scope, 224
- types, 447, 453

statistics, 344–346
`step`, 361
step size, 209
Stoer, J., 5, 160
stop, 32, *see also* .
`stopat`, 361, 363
`stoperror`, 363
stopping execution, 18, 231, 301, 361–363
`stopwhen`, 363
`storage`, 172
stretching, 81, 109
`string`
- in `convert`, 251, 252
- Maple type, 262, 301, 429

strings, 27, 28, 68–70, 401
- \ in, 38
- characters as, 41
- double quotes in, 39
- empty, 262, 430
- expanding, 262
- format, 134, 349, 436
- generating, 260
- I/O of, 346
- in C, 431
- indexing, 29
- joining, 37, 262
- newlines in, 39
- outputting, 436
- storage, 441

Stroustrup, B., 4
structural types, 300
structure, block, 224, 445
structured types, 295, 304–311, 314, 318, 466
structures
- C, C++, 446
- control, 35, 61, 87, 187–216, 222, 351, 357, 450
- data, 55, 61–63, 67, 82, 85, 86, 141, 159, 268, 330–337, 450, 467
- infinite, 256
- record, 448–450
- recursive, 244, 253, 255

`Sturm` (user proc), 373
Sturm sequences, 373–377
Sturm's theorem, 375
`style`, 83, 96
`SubMatrix`, 182
`submatrix`, 447, 455
subroutine, 219
`subs`, 47–49, 61, 64, 143, 149, 277, 374, 380, 416, 421
subscripting, 29, 69, 278, 457
subsets, 283, 284
subsidiary procedures, 338, 378, 379
`subsop`, 48, 61, 63, 64, 243, 261, 334, 393
substitution, 46–49, 53, 61, 143, 243
- local, 47
- multiple, 48–49

subtopic, help, 12
subtraction, *see* `-`, addition
`SubTree` (user proc), 254
`SubWords` (user proc), 261
`SUM`, 474
`Sum`, 58
`sum`, 40, 66–67, 243, 345

summation, 243, 341, 342, 345
  recursive, 247–248
sums, 243
  iterating over, 210
  mapping over, 330
  repeated, 53, 66, 474
  selecting from, 336
SunOS, 122
superimposing plots, 78–79, 90–93, 110, 118, 251
surface, 106, 107, 118
  plotting, 105, 107, 118
  types, 295, 297, 304, 305
`surfdata`, 126
SVD, 161, 178, 179, 182–184
`SYMBOL`, 93, 94
`symbol`, 83, 96
  Maple type, 298, 300, 301, 372
`symbol &+ integer` (Maple type), 309
`symbol &+ symbol` (Maple type), 308
`symbol &+ symbol &+ symbol` (Maple type), 308
symbols, 6
  font, 68, 93, 94
  indexed, 63
  mathematical, 68, 93
  special, 23–41, 68, 70, 83, 192, 238, 268, 269, 299, 301, 321, 427
`symmdiff`, 281, 282
symmetric
  difference, 281–282
  matrices, 170, 171, 174, 175, 178
  `mod`, 155
`symmetric`, 170
syntax, 23–41, 58, 445
  arrow, 25, 50, 220, 221, 239, 270
  block matrix, 470
  Boolean, 193
  complex, 131
  conversion, 19
  correction, 52
  errors, 51, 193
  functional, 268, 269, 272, 275, 301, 304, 323, 472
  invalid, 352
  Matrix, 161
  module, 224, 450
  new, xiii
  OO, 446, 451
  operator, 193, 267, 269, 279
  package, 88, 89
  plotting, 96, 105
  procedure, 221
  type, 298, 306–308

Tab key, 15, 17, 429
tab character, 69, 428–430
`table`, 334, 442
  Maple type, 300, 334
tables, 61, 88, 89, 160, 164, 165, 192, 212, 213, 243, 256, 297, 334, 337, 407, 432, 445, 448, 449
  displaying, 213, 215, 400
  empty, 442
  hash, 301, 441
  indexing, 29
  initializing, 442
  labelling, 214, 215
  mapping over, 330
  multiplication, 213, 404
  rectangular, 159
  remember, 238–241
  selecting from, 336
  truth, 191, 402
tabular
  input, 400
  output, 400
`tabular` (Maple type), 301
tabulator, horizontal, 428
tag, 143, 467
`tan`, 35, 140
Taylor series, 150, 151, 156, 203
`tdeg`, 382, 384, 392
teletypewriters, 69

term, 49, 150, 156, 288, 382, 384, 389
  leading, 382, 384, 386, 388, 389, 394
termcoeff (user proc), 389
terminal, 342, 343, 348
terminators, 35, 195, 222, 250, 350
terminology, 6, 24, 267, 299
  ambiguous, 6, 273
  Markov, 441
  OO, 446
ternary operator, 204
TEXT, 428
text, 8, 10, 11, 13–17, 36, 68, 70
  constant, 27, 28, 45
  export, 101
  file, 11, 346, 426–428, 433–444, 446, 465, 482, 486
  font, 93
  formatting, 436
  I/O, 346
  in spreadsheets, 401
  plain, 486
  plotting, 93, 112
  processing algorithms, 425
  regions, xv, 36
  types, 304
text_processing (user package), 431–433
textplot, 68, 93–94, 103, 112, 202, 259
textplot3d, 112, 125
then, 194–201
theorem, Sturm, 375
theta, 47, 79, 108, 109, 111
Theta0 (user proc), 421
THICK, 117
thickness, 103
throw
  control, 351
  dice, 349
  die, 348
  error, 351
  exception, 351
tickmarks, 103, 125
tilde, 6, 35–36, 68, 401, *see also* ~
time of archive, 480
TIMES, 103
times, *see* multiplication
title, 27, 39, 82, 84, 109, 111, 252
titlefont, 82
to, 207–209, 212
tolerance, 190
topic, help, 12, 13, 486
Trace, 89
trace, 359–360
tracing
  algorithms, 356
  execution, 359
  procedures, 359
transcendental, 128, 140, 145, 377
Transpose, 170, 175, 184
transpose, 170, 171, 175, 178, 179, 183
Tree (user proc), 255
tree, 305, 343
  binary, 253
trial-and-error, 2, 145, 418
triangle, 256
triangular, 170
triangular matrices, 161
tridiagonal matrices, 161
trigonometric, 95, 140
triple
  click, 13
  ditto, 34
  roots, 150
true, 85, 190–194, 283, 382, 402, 429, 438
  Maple type, 300
trunc, 141, 206
truth, 31, 55, 191
  tables, 191–193, 213, 402
try, 232, 351–354, 378
type, 50, 269, 283, 298, 301–323, 332, 334, 336, 378
typematch, 301
types, 6, 7, 27, 30, 35, 37, 55, 62, 63, 133, 139, 160, 166,

233, 247, 248, 251, 271, 273, 279, 295–323, 446
algebra of, 299
algebraic, 300
alternative, 298
Boolean, 300
checking, 452
combining, 304, 322
definitions, 481, 482
disjoint, 298, 299
dynamic, 447
identifiers, 301, 322
interface, 466
Maple
`*`, 300, 389
`**`, 300
`+`, 300, 307, 382
`+`({integer,symbol}), 309
`+`({symbol,integer}), 309
`+`(symbol), 307
`.`, 300
`.`(symbol), 309
`..`, 300
`<`, 300
`<=`, 300
`<>`, 300
`=`, 300
[constant,constant], 314
[integer,integer,integer], 307
[integer,integer], 307
[numeric,numeric], 314
{algebraic, procedure}, 378
{list,set,`+`,`*`,function}, 336
{list,set}, 316
{matrix,list(list),symbol}, 458
{name,procedure}, 332
{set,list}, 283, 302
`^`, 300
2, 318
algebraic, 300, 468
`and`, 300
And(fraction,negative), 304
And(integer,Not(prime)), 305
anything, 299, 475
Array, 300
array, 300, 334
boolean, 300
complex, 299
constant, 299, 316, 378
equation, 300
evaln, 311, 312
evaln(numeric), 312
even, 299
false, 300
float, 299, 320
fraction, 299, 318, 322
function, 300, 302, 310, 332
identical, 384
indexed, 300, 301
integer, 248, 299
integer &+ symbol, 309
list, 272, 300, 306
list(algebraic), 311
list(constant), 314
list(integer), 306
list(list(matrix)), 458
list(list), 458
list(name), 306, 382, 391
list(nonnegint), 452, 458
list(numeric), 314
list(polynom), 374, 376, 382
logical, 300
Matrix, 300
matrix, 300, 341, 452
matrix(matrix), 458
`module`, 301, 466, 467
name, 300–302, 313, 372
negative, 299
negint, 299
nonnegative, 299

nonnegint, 246, 289, 299, 318, 358
nonposint, 299
nonpositive, 299
'not', 300
numeric, 299, 313, 316, 374
odd, 299
'or', 300
Or(symbol,string), 304
polynom, 300, 372, 373, 382
posint, 234, 299, 338
positive, 299, 378
prime, 299
procedure, 50, 298, 301, 336, 378
range, 300
range(numeric), 376
rational, 299
ratpoly, 300
realcons, 299
relation, 300
rtable, 300
set, 285, 300, 391
set(numeric), 251
set(polynom), 391
specfunc, 475
string, 262, 301, 429
symbol, 298, 300, 301, 372
symbol &+ integer, 309
symbol &+ symbol, 308
symbol &+ symbol &+ symbol, 308
table, 300, 334
tabular, 301
true, 300
uneval, 311–313
Vector, 300
vector, 300, 338
module, 466–467
nested, 295, 304, 305, 307
new, 321–323
numerical, 299
of sequences, 310–311
of sets, 310
pattern, 307, 308, 318
primitive, 296–299, 305, 332, 466
procedural definition, 308, 322, 323
product, 308
programming, 301
pseudo, 311
sets of, 298, 304, 308
special argument, 311–313
static, 447
structural, 300
structured, 295, 304–311, 314, 318, 466
sum, 308
surface, 295, 297, 304, 305
syntax, 298, 306–308
text, 304
union of, 299
user
BlockMatrix, 466–468, 470
fracinteger, 323
halfinteger, 322
listset, 322
monord, 382, 384, 391
typography, xvi, 148

UCS, 427
unapply, 273, 274, 276, 278, 378
unary, 191, 267, 268, 270, 271, 304, 474
unassign, xvi, 29, 56, 284, 458
unassigning, 18, 42, 46, 76
unblock (user proc), 458
unblockIndex (user proc), 463
unbound, 35, 67
UNCONSTRAINED, 83
undebug, 359, 360
underflow, 134–135, 138
underscore, 6, 35, *see also* _
undo, 11, 14, 15, 357
assignment, 25, 42, 136
with, 473
uneval (Maple type), 311–313
unevaluation, 25, 26, 49, 65, 76, 201, 211, 227, 229, 277,

301, 303, 306, 349, 412, 457, 466, 479
Unicode, 68, 427
union
  of graphs, 80
  of types, 299
`union`, 269, 279–282, 284, 285, 391
unit
  imaginary, 129, 131
  matrices, 168, 169, 215
  step function, 198
unitary, 175
univariate, 286, 288, 289
  polynomial algorithms, 369–381
UNIX, xv, 8, 13, 15, 17, 227, 238, 343, 347, 427, 428, 433, 434, 436, 478, 481
unknown, 43, 129, 148, 149, 244
`unload`, 482
`unprotect`, 257
`unstopat`, 363
`unstoperror`, 363
`unstopwhen`, 363
until (repeat ... until), 211
`untrace`, 359
upper case letters, 52–53
`UseHardwareFloats`, 166, 178
UTF-8, 427

valid
  input, 301–302, 351, 352, 462
  numerical context, 76
`value`, 32, 53, 208
values, 25, 42–44, 54, 55, 363, 479
`Var` (user proc), 374
variables, 24–25, 27, 42–44, 54, 55, 131, 295, 363, 479
  ambiguity, 43
  clearing, 18, 25, 44, 76, 240
  control, 65, 67, 222, 224, 226
  dependent, 153–154
  environment, 136, 166, 178, 224, 268, 468
  free, 74–76, 78, 82, 89, 90, 141, 143, 145, 225
  global, 35, 44, 136, 239, 321, 388, 450, 455, 482
  in `RootOf`, 143
  independent, 75, 153, 223, 224
  local, 45–46, 244, 248, 340, 341, 361, 386, 445, 450, 468, 475
  loop control, 209
  polynomial, 289
  pseudo, 225
  reserved, 135
  scope, 222–227, 241–244
  with assumptions, 35
variation, 374–376
VCR, 90, 98
`VecAdd` (user proc), 338
`vectdim`, 339
`Vector`, 160, 166
  Maple type, 300
vector, 159–184, 256
  addition, 163
  algebra, 337–342
  as list, 257
  field, 117
  gradient, 117
  length, 167, 172, 175
  random, 166
  residual, 167, 180
  zero, 163
`vector`, 256, 382
  Maple type, 300, 338
`verboseproc`, 237, 238, 240
`version`, 85, 100
versions of Maple, xiii, 2, 8–10, 17–19, 27, 28, 30, 33, 45, 70, 195, 205, 246, 380, 445, 447, 469
vertical
  axis, 80
  bar, 37, 38, *see also* |
  coordinates, 79, 81, 93, 401
  plot range, 77, 82, 95
`view`, 95, 103, 109, 126

View menu, 16, 19, 69
Visual Basic, 260
Vivaldi, F., xiv
VRML, 121–123
`vrml`, 122, 123

watchpoints, 363
wave, 89
`wc`
    UNIX, 428
    user proc, 435
`wc_line` (user proc), 429
web, xiv–xvi, 347, 486
Weinberger, P., 436
`whattype`, 33, 296–298, 305–307, 309, 310, 332, 334, 336, 466
`while`, 204–212, 225, 226, 232, 234, 235, 303, 348, 372, 373, 380, 382, 391
while (repeat ... while), 211
white space, xvii, 269, 343, 428, 429, 436, 437, 441
Window menu, 17
Windows, Microsoft, xv, 8, 11, 15, 40, 68, 122, 141
    Explorer, 121, 433, 434
Winston, P. H., 4
`WIREFRAME`, 111
wizard, 1
word count, 428
Word, Microsoft, 9, 11, 16, 260, 436
    Picture, xvi
`Words` (user proc), 261, 262
wrapping lines, 16, 36
`WRITE`, 428
`writedata`, 346
`writefile` (user proc), 434, 437, 438, 440
`writeline`, 346, 434, 435
`writeto`, 342, 343, 348
`.wrl`, 122, 123

X Window System, 8, 14, 15
`xterm`, 347

yellow, *see* colour

zero
    contour, 119
    matrix, 163, 171, 172
    multiple, 150
    signed, 138
    singular value, 179, 183
    vector, 163
`ZeroMatrix`, 171, 172
zeros, finding, 142–146, 369–381, 418–422
`zip`, 273
zoom
    rtable, 165
    worksheet, 19